GREAT DEBATES IN AMERICAN ENVIRONMENTAL HISTORY

GREAT DEBATES IN AMERICAN ENVIRONMENTAL HISTORY

Volume 2

Brian Black and Donna L. Lybecker

GREENWOOD PRESS
WESTPORT, CONNECTICUT • LONDON

Library of Congress Cataloging-in-Publication Data

Black, Brian.

   Great debates in American environmental history / Brian Black and Donna L. Lybecker.

     p. cm.

   Includes bibliographical references and index.

   ISBN 978-0-313-33930-1 ((set) : alk. paper) — ISBN 978-0-313-33931-8 ((vol. 1) : alk. paper) — ISBN 978-0-313-33932-5 ((vol. 2) : alk. paper)

   1. Human ecology—United States—History. 2. Nature—Effect of human beings on—United States. 3. United States—Environmental conditions—History. I. Lybecker, Donna L. II. Title.

   GF503.B527 2008

   363.700973—dc22   2008002106

British Library Cataloguing in Publication Data is available.

Library of Congress Catalog Card Number: 2008002106

ISBN: 978-0-313-33930-1 (set)

      978-0-313-33931-8 (vol. 1)

      978-0-313-33932-5 (vol. 2)

First published in 2008

Greenwood Press, 88 Post Road West, Westport, CT 06881

An imprint of Greenwood Publishing Group, Inc.

www.greenwood.com

Printed in the United States of America

The paper used in this book complies with the Permanent Paper Standard issued by the National Information Standards Organization (Z39.48–1984).

10 9 8 7 6 5 4 3 2 1

CONTENTS

LIST OF ENTRIES

GUIDE TO RELATED TOPICS

Environmental Conservation

Pinchot Argues Conservation as a Development Strategy
Muir Argues for the Soul of Wilderness
Pinchot, Muir, and the Conservation Movement Meet at Hetch Hetchy
Public Learns Gardening and Rationing to Support the Cause
George Perkins Marsh Spurs Consideration of Industrialization
Discovering Alaska
Grazing Rights in the West and on Public Lands
Franklin D. Roosevelt Implements Conservation Policies
Debt for Nature and Development Swaps
Boone and Crockett Club Uses Virility to Attract Environmental Support
The Boy Scouts of America Involve Young Men in Outdoors
Progressives Demand Federal Action on Conservation
Yellowstone to Yukon Conservation Initiative

Influence and Effects of Aridity

Long-Term Implications of Sodbusting and Conversion of the American West for
    Agriculture
Using Chicago to Make the Great Lakes the Nation's Fifth Coast
Jefferson Argues for the Louisiana Purchase and American Expansion
American Leaders Create a Culture and System of Expansion
Mormons Create a Model for Interpreting the Aridity of the West
Settlement Systematizes and Simplifies the Ecology of the West
Powell and Efforts to Explore the Contours of the West

Urban Issues

City Beautiful, Urban Renewal, and the Effort to Reform the Modern City
Can the City of Boston Be Artificially Grown?
Alexander Hamilton Envisions an Industrial America
Modeling Public Works in Philadelphia
Bringing America's Nature Aesthetic to Life in Parks
The American Suburb, Sprawl Nation, and the Emergence
    of New Urbanism
Boston Harbor as Political Football
Can You Dig the Big Dig?
The Western United States, Urban Growth, and the New West
Olmsted and Vaux Design a Central Park for New York City
Olmsted Helps to Define the American Movement for Parks
Social Reformers Set Sights on Urban Problems
Russell Sage Studies Urban Problems in Pittsburgh

Exploration/Expansion to the West

Explorers Search the Resources of the American West
Jefferson Argues for the Louisiana Purchase and American Expansion
Lewis and Clark Seek to Know the Unknown Continent
The Land Ordinance of 1785-1787 Constructs the American Grid for
    Land "Disposal"
American Leaders Create a Culture and System of Expansion
Hitching the Nation's Future to the Railroad
Settlement Systematizes and Simplifies the Ecology of the West
Border Disputes and the Settlement of Texas
Gold Opens Up the West
Managing Resources in the Civil War
Powell and Efforts to Explore the Contours of the West
Frontier Thesis and American Meaning

Management of Resources

Pinchot Argues Conservation as a Development Strategy
Muir Argues for the Soul of Wilderness
Pinchot, Muir, and the Conservation Movement Meet at Hetch Hetchy
"Broad Arrow" Policy Conserves New England Forests and
    Foments Revolution
What Is the Importance of America's Natural History?
Internal Improvements in the Early Republic
Federalizing Forest Conservation
Turning to Fossil Fuels and an Electric Life
Foraging for Food in a Society of Mass Consumption
Energy Development on the Public Lands

Management of Water (Coasts, Canals, and Rivers)

Colonial and Indian Treatment of the Land and Rights

Agriculture

Politics and Social Reform

Ethics of Environmental Policies and Theories

Animals

Energy

Civil War

Miscellaneous

PREFACE

The perspective of these volumes is that of environmental history. It requires that we attach blinders to our view of history. As we selected events and issues for inclusion, we accentuated the changing idea of nature in American life. American definitions of terms such as nature or environment have changed a great deal over time. This book seeks to catalog such changes in definition, including romantics and extreme environmentalists who believed that God could be found in nature to wise users who saw no rationale for humans relenting on their use of natural resources. The issues covered in this volume seek to run the gamut of American ideas of nature and environment, including major policies that have been fashioned—and refashioned—to contend with social and environmental problems.

The essential common elements in each entry, then, are human users and the natural environment. In some of the topically organized entries, readers will be able to chart alterations in human approaches to the issue over time. In other entries, we have explored one event in depth and allow you to make the broader connections. Very often, government regulation or law making proves to be the outcome to changes in ideas about land use. We hope that these volumes will provide a valuable reference for policy makers as well as those involved in environmental law. Readers will see that some of these efforts have proven successful over time, whereas others have cleared the way for future adjustments. Throughout the volume, however, we have offered general themes with each entry, such as "Land Use," "Water Management," or "Planning." These, combined with the thematic table of contents, will assist readers who want to explore different dimensions of one issue in a variety of instances.

A clear theme in the entire volume is the need for an aware, educated public to speak out and take action when it deems it to be necessary, whether as a result of exploitation or inaction by others. We hope that these reference volumes will assist readers in approaching new environmental issues and challenges with a solid foundation of background knowledge.

Although the bibliography for this book will provide readers with a fine resource for further reading in environmental history, readers may also want to contact the American Society of Environmental History (http://www.h-net.org/~environ/ASEH/welcome_NN6.html).

We would like to thank the environmental studies students of Penn State Altoona for their unceasing interest and constant questioning of environmental topics. In addition, numerous colleagues and students at Gettysburg and Skidmore Colleges, Colorado State University, Penn State Altoona, and Idaho State University have taught, cajoled, and discussed many of these issues with us. We owe them an immense debt of gratitude for their contributions and inspiration. Finally, Chelsea Burket and Carole Bookhammer provided superior assistance in the preparation of this manuscript.

INTRODUCTION: WHAT LAY BELOW THE ARCTIC ICE CAP

That man is, in fact, only a member of a biotic team is shown by an ecological interpretation of history. Many historical events, hitherto explained solely in terms of human enterprise, were actually biotic interactions between people and land. The characteristics of the land determined the facts quite as potently as the characteristics of the men who lived on it.

—Aldo Leopold

Similar to other specialized approaches to studying the past, environmental history requires that we attach blinders to our view of history. In our survey of environmental historians, we can't be interested in every historical event; in fact, there are many essential events of the past that we will not discuss at all. Great characters of history will need to be entirely overlooked so that our focus might properly consider the specific factors that relate to environmental inquiry.

As the words of the 1930s biologist and naturalist Aldo Leopold declare, the perspective of environmental history accentuates the interaction between humans and nature over time—to explore events possibly hitherto considered to be of significance only to human actors. To do so, environmental historians overcome the limits of the human world by using scientific and biological concepts. Such concepts form the foundation of the historical stories that are told in the pages that follow.

At many junctures, these events elicited strong debate between alternative perspectives. The articles that follow are organized around these altercations, sometimes organized topically and at other times by specific events or time periods. Viewed chronologically, as they are organized in the two volumes, each of the debates dramatizes changing ideas about Americans' changing ideas of and relationship to the natural environment. Although they do not form a certain trajectory, these entries demonstrate changing ideas about free enterprise and about how active a role the federal government should take in American life. Following a clear high point of government involvement in resource management in the 1930s and again in the 1970s, the start of the twenty-first century brings new priorities and new possibilities. It is our contention that knowledge of past great debates of environmental history and policy will help us to better manage those that are certain to be around the next bend.

For instance, August 2007 may have witnessed the start of the next great resource debate. Eyeing future wealth (particularly energy resources), Russia sent two submersible ships two miles below the Arctic ice cap to perform a fairly simple task: planting a Russian flag. Similar to the United States planting its flag in the moon's soil a half century before, an enterprising nation has attempted to identify the next frontier of development in hopes of staking its claim ahead of any competitors. One reason for opportunity in the Arctic is global climate change, which is leading ice to recede at a record pace. Another reason is the region's appeal throughout human history: remoteness.

Five nations have cordoned off land in the Arctic (Denmark, Canada, the United States, Norway, and Russia). Frozen or not, crucial resources can be extracted there. Most appealing of all, there are few nongovernmental organizations or native residents to complain. Although activists, scientists, and health officials have taught us important lessons and helped to make the American living environment safer over the past few centuries, acquiring necessary resources and products has become somewhat more complicated. The remote, largely unspoken-for Arctic could be the next great frontier for development.

Of course, we can be certain that such a change will elicit debate and discussion, as will many other issues of importance to American land use. This book strives to construct a basic chronology of many debates in Americans' relationship with the natural environment. Some of the essays are organized very specifically on a moment at a specific time; in other essays, though, broader topics demonstrate the significance of specific events or policies. Although it is necessarily selective, these pages reveal a basic chronology in Americans' ability to consider some of the complications related to land use and development.

Brian Black, Hollidaysburg, PA
Donna Lybecker, Pocatello, ID

TIMELINE: THE STIRRING OF ENVIRONMENTAL ACTION IN THE 1960S– 1970S

1962   Rachel Carson writes *Silent Spring*.

1964   Barry Commoner's group, formed in the 1950s to oppose development of atomic energy, begins publishing the journal that will become known as *Environment*.

1965   Commoner and Ralph Nader lead a critique of the American economic system and the ability of science to question development.

1965   The Sierra Club files a lawsuit to protect Storm King Mountain in New York from a power project. The U.S. Supreme Court rules in favor of the club and of non-economic interests in a conservation case.

1966   The Sierra Club opposes the building of two dams that would have flooded the Grand Canyon and publishes newspaper ads that say, "This time it's the Grand Canyon they want to flood. The Grand Canyon." The dam plan is defeated.

1966   Congress passes the Endangered Species Act.

1968   Apollo 8 sends back the earth-rise pictures from space (see below, *Facing Up to American Energy Independence*).

1969   The Santa Barbara oil spill attracts public attention to polluted beaches.

1969   Congress passes the National Environmental Policy Act to mandate consideration of environmental issues before major public decisions.

1970   The first Earth Day is April 22.

1972   Congress passes the Clean Water Act, Coastal Zone Management Act, Federal Environmental Pesticide Control Act, Marine Mammal Protection Act, Ocean Dumping Act, and Federal Advisory Committee Act to require representation of public interest advocates on committees.

1972   Oregon passes the first bottle-recycling law.

1972   The Club of Rome publishes "Limits to Growth" in 1972.

1973   Congress passes the Endangered Species Act.

1974   Congress passes the Safe Drinking Water Act.

1977   Love Canal, New York, identified as a chemical waste site requiring evacuation of all inhabitants.

1979   Three Mile Island nuclear reactor near-meltdown in Pennsylvania.

1980   Congress passes the Superfund bill to help identify and pay for cleaning up abandoned toxic waste sites.

(Source: EPA)

FEDERALIZING FOREST CONSERVATION

Time Period: 1876 to 1920s
In This Corner: Progressive conservationists
In the Other Corner: Industrial interests
Other Interested Parties: Native peoples
General Environmental Issue(s): Forests, resource management

Conservation is one of the easiest environmental concepts to grasp. It is based, of course, on the premise that it is not in humans' best interest to extinguish or even to deplete the resources on which they rely. This reality, however, went against some of the basic principles on which the United States was established and, therefore, was not automatically accepted.

Colonial settlement of North America reinforced the concept that the continent teemed with such bountiful supplies of natural resources that it could never be depleted. With a fairly small population on the continent, little occurred that would dislodge this perception. After 1850, however, industrialization expanded the ability of Americans to harvest resources such as forests. In addition, economic success helped to create a wealthy class of Americans who began to think in alternative ways about America's natural resources. Therefore, an active debate began in the late 1800s over what responsibilities the federal government should take over managing the consumption of natural resources, such as North America's remarkable supply of trees.

Chopping Trees on the Plains and Arbor Day

On the Great Plains, where there were very few trees, the idea of development and conservation often faced off. An important fuel and building material, lumber played a crucial role in American development after 1850. With each press westward of European settlement, the supply of timber grew. The need for lumber, of course, grew as well. The boom in building, particularly in the American West, increased lumber development through 1900. For instance, Oregon doubled its population between 1890 and 1910. In the Pacific Northwest, timber production in the 1880s increased to 2.6 billion board feet per year. By 1910, Oregon and Washington ranked in the top three lumber producing states in the United States (Hays 1999, 55–57).

Forests, of course, could not replenish themselves as quickly as the trees were felled. In fact, clear-cutting often left forests with no ability to regenerate. This observation as well as timber shortages throughout Europe alerted the public to the need to conserve forests as early as the 1870s. In an effort to please all concerned, Congress passed the Timber Culture Act in 1873, which promoted tree planting as well as western settlement. This effort was partly spurred by the scientific theory of Joseph Henry of the Smithsonian Institution, who argued that planting trees would stimulate rainfall in the arid West. Clearly, settlers thought they could fell the forests and still have sufficient supply to support the nation's future.

With Henry's theory in mind, Nebraskans in 1872 even went as far as implementing the first celebration of Arbor Day, a holiday intended for tree planting. Many other western states followed. By 1907, however, Arbor Day was observed by all of the United States. Although the belts of trees did not bring additional rain to the West, they did help with settlement by breaking wind and providing shade.

Conserving American Forests

The concept of conservation in the United States began in reference to forest reserves, which trace their origin to an 1876 act of Congress that created the Forest Service. Based on the lessons of European nations, this act authorized the USDA to hire a forestry agent who would investigate the present and future supply of timber. In addition, he studied the methods used by other nations attempting to manage or conserve timber supplies. Although federal forest lands were not set aside until fifteen years later, a Division of Forestry was set up in 1881. The first head of the agency was a Prussian-educated forester named Bernhard E. Hough. Compiling many reports about the past and present use of American forests, Hough was very critical of American patterns of use. He urged the nation to adopt regulations to "secure an economical use" (Hays 1999, 28–30).

Although the division began primarily as a disseminator of information, by the 1890s, it provided forestry assistance to states and to private forest land owners. In 1891, Congress granted the president the authority to establish forest reserves from the existing public domain lands under the jurisdiction of the Interior Department. The Organic Administration Act of 1897 stipulated that forest reserves were intended "to improve and protect the forest within the reservation, or for the purpose of securing favorable water flows, and to furnish a continuous supply of timber for the use and necessities of the citizens of the United States." By the early 1900s, this office administered approximately fifty-six million acres. However, the agency was about to take a leading role in the broader American conservation movement (Hays 1999, 30–33).

Entrusting Forests to the Federal Forest Service

This interest in forest conservation arrived at the federal government during the 1870s. The formation of the Division of Forestry within the USDA during this decade proved to be a fairly insignificant political step. A corrupt federal bureaucracy overwhelmed any bona fide efforts at conservation. Instead, the division was used to more readily sell off timbered tracts of land. In hindsight, historians have noted that the Timber Culture Act allowed for twenty more years of unregulated clear-cutting. It was this intensification, however, that stirred public cries for reform (Hays 1999, 61–65).

When Bernard Fernow became the chief of forestry in 1886, he brought with him a commitment to the scientific management of American forests. Although the division continued its primary effort at disseminating public information, it soon added scientific experimentation to the division's responsibilities. President Harrison established the first timber land reserve on March 30, 1891 and placed it under the control of the General Land Office rather than the Division of Forestry. The president now had the authority to set aside public land in the West. By 1897, forty million acres in the Northern Rocky Mountains had been set aside. In this year, Congress moved the National Forests into the Department of the Interior's General Land Office.

Many of Fernow's scientific ideas were incorporated into the national forests by the first American trained in forestry, Gifford Pinchot. Pinchot replaced Fernow as director in 1898 and set out to force private timber firms to undertake cooperative forestry management on the reserves and reforestation programs on private lands (Hays 199, 62–63).

Gifford Pinchot, America's First Forester

Forest conservation was the business of Gifford Pinchot, America's first native professionally trained forester. He returned from forestry school in Germany to establish one of the United States' first model forests, which lay on the Biltmore estate of Cornelius Vanderbilt in North Carolina. He took over administration of the national forests in 1898 and became the first chief of the renamed Forest Service in 1905, when the agency took over the forest reserves from the Department of the Interior's General Land Office. The agency was renamed the U.S. Forest Service (USFS), and, ultimately, forest reserves were renamed national forests. Pinchot voiced a refrain of the conservation movement when he instructed Secretary of Agriculture James Wilson that the goal of forest administration needed to be based on "the greatest good for the greatest number in the long run" (Opie 1998, 389–90).

Pinchot's real fame came when his close friend Theodore Roosevelt became president in 1901. Together, they oriented the Forest Service toward the wise use of timber resources. This guideline was the forerunner of the multiple-use and sustained-yield principles that have guided forest management in recent years. These principles stress the need to balance the uses that are made of the major resources and benefits of the forests—timber, water supplies, recreation, livestock forage, wildlife and fish, and minerals—in the best public interest.

With Roosevelt's support, the 1905 act transferred the Forestry Division from the Department of the Interior to the USDA and combined it with the larger Bureau of Forestry. Roosevelt and Pinchot more than doubled the forest reserve acreage in the two years following the merger, to a total of 151 million acres by 1907. This total included sixteen million acres that were squeezed through after Congress placed limits on the president's authority to proclaim additional forest reserves. Without congressional approval, Roosevelt designated these areas in defiance of Congress. When he left office in 1909, there were 195 million acres of National Forest (Nash 1982, 152–59).

Pinchot's philosophies influenced the policy initiatives of the nation when his close friend, Theodore Roosevelt, became President in 1901. By 1905, Roosevelt had overseen the transferal of sixty-three million acres of reserves into Pinchot's domain and had renamed the agency as the U.S. Forest Service. An act on February 1, 1905 established the U.S. Forest Service and also transferred the national forest reserves from the General Land Office to the USDA.

Pinchot organized the Forest Service quickly. In 1907, the reserves were renamed national forests. During the following year, six district or regional offices were organized in the West for administering field work. These national forests, however, considered their priority to be ensuring that the nation would maintain an abundant timber supply in perpetuity. This conservation mandate concerned chopping wood, making roads, and had nothing to do with wilderness or preservation. National forests were similar to vaults holding a natural resource of vital importance to the nation and had little similarity to national parks (Pinchot 1998). The following is an excerpt from Pinchot's *A Primer of Forestry*:

THE SERVICE OF THE FOREST

Next to the earth itself the forest is the most useful servant of man. Not only does it sustain and regulate the streams, moderate the winds, and beautify the land, but it also supplies wood, the most widely used of all materials. Its uses are numberless, and the demands which are made upon it by mankind are numberless also. It is essential to the well-being of mankind that these demands should be met. They must be met steadily, fully, and at the right time if the forest is to give its best service. The object of practical forestry is precisely to make the forest render its best service to man in such a way as to increase rather than to diminish its usefulness in the future. Forest management and conservative lumbering are other names for practical forestry. Under whatever name it may be known, practical forestry means both the use and preservation of the forest....

FEDERAL FOREST RESERVES

When the President was given the power to make forest reserves, the public domain still contained much of the best timber in the West, but it was passing rapidly into private hands. Acting upon the wise principle that forests whose preservation is necessary for the general welfare should remain in Government control, President Harrison created the first forest reserves. President Cleveland followed his example. But there was yet no systematic plan for the making or management of the reserves, which at that time were altogether without protection by the Government. Toward the end of President Cleveland's second Administration, therefore, the National Academy of Sciences was asked to appoint a commission to examine the national forest lands and report a plan for their control. The academy did so, and upon the recommendation of the National Forest Commission so appointed, President Cleveland doubled the reserved area by setting aside 13 additional forest reserves on Washington's Birthday, 1897.

The Cleveland forest reserves awakened at once great opposition in Congress and throughout the West, and led to a general discussion of the forest policy. But after several years of controversy widespread approval took the place of opposition, and at present the value of the forest reserves is rarely disputed, except by private interests impatient of restraint.

The recommendations of the National Forest Commission for the management of the forest reserves were not acted upon by Congress, but the law of June 4, 1897, gave the Secretary of the Interior authority to protect the reserves and make them useful. The passage of this law was the first step toward a national forest service....

The forest reserves lie chiefly in high mountain regions. They are 62 in number, and cover an area (January 1, 1905) of 63,308,319 acres. They are useful first of all to protect the drainage basins of streams used for irrigation, and especially the watersheds of the great irrigation works which the Government is constructing under the reclamation law, which was passed in 1902. This is their most important use. Secondly, they supply grass and other forage for many thousands of grazing animals during the summer, when the lower ranges on the plains and deserts are barren and dry. Lastly, they furnish a permanent supply of wood for the use of settlers, miners, lumbermen, and other citizens. This is at present the least important use of the reserves, but it will be of greater consequence hereafter.... (Pinchot 1905)

Conclusion: Educating Foresters

By the model set by the new Forest Service, Americans learned to view forests very differently than they once had. Pinchot made sure that applicants to work for the Forest Service needed to demonstrate a mix of formal education and practical aptitude. These new types of foresters, he hoped, would approach old problems differently. Primarily, science would now be included in efforts to solve some of the service's most pressing problems, including fires, overgrazing by cattle and sheep, soil disturbance and stream pollution caused by mining, and insect and disease impacts.

In 1908, the Forest Service established its first experimental station near Flagstaff, Arizona. Other stations were later added throughout the West and eventually in other regions. The Forest Service also established its laboratory to experiment with new products that could be manufactured from forests, while also establishing the framework that would allow states to use the proceeds from any products made from their national forests to be used locally for new schools and other services. With science at its core, the Forest Service initiated some of the nation's first efforts to set aside wilderness and primitive areas. The most famous of these efforts was the Gila Wilderness in New Mexico, which was the first primitive area in the service set aside in 1930 (Hays 1999, 44–48).

Forest research got a big boost in 1928 through the McSweeney-McNary Act, which authorized a broad permanent program of research and the first comprehensive nationwide survey of forest resources on all public and private lands. Tree planting in national forests was expanded under the Knutson-Vandenberg Act of 1930. The Forest Service also operated more than 1,300 Civilian Conservation Corps (CCC) camps in national forests during the 1930s New Deal. More than two million unemployed young men in the CCC program performed a vast amount of forest protection, watershed restoration, erosion control, and other improvement work, including the planting of 2.25 billion tree seedlings. Although the administration of forests would shift with the political wind, a core ethic had crept into the Forest Service that emphasized conservation of lumber supplies and, at times, administration of national forests as complex ecological sites (Hays 1999, 264–68).

Sources and Further Reading: Fox, *The American Conservation Movement*; Hays, *Conservation and the Gospel of Efficiency*; Miller, *Gifford Pinchot and the Making of Modern Environmentalism*; Nash, *Wilderness and the American Mind*; Pinchot, *A Primer of Forestry*, http://www.forestry.auburn. edu/sfnmc/class/pinchot.html; Pinchot, *The Use of the National Forests*; Steen, *The U.S. Forest Service: A History*.

FIRE ON PUBLIC LANDS: TO BURN OR NOT?

Time Period: Late 1800s to the present
In This Corner: Administers of federal lands
In the Other Corner: Ecologists
Other Interested Parties: Residents near federal lands
General Environmental Issue(s): Fire, park management

Historian Stephen Pyne wrote the following:

> We hold a species monopoly over fire. With fire we claim a unique ecological niche: this is what we do that no other creature does. Our possession is so fundamental to our understanding of the world that we cannot imagine a world without fire in our hands. Or to restate that point in more evolutionary terms, we cannot imagine another creature possessing it.
>
> Yet while humans come genetically equipped to manipulate fire, we do not come programmed in its use. We apply and withhold it according to social institutions, cultural norms, perceptions of how we see ourselves in nature. Different people have created distinctive fire regimes, as they have distinctive literatures and architectures. In this way fire became both natural and cultural. If fire measures our ecological agency, so how we choose to apply and withhold it testifies to our understanding of that agency and the values, choices, and means by which we act. Fire enters humanity's moral universe, and thus into the scholarly realm of the humanities. (Pyne, *History with Fire in Its Eye*)

What is a forest fire? A necessary part of the natural processes of a forest? A scourge endangering anyone near forest? A waste of important resources? Part of the American fire regimen has been to interpret fire's meaning, particularly for places such as federal lands, including national parks. Changes in the definition of fire for such land elicits dramatic debate from many Americans, particularly residents living near such locales.

In the late nineteenth century, fires burned throughout the United States but particularly in the West. Lightning—and, therefore, nature—ignited some of these fires, but more and more were caused by human means, including the use of steam-powered trains throughout the countryside that left burning ash in their wake and settlers burning fields fallow for pasturage. Although seasonal, fire occurrence was tolerated. In fact, in many areas it was thought that lands were uninhabitable without it. By the late 1800s, progressive conservationists argued for government intervention to put a stop to the burning.

During the twentieth century, park administrators frequently debated the question: to burn or not to burn? In short, whether or not to allow fires in such locales to burn themselves, thereby seeing fire as an organic portion of nature or, conversely, to extinguish the fire and, thereby, perceive it as an artificial interloper in otherwise natural scenes. Pyne wrote:

> The problem with fires became one of maldistribution—too much of the wrong kind of fire, not enough of the right kind. Most fires burned as wildfires, set by lightning, accident, or arson. Many lands suffered from a fire famine, the shock of having a process to which they had long adjusted abruptly removed. On many sites, natural fuels

had ratcheted up to levels against which fire suppression stood helpless. Government fire agencies sought to reinstate fire, often at considerable cost and risk, even in the face of public skepticism. (Pyne, *History with Fire in Its Eye*)

Yellowstone has functioned as a case study for the changing approaches to fire. The nature that had been preserved in the national park was at least partly a product of fires, large-scale conflagrations sweeping across the park's vast volcanic plateaus, hot, wind-driven fires torching up the trunks to the crowns of the pine and fir trees at several hundred-year intervals. Park officials did not choose to view fire in this way. However, by the 1940s, ecologists recognized that fire was a primary agent of change in many ecosystems, including the arid mountainous western United States.

Controlled burns were used in Yellowstone, and, by the 1970s, the park had implemented a natural fire management plan that allowed the process of lightning-caused fire to be allowed to continue if it started. In the first sixteen years of Yellowstone's natural fire policy (1972–1987), 235 fires were allowed to burn 33,759 acres. Only fifteen of those fires were larger than one hundred acres, and all of the fires were extinguished naturally (Pyne, 1982).

The incredible destruction of fires in 1988, however, changed everything. By July 15, only 8,500 acres had burned in the entire greater Yellowstone area. The fires did not become noticeable to visitors, however, until the end of July when they grew to engulf nearly 99,000 acres. In August, fires reached across more than 150,000 acres.

Great public debate followed. A symbol of preservation, Yellowstone was now a symbol of the ecology of fire. NPS studies showed that 248 fires started in greater Yellowstone in 1988; fifty of those were in Yellowstone National Park. Only thirty-one of them were allowed to burn. Fighting the fires in the region was estimated to cost $120 million and to involve a total of 25,000 firefighters. Ecosystemwide, about 1.2 million acres were scorched; 793,000 (about 36 percent) of the park's 2,221,800 acres were burned. Subsequent studies, however, have also demonstrated nature's remarkable ability to recover.

Therefore, national parks and forests across the United States suspended and updated their fire management plans. The 1992 Yellowstone revised plan continued a wildland fire management plan but with stricter guidelines under which naturally occurring fires may be allowed to burn. Through continued public education, scientific research, and professional fire management, Yellowstone hopes to preserve the process of natural fire in the park while minimizing adverse effects on park visitors and neighbors, recognizing the inevitability of this force to continue shaping the landscape as it has for centuries.

Throughout the United States, after a terrible fire season in 1994, a revised federal policy emerged in December 1995 that was organized by controlled burning, which could help thin forests of dead wood and, thereby, eliminate excessive fuel for naturally caused fires. Under this new approach to fire, the NPS attempted burns and lost control of two in 2000. As a result, the secretary of the interior placed the NPS program under a moratorium.

Currently, federal administrators extinguish fires that threaten facilities and surrounding homes and businesses. Busy fire seasons in the far West in 2006–2007, however, have made each side of the fire argument flare up.

Sources and Further Reading: Hays, *Conservation and the Gospel of Efficiency*; Hirt, *A Conspiracy of Optimism: Management of the National Forests since World War Two*; Miller, *Gifford Pinchot and the Making of Modern Environmentalism*; Nash, *Wilderness and the American Mind*; National

Interagency Fire Center, www.nifc.gov; Pinchot, *The Use of the National Forests*; Pyne, *Fire in America: A Cultural History of Wildland and Rural Fire*; Pyne, *The Ice: A Journey to Antarctica*; Pyne, *Burning Bush: A Fire History of Australia*; Pyne, *World Fire: The Culture of Fire on Earth*; Pyne, *Vestal Fire: An Environmental History, Told through Fire, of Europe and Europe's Encounter with the World*; Pyne, *Fire: A Brief History*; Pyne, *History with Fire in Its Eye: An Introduction to Fire in America*, http://www.nhc.rtp.nc.us/tserve/nattrans/ntuseland/essays/fire.htm; Steen, *The U.S. Forest Service: A History*; Williams, *Deforesting the Earth*.

PROTECTING THE WILD IN THE ADIRONDACKS

Time Period: 1890s to 1910s
In This Corner: Preservationists, many Adirondack residents
In the Other Corner: Developers, private land owners
Other Interested Parties: Consumers of water supply, such as residents of New York City
General Environmental Issue(s): Preservation, parks, wilderness

As the mechanization of American life increased at the close of the 1800s, it spurred a contrary reaction in American culture. For many Americans, the growing intensity with which nature was put to use made them increasingly concerned with the wilderness that remained untouched. In addition, many Americans who appreciated roller coasters and other new leisure activities began also to better appreciate the antitechnological, the natural world. Although this movement occurred throughout the nation, its nexus grew from metropolitan New York and reached directly into one of the most accessible wild areas: the Adirondacks.

In the late 1800s, the Adirondacks became the centerpoint in an intellectual reevaluation of nature's role in American life. Part of this was chance: with its proximity to New York City, the Adirondacks served as one of the most accessible vestiges of raw nature for urbanites. It was these upper-middle-class city dwellers who, by the 1890s, had created a reactionary cultural ripple to the massive use of technology seen throughout society at large. A major portion of this reaction was a rejection of mechanical progress and a new celebration of nature in its rawest forms.

The Adirondacks became the setting for many of these wealthy Americans to express their taste and civility by "roughing it," at least briefly. Often, their ideas derived from romantic and transcendental leanings. However, as the wealthy made retreating to nature trendy and tasteful, they also defined a grander vision than the tents and the hunting parties that occupied them in the 1870s and 1880s. By the late 1890s, the Adirondacks had reached its era of "Great Camps" (Terrie 1994, 10–14).

Distinctive to New York State, the Adirondack camps illustrate a style of architecture that was meant to mimic nature. Most often, these camps were located on vast tracts of forested land in the Adirondack Mountains. The camps allowed the elite of New York City to use and enjoy the region's lakes, streams, and forests. The interest of upper-class Americans in the outdoors emerged at the end of the 1800s, which also marks the establishment of many of the Great Camps. Although their location was remote, the camps often made leisure and amenities available in a rustic setting.

Architectural historians use three characteristics to define the Adirondack camp: a distinctive compound plan consisting of separate buildings for separate functions; the close integration of camp buildings with existing natural features; and a rustic aesthetic of decoration, design, and building.

Perfecting the American Idea of Preservation

Although unclear in its original ethic or motives, the preservation effort in the Adirondacks eventually became more formal. By the mid-1890s, New York had created a model of legislative discussion that would have national implications. The movement to designate a park in the Adirondacks was spurred by the writer Verplank Colvin, who wrote in 1885, "Had I my way, I would mark out a circle of a hundred miles in diameter, and throw around it the protecting aegis of the constitution. I would make it a forest forever. It would be a misdemeanor to chop down a tree and a felony to clear an acre within its boundaries" (Graham 1978, 70–78).

The effort to argue for a park in the Adirondacks did not make headway when it was based purely on romantic arguments about natural beauty. The key came when preservationists tied their argument to watershed preservation, especially that of New York City, and the state government paid attention (Nash 1982, 116–21). In short, preservation for its own sake was not attractive to nineteenth-century Americans. Ensuring good water supplies for the nation's most important urban area, however, was a tangible outcome that interested preservationists (many of whom lived in the city). For this purpose, all the Adirondack land owned by the state (approximately 681,000 acres) was designated a forest preserve in 1883–1885. To make it simpler to acquire additional land, Governor David B. Hill urged the legislature to create an Adirondack park in 1890 (Terrie 1994, 100–105).

The vote to establish a preserve came in 1892. The New York legislature voted to place a blue line on the map to denote the parts of the region that it hoped to acquire and include in the park. The total area covered more than 2.8 million acres. Many conservationists, however, immediately felt that the law was a mixed blessing. Although it created the park, it also weakened some earlier policies. The political winds continued to alter the park in 1893 when Governor Roswell P. Flower proposed a bill that authorized the park to sell trees from any part of the forest preserve. This idea, of course, undercut the whole idea for the forest preserve.

The reaction came at a convention held in 1894. Nearly halfway through the meeting, David McClure proposed what became known as the "Forever Wild" amendment. After witnessing the threat of the 1893 Cutting Law, McClure and others wanted to create a constitutional barrier against any similar efforts in the future. A committee convened to establish whether or not the Adirondacks merited amending the constitution. On the last day of the convention, the "Forever Wild" clause came to a vote. By a margin of 112 to 0, Article VII, Section 7 (which became Article XIV, Section 1 in 1938) was adopted into the New York State Constitution. Approved by New York state voters, on January 1, 1985, the new constitution went into effect (Terrie 1994, 95–100).

By 1900, the area of the forest preserve was more than 1.2 million acres. Today, the Adirondack Preserve is composed of nearly three million acres. It remains a primary recreation area for the northeastern United States and a critical mainstay for the preservation of wilderness.

Sources and Further Reading: Hays, *Conservation and the Gospel of Efficiency*; Nash, *Wilderness and the American Mind*; Price, *Flight Maps*; Reiger, *American Sportsmen and the Origins of Conservation*; Steinberg, *Down to Earth*; Runte, *National Parks: The American Experience*; Sellars, *Preserving*

Nature in the National Parks: A History; Terrie, *Forever Wild: A Cultural History of Wilderness in the Adirondacks.*

WORLD WAR I PERFECTS KILLING TECHNOLOGY

Time Period: 1910s
In This Corner: Soldiers, inventors, political leaders of the Allies
In the Other Corner: Soldiers, inventors, political leaders of the Germans
Other Interested Parties: 1920s consumers
General Environmental Issue(s): Technology, toxic waste, weapons and the environment

Swift changes in technology could be seen in many portions of American society in the early 1910s, from entertainment to transportation. The first real demonstration of the capabilities of new ideas and technologies, however, arrived when the great powers of the world squared off in the first large-scale war of the modern era. Critics argued that the battlefield was a place of honor that demanded each side to use consistent strategies and tactics. However, just as the modernizing technologies impacted American life and leisure, so did they change the way war was fought. The horror of it could be ghastly.

Events and agreements between nations brought Europe, the United States, and other nations to war in 1915. As each side used technology to attempt to win the war, World War I emerged as a transitional war in which new weapons were used in old forms of battle. The outcome was horrific trench warfare and unbelievable carnage. All totaled, approximately ten million soldiers died worldwide, as did an estimated fifty million civilians. Machines had been made to kill efficiently, just as they were also made to make car parts or stuff sausage casings.

The new era of killing technologies was defined by a gun that began with no use for hunting, the primary use of guns up to this point. Although versions of the machine gun had been used previously, World War I brought the machine gun into widespread application. The 1914 machine gun was quite difficult to use. Each gun had to be placed on a flat tripod and required a four- to six-man crew. Once in place, however, there had never been a weapon quite like it. Models at the start of World War I fired 400 to 600 rounds per minute; by the end of the war, this capacity had doubled. High technology came at a price, however. Machine guns rapidly overheated. Without a cooling mechanism, the gun quickly became unusable. For this reason, the early guns were used in short bursts and then allowed to cool. Strategically, armies worked around this limitation by grouping machine guns together and having them fire in shifts. Officials estimated that a properly positioned machine gun was worth approximately sixty to one hundred rifles. They began the war as a defensive weapon, but, during the war, they were adapted for use on tanks, aircraft, and warships by 1915.

New technologies also expressed the desperation of what was referred to as "trench warfare." Faced with lengthy stalemates created by armies holding out in underground bunkers and trenches, World War I armies used poison gas to force soldiers out from hiding (Russell 2001, 146–50). Chlorine was used by the Germans on April 22, 1915 at the start of the second battle of Ypres. Within seconds of inhalation, the gas destroyed soldiers' respiratory organs. The Germans' use of chlorine gas was criticized worldwide. Just as it condemned the Germans, however, Britain made plans to also put gas to use on the battlefield. The Allies' retaliation began later in 1915. By 1917, each army had turned primarily to the use of mustard gas.

Although many new technologies emerged in World War I, the strategic fighting of war remained primitive in many ways. The most striking of this was the use of animals for transportation and other tasks. Although horses had been important to many wars in human history, World War I put more to work than any previous conflict. The British army acquired approximately 200,000 horses at the start of the war and reportedly added approximately 15,000 per month throughout the war. These animals were brought together from the United Kingdom, South Africa, New Zealand, India, Spain, and Portugal. Horses and mules carried ammunition, artillery, guns, and shells, as well as soldiers. In World War I alone, around eight million horses, mules, and donkeys died.

Finally, possibly the most famous animal participant in World War I was the messenger pigeon. Although radios began to be used during the war, soldiers sent most messages by messenger pigeon. Possibly, this transition in information transfer best represents the watershed change in technology by the end of the war.

New technologies wrought more damage and death than in any previous war. The toleration of humans to live in a world with technologies capable of killing masses set the stage for later developments, including nuclear weapons. As Thomas Edison and others lent their technical abilities to the war effort, however, a crucial connection was permanently forged between technology, engineers, and the military (Hughes 1989, 96–100). Technical innovation became indelibly linked to national security by the end of World War I, whether it was honorable or not.

Sources and Further Reading: McNeil, *Something New Under the Sun: An Environmental History of the Twentieth-Century World*; Russell, *War and Nature: Fighting Humans and Insects with Chemicals from World War I to Silent Spring*; Russell and Tucker, *Natural Enemy, Natural Ally: Toward an Environmental History of War*.

TURNING TO FOSSIL FUELS AND AN ELECTRIC LIFE

Time Period: 1890s to 1910
In This Corner: Rockefeller and petroleum production lobby
In the Other Corner: Edison, Insull, and electricity pioneers
Other Interested Parties: Consumers, federal interests
General Environmental Issue(s): Energy, resource management, petroleum, electricity

The rapid technological innovations of the late nineteenth century came directly into American homes through changes in the methods for making light. Whale oil had been used as an illuminant in lamps throughout the nineteenth century. When petroleum emerged in significant supply during the 1860s, it became a viable replacement for whale oil in the form of kerosene. In fact, kerosene was so much more affordable that it made illumination much more widely available.

Petroleum's infrastructure grew, particularly under the influence of Rockefeller's Standard Oil Trust. However, by the 1890s, the pace of innovation moved so quickly that petroleum-based kerosene was about to be outmoded.

Electricity Becomes Vital

By the 1890s, humans accepted their ingenuity and its ability to remake their place in the world. Through engineering, Americans defied the limitations of previous eras. Both fueling and helping to cause this transition were new ways of viewing energy.

The change in energy use after the Civil War was an expression of a new culture of industry. In 1860, there were fewer than a million and a half factory workers in the country; by 1920, there were 8.5 million. In 1860, there were about 31,000 miles of railroad in the United States; by 1915, there were nearly 250,000 miles. Such infrastructure demanded energy to power it.

In the nineteenth century, energy defined industry and work in America but did not necessarily impact everyday cultural life. This would change dramatically by the end of the 1800s with the development of technology to create, distribute, and put to use electricity. Although electricity is the basis for a major U.S. energy industry, it is not an energy source. It is mostly generated from fossil fuel (coal, oil, natural gas), hydroelectric (water power), and nuclear power. The electric utilities industry includes a large and complex distribution system and, as such, is divided into transmission and distribution.

Following experiments in Europe, the the electric future of the United States fell to the mind of Thomas Edison, one of the nation's great inventors. In 1878, Joseph Swan, a British scientist, invented the incandescent filament lamp, and, within twelve months, Edison made a similar discovery in America. Edison used his direct current generator to provide electricity to light his laboratory and later to illuminate the first New York street to be lit by electric lamps in September 1882. From this point, George Westinghouse patented a motor for generating alternating current. Society became convinced that its future lay with alternating current generation. This, of course, required a level of infrastructural development that would enable the utility industry to have a dominant role over American life.

Once again, this need for infrastructural development also created a great business opportunity. George Insull went straight to the source of electric technology and ascertained the business connections that would be necessary for its development. In 1870, Insull became a secretary for George A. Gourand, one of Thomas Edison's agents in England. Then, he came to the United States in 1881 at age twenty-two to be Edison's personal secretary (Hughes 1989, 226–30).

By 1889, Insull became vice president of Edison General Electric Company in Schenectady, New York. When financier J. P. Morgan took over Edison's power companies in 1892, Insull was sent west to Chicago to become president of the struggling Chicago Edison Company. Under Insull's direction, Chicago Edison bought out all its competitors for a modest amount after the Panic of 1893. He then constructed a large central power plant along the Chicago River at Harrison Street. The modest steam-powered, electricity-generating operation would serve as Insull's springboard to a vast industrial power base.

By 1908, Insull's Commonwealth Edison Company made and distributed all of Chicago's power. Insull connected electricity with the concept of energy and also diversified into supplying gas. Then he pioneered the construction of systems of dispersing these energy sources into the countryside. The energy grid was born. It would prove to be the infrastructure behind each American life in the twentieth century. Through the application of this new technology, humans now could defy the limits of the sun and season (Hughes 1989, 234–40).

In the process of electricity's emergence, petroleum had become an illuminant of the past. Simultaneously, however, a new use had emerged for it.

Sources and Further Reading: Hughes, *American Genesis*; Nye, *Electrifying America: Social Meanings of a New Technology*; Nye, *Technological Sublime*.

FORAGING FOR FOOD IN A SOCIETY OF MASS CONSUMPTION

Time Period: Late 1800s
In This Corner: Food packagers, farmers
In the Other Corner: Farmers, some previously existing food suppliers
Other Interested Parties: Consumers
General Environmental Issue(s): Food, resource management

Labor-saving technologies influenced many aspects of American life in the nineteenth century, but none was more basic than the changes on what and how humans ate. As a species, resupplying our bodies is one of the most essential functions of our everyday life. Although many Americans marveled at innovations that helped to create food with more ease, others criticized these huge changes in everyday life.

Canning a Stew

In this new era, Americans learned a new comfort with technology and modern innovations across the board. Canned food, for instance, had been packaged in the United States since the early 1800s, but, during the late 1800s, many manufacturers took the process of canning to an industrial level. By 1863, Chicago had at least forty-five packing houses, several of which killed more than 100,000 hogs and cattle per year.

Canning was also used to create new markets for an old favorite: sodas. Soft drinks had been popular in the United States since the start of the nineteenth century, and soda fountains had become very popular by the 1890s. Often, drinking such sparkling beverages was considered healthful, and pharmacists made the drinks for their customers. Beginning with mineral water, many pharmacists started to add medicinal and other flavorful herbs to the unflavored beverage, including birch bark, dandelion, sarsaparilla, and fruit extracts. In the early drug stores sprouting in urban areas, the soda fountain was one of the most popular destinations. This arrangement, however, still required that customers travel to the soda fountain to enjoy sodas. Of course, customers wanted to take the drinks home with them, and this desire gave birth to the soft-drink bottling and canning industries.

One of the most important innovations in this process arrived in 1892 when William Painter, a Baltimore machine-shop operator, perfected the "crown cork bottle seal." The bottle seal was the first successful device for keeping the drink carbonated while it was in the bottle. Next, production of glass bottles surged in 1899 when the first patent was issued for a glass-blowing machine that could automatically make glass bottles. Soon glass-bottle production increased from 1,500 bottles a day to 57,000 bottles a day.

The most popular drink in America quickly became the artificially made Coca-Cola, which was originally invented by Dr. John Stith Pemberton in 1886. Using coca leaves and the kola nut as a basis, Pemberton created a drink to imitate French Coca-Wine. The recipe was bought and sold until the 1890s, when it came into the possession of Atlanta pharmacist Asa Candler. Candler increased syrup sales by over 4,000 percent between 1890 and 1900. He began massive ad campaigns and sold the drink across the United States and Canada. In addition, Coca-Cola began selling syrup to independent bottling companies that were licensed to sell the drink. By the turn of the century, Candler would become one of the wealthiest men in Atlanta, and Coca-Cola would become the most popular soft drink in America.

Such uses of technology brought impressive changes to Americans' everyday life. Primarily, canned and bottled foods and drinks enabled Americans a flexible lifestyle that would define the twentieth century. The list of new products in this era is remarkable, including:

1872, Blackjack chewing gum
1876, Premium soda crackers (later Saltines)
1881, Pillsbury flour
1886, Coca-Cola
1887, Ball-Mason jars
1888, Log Cabin syrup
1889, Aunt Jemima pancake mix
1889, Calumet Baking Powder
1889, McCormick Spices
1889, Pabst Brewing Company
1890, Knox gelatine
1890, Lipton tea
1891, Fig Newton
1891, Quaker Oats Company
1893, Cream of Wheat
1893, Juicy Fruit gum
1895, Triscuits
1896, Cracker Jack
1896, Michelob beer
1896, Tootsie Roll
1897, Campbell's condensed tomato soup
1897, Grape Nuts
1897, Jell-O
1898, Nabisco graham crackers
1898, Shredded wheat cereal
1899, Wesson oil

Were Americans better off because of Jell-O? Gathering food had certainly become more simple.

Making Meat for the Masses

In terms of food production, meats faced particularly significant changes at the end of the nineteenth century. Conceived by George Henry Hammond and Gustavus F. Swift, the refrigerated train car had widespread implications for the nature of American eating. Up to this point, meat had been available from local butchers or private livestock supply. Preservation techniques such as salting, pickling, or smoking were considered to be technological advances earlier in the 1800s. The first refrigerated railroad car was used in 1867. When Swift began the widespread use of the refrigerated cars in the 1880s, however, he enabled meat from the Chicago stockyards to reach markets throughout the nation.

Chicago's Union Stockyards had been constructed in the 1860s. By the late 1880s, Chicago's meat packers, particularly Swift and Philip D. Armour, controlled the nation's market. Their specialty, however, was not fresh meat; instead, they popularized processed meat. By mixing gelatin or other materials with chopped meats, these entrepreneurs made them capable of being packaged into cans. For the first time, meat could be sold in a form that would not spoil.

Ironically, the use of the refrigerated car in the late 1800s made fresh meat also able to be more easily transported and sold throughout the nation. With the use of the refrigerated car, however, the Union Stockyards by 1883 processed 1.9 million cattle and more than 5.6 million pigs annually.

Together, these innovations meant that there were new, large-scale markets for meat that entirely changed the history of the American West. The stockyards, disassembly lines, railroads, cattle drives, and cowboys became vital cogs in feeding the nation. Each acre of range land in the West now could be profitable if it hosted or fed animals for market. Cattle, but especially pigs, became the most profitable way for farmers to transform fields of corn into cash.

Foraging in the Cities and Suburbs

Packaged food altered one of the most basic portions of American life. As the new century unfolded, Americans would be more liberated than ever from the need to raise their own food. In addition, they would readily be freed from the support network of urban butchers, bakers, and other specialized preparers of food. Packaged food in general stores and eventually in grocery stores would provide people with the freedom to live outside of urban areas without having their own farms. This is one of the foundational planks of suburbanization.

The first grocery stores, of course, began as small, privately owned markets in the early 1800s. By 1850, the Great American Tea Company opened a store in New York City and began selling tea, coffee, and spices at inexpensive prices. The stores spread throughout New York and into New England. In 1870, the company was renamed the Great Atlantic and Pacific Tea Company (A & P) in honor of the first transcontinental railroad. The name, of course, belied the company's hope to expand across the continent.

By the end of the 1800s, A & P grocery stores appeared throughout the United States. In 1880, A & P introduced the first private-label product, baking powder. From that point forward, A & P specialized in its own privately manufactured products. From this base, the grocery industry revolutionized American foodways. Gradually, transportation took on a primary role in structuring the patterns by which Americans acquired the food that they needed to survive.

Although bicycles and streetcars became the nation's most popular forms of transportation in the 1890s, it was in 1896 that the Duryea Motor Wagon Company of Springfield, Massachusetts, sold thirteen identical gasoline-powered vehicles. Once automobiles became available on a mass scale, grocery markets were liberated from urban merchants and began to spread out of the cities into the suburbs that were springing up by the 1920s.

Conclusion: Drive-Thru Nation

By prioritizing speed and ease of food availability, the slippery slope to the "fast-food nation" of the twenty-first century had begun. Just as grocery stores made themselves more available to automobile consumers, restaurants were soon to follow. As cars became more familiar in everyday Americans' lives, planners and developers formalized refueling stations for the human drivers as well.

Food stands informally provided refreshment during these early days, but soon restaurants were developed that used marketing strategies from the motel and petroleum industries. Diners and family restaurants sought prime locations along frequently traveled roads; however, these forms did not alter dining patterns significantly. In 1921, White Castle hamburgers combined the food stand with the restaurant to create a restaurant that could be put almost anywhere. Drive-in restaurants would evolve around the idea of quick service, often allowing drivers to remain in their automobile.

Fast food as a concept, of course, derives specifically from Ray Kroc and the McDonald's concept that he marketed out of California beginning in 1952. By the twenty-first century, McDonald's annual sales had topped $40 billion.

Was this convenience worth it? Most medical findings suggest that Americans are more obese today than they were in the late nineteenth century.

Sources and Further Reading: Hughes, *American Genesis*; Pollan, *Omnivore's Dilemma*; Schlosser, *Fast-Food Nation*; Trachtenberg, *The Incorporation of America*.

JUDGING THE HUMAN SPECIES: FROM SOCIAL DARWINISM TO THE MOVEMENT FOR CIVIL RIGHTS

Time Period: 1870s to 1960s
In This Corner: Reform-minded Civil Rights workers, members of minority populations
In the Other Corner: American norms, white supremacists
Other Interested Parties: Politicians, government officials
General Environmental Issue(s): Interpreting the human species

Bias and bigotry have played an important role in American history, although we may wish this was not so. As humans have attempted to make sense of their role in the world and within their own species, there have been glitches. At times, convictions of quasi-science or religion have resulted in institutionalized bias and bigotry.

Most often, variations in the view of other humans have begun with physical distinctions that exist between humans. From this starting point, however, different cultural rationale takes over to possibly form bias. The basis of such bias or bigotry has most often been one's perception of a specific race's particular strengths and weaknesses, particularly its intellectual capabilities. Although it is invalid to base generalizations about individual capabilities into the logic of race, such opinions can often form very firm convictions for many individuals. For many people, opinions based in biases serve as a subconscious way of establishing their own standing in society. In American history, though, such bias was also, at times, the rationale for offering aid and assistance to minority races.

Class and Social Darwinism

The starting point for exploring bias and bigotry in U.S. history is the concept of social Darwinism. In the late 1800s, well-intentioned wealthy Americans sought to offer assistance to groups in need. The massive influx of immigrants at the turn of the century forced reformers to make assumptions about specific ethnic groups. Reformers had to make the assumption that the poverty stricken were somehow morally and, by extension, civically deficient. During the Gilded Age, this concept had combined with the popularity of Charles Darwin's theories of survival of the fittest and Edward Spenser's translation of these ideas into the social realm. "Common to almost all the reformers ... was the conviction—explicit or implicit—that the city, although obviously different from the village ... should nevertheless replicate the moral order of the village. City dwellers, they believed, must somehow be brought to perceive themselves as members of cohesive communities knit together by shared moral and social values" (Boyer 1994, vii).

These beliefs were given quasi-scientific status (particularly among upper-class Americans) under the term "Social Darwinism." Although they applied the concepts to early 1900s American life, social Darwinists legitimized social inequality by reaching back to the writings of Herbert Spencer. In *Progress: Its Law and Cause* (1857), Spencer wrote the following:

> ... this law of organic progress is the law of all progress. Whether it be in the development of the Earth, in the development of Life upon its surface, the development of Society, of Government, ... this same evolution of the simple into the complex, through a process of continuous differentiation, holds throughout.

While such notions of unidirectional progress are not identical to those described in Darwin's *Origin of Species*, many Americans made a connection. Particularly among the wealthy white Americans who had benefited from the economic disparity of the Gilded Age, social Darwinism helped to assuage any guilt that they might feel about their opulence. In the social form of Darwinism, different types of humans (cultures and ethnicities) were more fit to succeed than others. The less desirable types of humans, went the dangerous logic, including native peoples and Africans, who were less capable of succeeding in society and should be given menial labor tasks to perform. Such loose social rankings were also applied to white-skinned, European immigrants based on their country of origin. This allowed informal, and at times formal, ranking systems for those hiring factory workers.

In an era when technological change occurred so rapidly, many social ideas lagged behind modern sensibilities. In fact, examples such as social Darwinism demonstrate a particularly alarming misuse of scientific knowledge to perpetuate archaic ideas. Ideas of racial difference are a primary example of this.

Chinese Exclusion Act

One of the first examples of these ideas being transferred into policy concerned labor supplies in California. In the spring of 1882, Congress passed one of the first openly racist pieces of legislation in its history. After Californians had grown more and more convinced that Asian immigrants were taking too many of the available jobs, the Chinese Exclusion Act provided an absolute ten-year moratorium on Chinese labor immigration. For the first time, federal law proscribed entry of an ethnic working group on the premise that it endangered the good order of certain localities.

The Chinese Exclusion Act required the few nonlaborers who sought entry to the United States to obtain certification from the Chinese government that they were qualified to immigrate, but this group found it increasingly difficult to prove that they were not laborers because the 1882 act defined excludable as "skilled and unskilled laborers and Chinese employed in mining." With this law in place, very few Chinese could enter the country.

In addition, the act placed new requirements on Chinese who had already entered the country. If they left the United States, they had to obtain certifications to reenter. Congress, moreover, refused state and federal courts the right to grant citizenship to Chinese resident aliens, although these courts could still deport them. In 1892, the Geary Act extended these terms for ten more years. The Geary Act regulated Chinese immigration until the 1920s.

Formalizing Racial Distinctions: *Plessy v. Ferguson*

Although there were regional examples of this such as the Chinese Exclusion Act, national biases were most obvious in the area of race. In conceiving of racial difference, most Americans remained committed to stereotypes about mental inabilities of different types of humans, particularly African Americans. For African Americans, the policies often took the form of local ordinances or state laws. In the South after the Civil War, these Black Codes or grandfather laws were used to construct a line of separation between the races. After Reconstruction, the nation continued in an era of segregation when the separation of races was simply an accepted part of life. Of course, at the root of segregation lay the nation's very ideas of the nature of racial differences within the human race.

Although countless examples were available to demonstrate that African Americans had equal or better abilities and talents than many white Americans, the institutionalization of bigotry reached a high point at the close of the nineteenth century. In 1896, the Supreme Court ruling *Plessy v. Ferguson* showed just how deeply ran the biases of the era. In an era of such wildly innovative new ideas, many of the attitudes toward racial and ethnic differences remained primitive.

After Reconstruction ended in 1876, every southern state enacted Jim Crow laws that mimicked the Black Codes that had been enacted immediately after the Civil War. In an 1878 case, the Supreme Court ruled that the states could not prohibit segregation on common carriers, such as railroads, streetcars, or steamboats. Twelve years later, it approved a Mississippi statute requiring segregation on intrastate carriers. In doing so, it acquiesced in the South's solution to race relations.

In *Plessy v. Ferguson*, Justice Billings Brown asserted that distinctions based on race ran afoul of neither the Thirteenth nor Fourteenth Amendments, which were two of the Civil War amendments passed to abolish slavery and secure the legal rights of the former slaves. The Supreme Court's decision essentially argued that separate could be equal.

Educating African Americans at the Tuskegee Institute

Working within the confines of segregation, there were some great things going on in the South and elsewhere. For instance, after Reconstruction, Booker T. Washington worked against many of these stereotypes to create the Tuskegee Institute in Mississippi. In a

The meeting of a history class at Tuskegee Institute, Tuskegee, Alabama, in 1902 demonstrates that Washington's school was not exclusively vocational in its intent. Library of Congress.

post-Reconstruction era marked by growing segregation and disfranchisement of blacks, this spirit was based on what realistically might be achieved in that time and place. "The opportunity to earn a dollar in a factory just now," he observed, "is worth infinitely more than the opportunity to spend a dollar in an opera house." Prioritizing marketable skills, Tuskegee instructed its students in trades, including bricklaying, farming, home building, and small-scale manufacturing.

Tuskegee prospered as it did in part because Washington won widespread support in both the North and South. His efforts tapped into many Americans' growing willingness to view the world in a modern way, unfettered by social Darwinism. By the time of Washington's death in 1915, Tuskegee had become an internationally famous institution. The main campus has since grown to include 161 buildings on 268 acres and an academic community of nearly 5,000 students, faculty, and staff.

Tuskegee's success was not entirely greeted with acclaim. Many felt that vocational training for blacks would tend to keep them in a subordinate role. Instead, critics argued for a greater emphasis on traditional higher education. Most notably, the sociologist and writer W. E. B. Du Bois argued for African Americans to foster their own "talented tenth" of artists and creative people. Although each side in this debate recognized the need for both kinds of education, the concern was with the disproportionate emphasis on vocational training that Washington's approach and Tuskegee's popular success were fostering. Growing racial discrimination heightened the urgency of the debate. Although Washington combated racial injustice behind the scenes, his critics knew little or nothing of his activity and criticized what they saw as inaction.

At least one of Tuskegee's faculty members, however, used his expertise to reach out and to transcend the color line on a national level. Born a slave of Moses and Susan Carver,

George Washington Carver had an unwavering interest in science and natural curiosity. After working his way through school and earning a master's degree in agriculture from Iowa Agricultural College (later Iowa State University), Carver chose to commit himself to educating other black southerners. There was, of course, no better place to do so than at Washington's Tuskegee Institute.

While heading Tuskegee's Department of Agriculture, Carver conducted research that he hoped would provide African Americans with a unique niche in southern agriculture. He published bulletins and gave demonstrations on using native clays for paints, increasing soil fertility without commercial fertilizers, and growing alternative crops along with the ubiquitous cotton. These alternative crops included cow peas, sweet potatoes, and peanuts. Particularly with peanuts, what he called goobers, Carver attempted to develop new uses that would make them more enticing to grow.

Carver's work with peanuts drew the attention of a national growers' association, which invited him to testify at congressional tariff hearings in 1921. That testimony as well as several honors brought national publicity to the "Peanut Man." A wide variety of groups adopted the professor as a symbol of their causes, including religious groups, New South boosters, segregationists, and those working to improve race relations.

Carver revolutionized the southern agricultural economy by showing that 300 products could be derived from the peanut. By 1938, peanuts had become a $200 million industry and a chief product of Alabama. Carver also demonstrated that one hundred different products could be derived from the sweet potato.

Conclusion: The Civil Rights Era

Despite the strong statement made by *Plessy v. Ferguson*, the 1890s had bred a spirit of reform that would continue to grow during the twentieth century. The ideas of Jacob Riis, Jane Addams, Ida Tarbell, and others would grow into Progressive reform in the first decades of the 1900s and then into full-blown social movements by the 1960s.

These reformers marked a crucial point in the nature of America's view of its own residents. Was the human body a resource to be mined and exploited? Or did every type of person of every economic class merit a certain standard of treatment and concern? Slowly, these concepts were given scientific verification and political support. By the late 1950s, the great American movement for Civil Rights demanded federal laws that would ensure the rights of all Americans and press back the influence of bias and bigotry. Realizing the vision of equality that was expressed in the founding ideas of the United States, Civil Rights legislation of the mid-1960s ensured that basic rights such as voting would not be influenced by racial and ethnic differences.

Sources and Further Reading: McNeil, *Something New Under the Sun: An Environmental History of the Twentieth-Century World*; Opie, *Nature's Nation*; Steinberg, *Down to Earth*.

DISCOVERING ALASKA

Time Period: 1870s to late 1900s
In This Corner: Alaskan residents, proponents of development
In the Other Corner: Americans fearing overexpansion, Americans in the lower 48, environmentalists

Other Interested Parties: Most Americans, petroleum companies, native Alaskans, political officials

General Environmental Issue(s): Preservation, wilderness, petroleum

Getting to know a strange place can have many stages, even when it is part of your own nation. In the case of Alaska, since it became part of the United States, it has been the focus of debate by Americans for various reasons. This is at least partly because, for most Americans, Alaska exists less as a "real" place and more as a symbol. This symbolic value has also taken shape over time.

What does Alaska mean to Americans living elsewhere? How has this changed over time? The answer to these questions shows that Alaska has been the focus of ongoing debate even before it was established as part of the United States. Rarely allowed to function with the autonomy of most states, Alaska has been pulled in a variety of directions over the last century. Functioning almost as a measuring stick, opinions and approaches to Alaska's use and management provide a way of understanding Americans' changing approaches to the environment.

Seward's Folly

As the Russian American Company found profits from fur sales dropping, the Russian government, embroiled in a number of conflicts in Europe, lost interest in Alaska. In 1859, the government authorized Edoard de Stoeckl, a Russian diplomat in the U.S. delegation, to broach the subject of selling Alaska to the United States.

William H. Seward, who was secretary of state in both the Lincoln and Johnson administrations, reached agreement in March 1867 to transfer Alaska to the United States in return for a payment of $7.2 million. This price worked out to approximately 0.025¢ per acre to acquire an area twice the size of Texas. Even so, when Americans considered the frozen, seemingly worthless state of much of the terrain in the new purchase, they dubbed it "Seward's Folly." Ultimately, the purchase brought to an end Russian efforts to expand trade and settlements to the Pacific coast of North America. Coordinated with the development of the U.S. Navy, the Alaskan purchase marked an important step in the United States's rise as a great power in the Asia-Pacific region.

Alaska had been a point of ongoing settlement efforts by Russia as early as 1725, when Russian Czar Peter the Great dispatched Vitus Bering to explore the Alaskan coast. It was natural resources and not settlement that empowered the Russians. As American settlers moved steadily westward, they found themselves competing with Russian explorers and traders. With neither the financial ability nor a nearby military to support large-scale settlement, Russia first offered to sell Alaska to the United States in 1859. After the delay of the American Civil War, Seward took the Russians up on their offer.

Few Americans thought about Alaska over the next few decades. Indeed, Alaska was governed under military, naval, or treasury rule. At other times, it appeared there was no one in charge of the area. Seeking a way to impose U.S. mining laws, the United States constituted a civil government in 1884. When a major gold deposit was discovered in the Yukon in 1896, Alaska became the gateway to the world-famous Klondike gold fields.

Given this beginning in American history, the growing interest in Alaska at the dawn of the twentieth century was a signal of larger changes in American attitudes toward nature, particularly toward the few remaining wilderness areas. In what would become a pattern, however, Alaska's wildness seemed threatened even in the late 1890s. By the end of the nine-teenth century, Alaska was rapidly becoming a resource for industrial development. The Gold Rush was in full swing, salmon canneries hummed round the clock, and fur seal rook-eries exported thousands of skins every year. Because of these new activities, many native Alaskan communities had collapsed. A few remaining villages became tourist attractions.

Alaska as the Last Frontier

In this context, scientists organized a few expeditions to explore and catalog this wilderness before it was lost forever. One of the best-known expeditions was organized by Edward Har-riman, the nation's most powerful railroad magnate. It included prominent scientists as well as John Burroughs, the best-selling nature writer of the era, and John Muir, a writer and leader of the emerging American conservation movement. Their boat, the *Elder*, left Seattle on May 31, 1899 and traveled almost 9,000 miles along the coasts of British Columbia and Alaska over the next two months. They made some fifty stops, sometimes brief visits that lasted an afternoon, sometimes longer excursions.

The businessman Harriman's decision to turn a wilderness trip into a serious exploration of the coast indicates the growing interest in gathering knowledge by a generation that feared that the last bastions of wildness were being lost. The expedition returned with more than one hundred trunks of specimens and more than 5,000 photographs and colored illustra-tions. The scientists produced thirteen volumes of data that took twelve years to compile. There were 8,000 insects, 344 of which had been unknown previously to scientists. The col-lections included thousands of shellfish, birds, and small mammals and even a small number of large mammal specimens (Nash 1982, 279–84).

Alaska Helps to Create a Preservation Ethic at the Century's Close

The overall interest in Alaska, however, showed a growing appreciation for science and an acknowledgement that human activity was causing the loss of something worthwhile. This same sentiment could be seen in efforts to preserve nature but also in new efforts to develop and maintain elements of nature, such as rivers. One example of this growing sentiment was the public interest in the writings of the travelers taking part in the Harriman Expedition. Here is an excerpt from *Travels in Alaska* by John Muir, Chapter X:

> Looking southward, a broad ice-sheet was seen extending in a gently undulating plain from the Pacific Fiord in the foreground to the horizon, dotted and ridged here and there with mountains which were as white as the snow-covered ice in which they were half, or more than half, submerged. Several of the great glaciers of the bay flow from this one grand fountain. It is an instructive example of a general glacier covering the hills and dales of a country that is not yet ready to be brought to the light of day—not only covering but creating a landscape with the features it is destined to have when, in the fullness of time, the fashioning ice-sheet shall be lifted by the sun, and the land become warm and fruitful. The view to the westward is bounded and almost filled by the glorious Fairweather Mountains, the highest among them springing aloft in sublime beauty to a height of nearly sixteen thousand feet, while from base to sum-mit every peak and spire and dividing ridge of all the mighty host was spotless white,

as if painted. It would seem that snow could never be made to lie on the steepest slopes and precipices unless plastered on when wet, and then frozen. But this snow could not have been wet. It must have been fixed by being driven and set in small particles like the storm-dust of drifts, which, when in this condition, is fixed not only on sheer cliffs, but in massive, overcurling cornices. Along the base of this majestic range sweeps the Pacific Glacier, fed by innumerable cascading tributaries, and discharging into the head of its fiord by two mouths only partly separated by the brow of an island rock about one thousand feet high, each nearly a mile wide.

Dancing down the mountain to camp, my mind glowing like the sunbeaten glaciers, I found the Indians seated around a good fire, entirely happy now that the farthest point of the journey was safely reached and the long, dark storm was cleared away. How hopefully, peacefully bright that night were the stars in the frosty sky, and how impressive was the thunder of the icebergs, rolling, swelling, reverberating through the solemn stillness! I was too happy to sleep....

In the evening, after witnessing the unveiling of the majestic peaks and glaciers and their baptism in the down-pouring sunbeams, it seemed inconceivable that nature could have anything finer to show us. Nevertheless, compared with what was to come the next morning, all that was as nothing. (Muir)

When Muir and the rest of the Harriman Expedition visited Alaska, it was just approximately twelve years after Seward purchased the area for the United States. Muir was the first to chip away at the Americans' view that this purchase was folly (Nash 1982, 281). He introduced Americans to this place and to a vision of nature that found value in places other than resources to be used by humans.

Through his writings in magazines such as *Century*, Muir literally introduced an ethic. His ideas were not entirely new; thinkers such as Thoreau had expressed them earlier in the century. However, Muir's viewpoint demonstrated that the idea of preservation and valuing nature for its own sake had endured the era of massive industrialization.

Alaska represents the ongoing nature of America's dynamic view of nature. Of course, contemporary Americans continue to use Alaska as a measuring stick for our values toward nature. The need for petroleum complicates the matter considerably. In Muir's time, however, Alaska represented possibly the most useless natural aesthetic. To paraphrase Frank Sinatra, "If you could appreciate nature's beauty there, you could appreciate it anywhere."

Through the Harriman Expedition, Americans learned about this strange place. Writing in *National Geographic* in 1901, one of the other members of the expedition, Henry Gannett, may have described the preservation mandate best when he wrote this of Alaska:

If you are old, go by all means, but if you are young, stay away until you grow older. The scenery of Alaska is so much grander than anything else of the kind in the world that, once beheld, all other scenery becomes flat and insipid. It is not well to dull one's capacity for such enjoyment by seeing the finest first. (Nash 1982, 283)

Conclusion: A Place Like No Other

The strategic importance of Alaska was finally recognized in World War II. Alaska became a state on January 3, 1959.

By the end of the twentieth century, Alaska had also become well known for its reserves of petroleum and natural gas. Even this image, however, was influenced by Alaska's role as the nation's bastion of wilderness. For instance, when oil companies and most state residents wanted to access petroleum in the frozen northern slope of Alaska, national concerns were voiced. Even after the attempt to compromise and construct the Trans-Alaskan Pipeline and to establish the Alaska National Wildlife Refuge (ANWR), the debate over further drilling extended into the twenty-first century.

Sources and Further Reading: Coates, *Trans-Alaskan Pipeline Controversy: Technology, Conservation, and the Frontier*; Muir, *Travels in Alaska*; Nash, *Wilderness and the American Mind*; Strohmeyer, *Extreme Conditions: Big Oil and the Transformation of Alaska*; Wheelwright, *Degrees of Disaster: Prince William Sound, How Nature Reels and Rebounds*; Yergin, *The Prize: The Epic Quest for Oil, Money & Power*; Library of Congress, *Treaty with Russia for the Purchase of Alaska*, http://www.loc.gov/rr/program/bib/ourdocs/Alaska.html; U.S. Department of State, *Purchase of Alaska, 1867*, http://www.state.gov/r/pa/ho/time/gp/17662.htm.

THE U.S. CORPS OF ENGINEERS CONSERVES NATION'S RIVERS AND COASTS

Time Period: 1870s to the present
In This Corner: Federal government, riparian engineers
In the Other Corner: Some state interests, environmentalists
Other Interested Parties: Citizens living in riparian zones
General Environmental Issue(s): Rivers, engineering

Particularly in the nineteenth century, riverways were considered to be vital portions of the national infrastructure. The security and stability of these trade corridors became one of the first activities carried out by the federal government to conserve natural resources. Despite Congress's continued reservations about federal involvement in infrastructural development, some of the first efforts dubbed "conservation" related to river management and control. Most of these efforts in the late nineteenth century were carried out by the federal Corps of Engineers (Hays 1999, 100–102).

When the Corps of Engineers was created by Congress in 1867, it was put in charge of public parks and monuments. By the end of the century, however, it had taken on many additional projects in the Washington, DC area, including the improvement of navigation on the Potomac River and its tributaries, the expansion of the local water-supply system, completion of the Washington Monument, and the construction and design of many government structures, including the Lincoln Memorial, the Library of Congress, and the Government Printing Office. The corps defined much of its future work, however, with its efforts to reclaim the swamps of the tidal basins along the Potomac. After the Civil War, the corps also became involved in nationwide flood control. The major emphasis was on large rivers such as the Mississippi, where floods impaired commerce. In 1879, Congress created the Mississippi River Commission, composed of seven people, three from the corps including the commission president, three from civilian life including at least two civil engineers, and one from the U.S. Coast and Geodetic Survey. Congress created the commission to ensure that the best advice from both the military and civilian communities was heard on the subject of improving the Mississippi River for navigation and flood control.

After much debate, the commission decided to rely principally on levees to protect the lower Mississippi Valley. Cooperating with local levee districts, the Mississippi River Commission oversaw the construction of many levees along the river. Later, this levee construction was supplemented with considerable dredging on the river. The commission also attempted to stop the erosion of banks by constructing willow mattresses.

Before the era of reclamation, the corps also got involved in western water issues. Beginning in 1893, the corps established a California Debris Commission that regulated the streams of California that had been devastated by the sediment washed into them from mining operations. Throughout the nation, however, riparian science and engineering were put to use to make rivers function better for the use of American commerce (Hays 1999, 91–95).

The other emphasis of early conservation ideas was rivers. These efforts, however, clearly stretched modern environmental interpretations of the term conservation even more than did forest management. Most typically, river management emphasized development of human communities and how best to control and limit their impact on rivers. Possibly no organization influenced nature in flood-prone areas more than the Army Corps of Engineers. Although the organization was most active in the mid-1900s, it has a long history that began in 1879 when Congress created the Mississippi River Commission. This group of engineers mixed advice from both the military and civilian communities to improve the Mississippi River for navigation and flood control (Opie 1998, 308–10).

Cooperating with local levee districts, the Mississippi River Commission oversaw the construction of many levees along the river. Later, dredging would become an additional duty to prolong the effectiveness of the levees. To slow or stop erosion along this portion of the river, commission used mattresses made from willow branches. Experiments with concrete mattresses eventually helped the corps to develop the articulated concrete revetment that has been used for several decades to protect the banks of the lower Mississippi River (Hays 1999, 96, 212).

Defining a Mandate: The 1927 Mississippi Flood

Repeated flooding along the Mississippi forced the corps to try a variety of new measures; however, the floods of 1912 and 1913 paled compared with what the region endured in 1927. The 1927 flood displaced at least 700,000 and permanently ended any hope of truly controlling the river. This event left a lasting imprint on American politics, society, and management strategies for the Mississippi and other U.S. rivers (Barry 1998, 13–17). Some planners and historians have argued that the federal effort to recover from the 1927 flood will provide a template for any effort to help the region recover after Hurricane Katrina in 2005. Ironically, however, a significant number of the decisions made as a response to the 1927 flood resulted in the flooding of New Orleans in 2005.

Until 1927, the U.S. Army Corps of Engineers bypassed secondary channels and outlets and attempted a heavy-handed effort to steer the river where it wanted. The primary mechanism was the levee. In August 1926, the Mississippi began rising, and, by January 1, 1927, the river had passed flood stage at Cairo, Illinois. It is estimated that the river remained in its flood stage for 153 consecutive days. The flood shattered most of the levee system from

Illinois to the Gulf of Mexico. The flood inundated approximately 27,000 square miles of land in the midwestern United States. The corps managed to keep New Orleans from flooding by dynamiting levees and creating intentional floods at other points. Until that time, the U.S. federal government left relief from natural disasters in local and private hands. The flood of 1927 was so severe, however, that the government was forced to step in, ushering in the subsequent era of increasing federal involvement in disaster relief and recovery.

The flood of 1927 was most disastrous in the lower Mississippi valley. An area of about 26,000 square miles was inundated. Levees were breached, and cities, towns, and farms were laid to waste. Crops were destroyed, and industries and transportation were left at a standstill. The relief effort was massive but uneven, with inequities largely falling along racial lines. Property damage amounted to about $1.5 billion at today's prices. More than 200 lives were lost and more than 600,000 people were displaced. Out of it grew the Flood Control Act of 1928, which committed the federal government to a definite program of flood control. This legislation authorized the Mississippi River and Tributaries Project, the nation's first comprehensive flood control and navigation act. With these new policies came a change in the strategy for federal flood control. With the old "levees only" policy definitely swept away, there gradually emerged the multifaceted structural approach that remains in place today (Barry 1998, 403–8).

Remaking the Mississippi

With the Army Corps of Engineers functioning as the nation's beavers, engineers set out to "solve" the problematic Mississippi River. The four major elements of the Mississippi River and Tributaries Project are as follows: levees for containing flood flows; floodways for the passage of excess flows past critical reaches of the Mississippi; channel improvement and stabilization for stabilizing the channel to provide an efficient navigation alignment, increase the flood-carrying capacity of the river, and protect the levees system; and tributary basin improvements for major drainage and for flood control, such as dams and reservoirs, pumping plants, auxiliary channels, and the like (Colten 2001, 84–86).

The Mississippi River levees are designed to protect the alluvial valley against projected floods by confining flow to the leveed channel, except where it enters the natural blackwater areas or is diverted purposely into the floodway areas. The main stem levee system spans 2,203 miles and comprises levees, floodwalls, and various control structures. This system reaches 1,607 miles along the Mississippi River and then along 596 miles of the Arkansas and Red Rivers. During the twentieth century, construction of the levees was financed by the federal government. Once constructed, however, the levees were maintained by local interests with government assistance during major floods. The U.S. Army Corps of Engineers is responsible for inspecting the system in cooperation with local authorities.

Conclusion: The Best Laid Plans

Communities from Cairo, Illinois, to New Madrid, Missouri, have grown around the massive banks and levees that are used to protect them from seasonal floodwaters. Their relationship with the Corps of Engineers was almost paternalistic: the assumption being that the corps would provide the technology and know-how to keep the rivers at bay and the trade corridors passable. In many communities, such as New Orleans, residents have grown to take these barriers for granted.

The failure of this protection during Hurricane Katrina in 2005 demonstrated many different lessons for residents of New Orleans and other areas. Separate from the impact of the hurricanes that frequent the Gulf Coast region, 2005 showed the fallibility of the efforts of the Corps of Engineers. When the system of levees securing residential areas such as New Orleans' 9th Ward failed, the corps received the blame. Currently, some residents debate whether or not the corps can be trusted to ensure their futures.

In regions such as New Orleans, however, the corps engineers the nearly impossible. Whereas this federal agency attempts to keep nature at bay, residential development often works directly against it. In the case of the Gulf Coast, scientific experts point to the rapid disappearance of wetlands as the primary culprit for the intensified impact of hurricanes such as Katrina. The wetlands of this region are the buffer that separates land from the fluctuations in water level that are part of life here. Many planners argue for efforts to protect the remaining wetlands and to find ways to recreate some of what has been taken. If they get the go ahead to do so, they will, no doubt, be working with the Corps of Engineers (Colten 2001, 100–101).

Sources and Further Reading: Barry, *Rising Tide*; Colten, *Transforming New Orleans and Its Environs*; Reuss, *Water Resources Administration in the United States: Policy, Practice, and Emerging Issues*.

RIO GRANDE/RIO BRAVO

Time Period: 1800s to the present
In This Corner: Texas Republic, United States
In the Other Corner: Mexico
Other Interested Parties: Farmers, ranchers
Major Environmental issue(s): Water use, watersheds

The fifth longest river in North America is known as the Rio Grande in the United States and Rio Bravo (officially Rio Bravo del Norte) in Mexico. This river, which begins in the San Juan Mountains in Colorado, flows for 1,885 miles (3,034 kilometers). The path of the river goes from Colorado, south through the middle of New Mexico, and then southeast around Texas, giving the name to Big Bend National Park. Along the way, the river forms nearly 1,300 miles of the United States–Mexico border before finally emptying into the Gulf of Mexico at the twin cities of Brownsville, Texas, and Matamoros, Mexico.

The Rio Grande has marked the boundary between the United States and Mexico, from El Paso, Texas/Ciudad Juarez, Chihuahua, to the Gulf of Mexico since 1848. Shifts in the river's channel led to numerous border disputes between the United States and Mexico, leading to canalization of parts of the river to stabilize the river itself. The most complicated of these border controversies, over the location of the border at El Paso, was settled in 1968 when the Rio Grande was diverted into a concrete channel.

Even before the dispute over the river's shifting channel, the Rio Grande has long been the center of controversy within the southwest desert. In the first half of the 1800s, the Texas Republic claimed the Rio Grande as the southern and western borders of its territory; when the United States annexed Texas in 1845, it too maintained the claim of these boundaries. Mexico did not agree with these borders, believing the boundary to be farther north and thus claiming the Rio Grande/Rio Bravo River as its own. This disagreement was one of the causes of the Mexican-American War. Eventually, the war was brought to a close with the Treaty of Guadalupe Hidalgo, which recognized the Rio Grande/Rio Bravo River as the international border.

In addition to border wars, the Rio Grande/Rio Bravo has also been the center of controversy concerning the use of the river water. The waters of the Rio Grande have long been important for farming in the arid southwest. Prehistoric cultures used the river via irrigation systems, and the practice continues today as the Rio Grande supports the region's commercially important agricultural sectors in both the United States and Mexico. After years of both cooperation and conflict over the Rio Grande/Rio Bravo's water, water from the river became internationally regulated in 1944 when the United States and Mexico signed the Treaty for the Utilization of Water of the Colorado and Tijuana Rivers and of the Rio Grande/Rio Bravo (The Water Treaty of 1944). This treaty allocates future distribution of the river's water between the two countries. The Rio Grande/Rio Bravo river water is, however, overappropriated. Because the treaty was drawn up during a wet period, the amount of water allotted to the United States and Mexico exceeds the normal flow of the river. This is particularly problematic during drought years when both countries often accuse the other of overusing the water resources.

In recent years, the water level in the Rio Grande/Rio Bravo has dropped substantially as a result of drought conditions and increased consumption by both municipal and agricultural users. The problem is so acute that, in the summer of 2001, the Rio Grande/Rio Bravo failed to reach the Gulf of Mexico. Although spring rains at times supply enough water to allow the river to reach the gulf, ecologists fear that, without strict water conservation measures, the Rio Grande may become extinct, causing problems for citizens of the United States and Mexico, along with wildlife that depends on the river water for existence.

Sources and Further Reading: Burke, *Mestizo Democracy: The Politics of Crossing Borders*; Eaton and Anderson, *The State of the Rio Grande/Rio Bravo: A Study of Water Resource Issues along the Texas/Mexico Border*; Horgan, *Great River: The Rio Grande in North American History*: Volume 1, *Indians and Spain*. Volume 2, *Mexico and the United States*; Peschard-Sverdrup. *U.S.–Mexico Transboundary Water Management: The Case of the Rio Grande/Rio Bravo*; Reid, *Rio Grande*; Rivera, *Acequia Culture: Water, Land, and Community in the Southwest*; Sixeas, "Saving the Rio Grande"; Taylor, *Bloody Valverde: A Civil War Battle on the Rio Grande, February 21, 1862*.

UNITED STATES–MEXICO BORDER

Time Period: 1819 to the present
In This Corner: United States
In the Other Corner: Mexico
Other Interested Parties: Tribal groups, environmentalists, human rights groups
General Environmental issue(s): Habitat reduction, pollution

The United States–Mexico border spans 1,954 miles (3,141 kilometers) from the Pacific Ocean at San Diego, California, and Tijuana, Baja California, to the Gulf of Mexico in Brownsville, Texas, and Matamoros, Tamaulipas (International Boundary and Water Commission 2007). Between these two points, the border passes through four states in the

United States (California, Arizona, New Mexico, and Texas) and six states in Mexico (Baja California, Sonora, Chihuahua, Coahuila, Nuevo Leon, and Tamaulipas). The border delineates the divide between eleven sets of twin cities, cities that meet along the border and share some resources. The twin cities include the following: San Diego-Tijuana; Calexico, California-Mexicali, Baja California; San Luis, Arizona-San Luis Rio Colorado, Sonora; Nogales, Arizona-Nogales; Sonora; Douglas, Arizona-Agua Prieta, Sonora; El Paso, Texas-Ciudad Juarez, Chihuahua; Eagle Pass, Texas-Piedras Negras, Coahuila; Del Rio, Texas-Ciudad Acuna, Coahuila; Loredo, Texas-Nuevo Laredo, Tamaulipas; McAllen, Texas-Reynosa, Tamaulipas; and Brownsville, Texas-Matamoros, Tamaulipas. The border cuts through the Sonoran desert and the Chihuahuan desert, crosses the Colorado River Delta, and follows the Rio Grande (Rio Bravo del Norte). Going though diverse ecosystems, drawing a line between two diverse countries, the United States–Mexico border is the line at which many diverse issues meet.

The boundary between today's United States and Mexico was initially set by the 1819 Adams-Onis Treaty, also called the Transcontinental Treaty of 1819. This document defined the border between the American territory and the Spanish-controlled colonial lands. The document set a definite border between Spanish lands and the Louisiana Purchase; Spain retained control of most of modern-day Texas, New Mexico, California, Nevada, Utah, Arizona, and parts of Wyoming and Colorado. This treaty was not ratified by the new republic of Mexico and the United States until 1831.

The border was changed, in the eyes of the United States, in 1845 at the end of the war between Texas and Mexico when Texas became the twenty-eighth state of the union. However, the formal change to include Texas as a part of the United States, in addition to California, Arizona, New Mexico, and parts of Colorado, Nevada, and Utah, came in 1848 with the Treaty of Guadalupe Hidalgo at the close of the Mexican-American War. With this treaty, Mexico gave up half of its territory and agreed to the Texas-Mexico border at the Rio Grande/Rio Bravo.

Final changes to the United States–Mexico border came in 1853 with the Gadsden Purchase. The Gadsden Purchase consisted of approximately 30,000 square miles of land south of the Gila River to El Paso and west to California. The United States paid $10 million for this land and the right to move the United States–Mexico border from the main fork of the Gila River to where it is today.

The United States–Mexico border is, by definition, a division between two countries. However, just how solid the division should be is a discussion that brings many diverse answers from many people with diverse points of view. Issues brought up in this discussion include trade, security, environmental concerns, and human rights.

The United States–Mexico border is one of the most, if not the most, frequently crossed international borders in the world. There are nearly thirty commercial border crossing points along the border, nine of which handle the majority of the trade and six points that manage the rail transport between the two countries. The additional points handle some commerce and many individuals crossing between the two countries. Trade between the United States and Mexico has always been important to the economy of both countries and has become even more important since the initiation of the North American Free Trade Agreement (NAFTA) in 1994. This agreement expanded trade and trade-related transportation between the countries, increasing the numbers of both people and goods that cross the border. This change has put more stress on the United States–Mexico border region both concerning security (i.e., who and what is crossing the borer) and environmentally (i.e., how is the border's sensitive desert ecosystem dealing with increasing congestion).

Security along the United States–Mexico border was always a concern but draws much more attention today than it did before 9-11. The U.S. Border Patrol along the United States–Mexico border was established in 1904. It has maintained a presence in the area, in greater and smaller numbers, since that time, providing protection to not only the people of the United States but also the people crossing the border and the natural resources in the area. Some argue that the job of the U.S. Border Patrol could be made easier with the construction of a wall to separate the two countries, an idea that has been posed and questioned frequently throughout the last decades.

In 1990, beginning with Operation Gatekeeper in California, Operation Hold-the-Line in Texas, and eventually Operation Safeguard in Arizona the United States began constructing barriers on the southern border with Mexico. These operations were designed to focus on the major metropolis areas along the border and to limit the number of illegal immigrants coming into the United States through these areas. Although the operations did limit the number of immigrants crossing into the major metropolis, the overall number of illegal immigrants did not drop. The construction of barriers along the United States–Mexico border pushed the illegal immigrants into other areas of the border, the so-called difficult lands, the areas in the desert that are difficult for people to travel through and are ecologically delicate.

The minimal success of constructing the barriers (they slowed illegal immigration but only in the areas where they existed), in addition to added security concerns stemming from 9-11, led some people in the United States to support an effort to construct a wall along the entire length of the United States–Mexico border, effectively sealing off the two countries. Supporters of the construction of a wall suggest it would make the United States more secure, it would save money by minimizing the services for immigrants, and it would lessen environmental damage within the border region. Critics of the border wall include the Mexican government, which believes the wall would be isolationist and detract from neighborly interactions, a variety of industries in the United States that rely on Mexican workers and thus want to maintain open channels of cooperation between the two countries, and numerous environmental and human rights organizations who believe the wall will damage shared ecosystems and create further civil rights problems for people of Mexican descent along the border (Immigrant Solidarity Network 2006). The governor of Texas, Rick Perry, also rejected the idea of building a barrier between the United States and Mexico, suggesting technology is a safer, less expensive, and less environmentally damaging way to monitor the border (Harris 2006).

Despite criticism about a border wall separating the United States and Mexico, in December 2005, the U.S. House of Representatives voted to construct a barrier along part of the United States–Mexico border. The House vote called for mandatory fencing along 698 miles of the border. The Senate's vote on the same issues, which occurred in May 2006, included a plan to construct a blockade along 860 miles of the border, including vehicle barriers and triple-layer fencing. In September 2006, "Secure Fence Act of 2006" was passed by the U.S. House of Representatives and confirmed by the Senate. The Senate authorized and partially funded the possible construction of nearly 700 miles of barrier between the United States and Mexico. President George W. Bush signed it in October 2006. Possibly because a poll found that most Americans would prefer more Border Patrol agents to a physical barrier (CNN 2006a), Secretary of Homeland Security Michael Chertoff stated that there would be an eight-month test of virtual fencing that would precede construction of the physical barrier.

Supporters of the act suggest that the barrier will drastically reduce illegal drug and arms smuggling and illegal immigration. Critics note that the wall would do extensive damage to sensitive ecosystems, would further damage relations between the United States and Mexico, and is not likely to stop illegal activities—the activity would simply find a new path into the United States. Native Americans are also opposed to the act and construction of a barrier between the two countries, because this would divide their historic lands. Critics also note that the border wall will not be subject to any laws because in the Real ID Act of 2005 (H.R. 1268, "Emergency Supplemental Appropriations Act for Defense, the Global War on Terror, and Tsunami Relief, 2005"), a rider to a supplemental appropriations bill funding the wars in Iraq and Afghanistan, stated that "Notwithstanding any other provision of law, the Secretary of Homeland Security shall have the authority to waive all legal requirements … determine[d] necessary to ensure expeditious construction of the barriers and roads" (Real ID Act of 2005). Additionally, the Real ID Act states that the secretary of Homeland Security's decisions are not subject to judicial review. Secretary of Homeland Security Michael Chertoff used this provision to waive the Endangered Species Act (ESA), the Migratory Bird Treaty Act, the National Environmental Policy Act (NEPA), the Coastal Zone Management Act, the Clean Water Act, the Clean Air Act, and the National Historic Preservation Act to extend triple fencing through the Tijuana River National Estuarine Research Reserve near San Diego.

There is consistent controversy surrounding the United States–Mexico border. The issues of importance on the border, sovereignty, security, environmental sustainability, and human rights, are all complex issues that will not be easily resolved. Furthermore, main issues of concern along the border are often set up as incompatible with one another. Thus, the combined effect of problems with the United States–Mexico border is one that will not be easily resolved.

Sources and Further Reading: Andreas, *Border Games: Policing the U.S.–Mexico Divide*; Clough-Riquelme and Bringas, *Equity and Sustainable Development: Reflections from the US–Mexico Border*; Herzog, *Shared Space: Rethinking the U.S.–Mexico Border Environment*; Immigrant Solidarity Network, 2006; International Boundary and Water Commission, "International Boundary & Water Commission: United States and Mexico, United States Section"; Lorey, *The U.S.–Mexican Border in the Twentieth Century*; Milligan, "US Senate Passes Bill to Build Mexican Border Fence"; Rush, *Annexing Mexico: Solving the Border Problem through Annexation and Assimilation*.

ENERGY DEVELOPMENT ON THE PUBLIC LANDS

Time Period: 1920 to the present
In This Corner: Energy companies
In the Other Corner: Land owners, environmentalists
Other Interested Parties: Consumers, political officials
General Environmental Issue(s): Energy, resource management, public lands

When Woody Guthrie wrote "This Land Is Your Land," his rough-toned voice instructed Americans to recall that, "This land was made for you and me." Nowhere is this American ideal more obvious than in the nation's publicly owned lands, or so it would seem, yet the spirit of Guthrie's idealism has rarely penetrated the administration of the more than 600 million acres of land that belong to each citizen. One-third of all national land is administered by the federal government and owned collectively by the people of the nation. For a variety of reasons, including climate and the late entry of Euro-American settlement, most of the federal land can be found in the American West.

The public lands include parks, monuments, wild areas, refuges, underground mineral reserves, marine sanctuaries, historic parks, forests, and seashores. Throughout American history, the administration of this great national resource has been tied to politics and the powerful elements of economic development. The use of these sites and particularly the harvest of natural resources existing on them have been consistently debated in recent years. Should these lands be viewed as vast storehouses of resources on which our nation depends? Or are federal lands intended more for preservation, regardless of valuable resources that might be found within their borders? And, of course, when development takes place, who should gain the financial benefit?

As a matter of existing law, mineral resources on the public lands exist as national property primarily as a result of specific strategies to foster development in the western United States. Outside variables, including the needs of war or domestic crisis or even presidential preference, influence the perceptions of mineral resources, particularly those related to energy production. Although the use and availability of such resources varies with the philosophy of each presidential administration, each American leader has viewed these energy sources as an important tool to economically develop communities in the western United States. The harvest of such resources carries economic benefits locally and nationally.

Although wood must be included as an energy resource that was harvested from federal lands, the major sources in the twentieth century have included coal, oil, natural gas, shale, uranium, and geothermal. In recent years, there has additionally been a concerted effort to develop alternative energy resources (such as solar and wind) on public lands. By the end of the twentieth century, the energy industry estimated that Americans received thirty percent of their fuels for energy production from public lands.

Energy for All?

The remaining oil, natural gas, and mineral deposits in the continental United States are concentrated in the American West, where they were created between forty and a few hundred million years ago. The abundance of such resources on public lands is a historical confluence of variables: lands that were unsettled and therefore taken by the federal government happen to have been the site, thousands of years ago, of ecosystems that today result in abundant energy resources. Western sites such as Fossil Butte National Monument (Wyoming) and Dinosaur National Monument (Utah) contain the fossilized remains that demonstrate abundant life many years ago, both in and out of the sea. Of course, as these ancient plants and animals decomposed over millions of years, extreme heat and pressure transformed them into resources that can be mined and burned to provide energy.

Such nonrenewable sources are just a fraction of the available energy resources. Today, the open space available in much of the federal lands has also become a resource for the development of another type of energy: renewable, including wind, solar, and water power. The U.S. Department of the Interior oversees the leasing of this development, including mining. Within the Department of the Interior, the Bureau of Land Management (BLM) manages the development of all fossil fuels and minerals on nearly all of the land, excluding areas of special jurisdiction, including Naval Petroleum Reserves and hydroelectric watershed development areas that are administered by the Army Corps of Engineers.

The sheer amount of energy resources on federal lands is staggering. Of the 75.6 million acres of coal-bearing public lands, only 1 percent is currently being mined. As for oil and gas, energy companies are actively working between 18,000 and 25,000 wells on public lands, with tens of thousands of more sites held in reserve. A rarer mineral, uranium, is processed to create nuclear energy, which provides electricity to millions of homes, schools, and businesses. About one-third of all oil and gas comes from public lands. In addition, public lands are estimated to contain about 68 percent of all undiscovered oil reserves and 74 percent of all undiscovered gas reserves in our country. Nearly all reserves of uranium are found on federal land. Public lands are also used for harvesting hydroelectric power, although these areas are administered by the Army Corps of Engineers.

The story of the harvest and administration of this energy supply, however, reveals a theme that recurs in many publicly owned areas: public debate and contest. The use and management of the energy resources on public lands demonstrates important changes in the ethics guiding the management of federal lands. In addition, however, a survey of public lands' use demonstrates the primary components of a public controversy that promises to rage deep into the twenty-first century.

General Phases of Public Lands Administration

Broader philosophical ideas, called ethics, toward land use inform the administration of public lands. For the purposes of this essay, the administration of public lands can be separated into four distinct eras or shifts in these larger, governing ethics. Such ethics, of course, are formed by a complex set of cultural, social, and political variables. As a publicly administered entity, public lands are subject to the influence of larger patterns within American society. In the case of energy development on the public lands, the predominant cultural variable has been the ideological shifts caused by the evolution of the modern environmental movement in the twentieth century. The ethic of the environmental movement has provided a non-utilitarian impulse toward the administration of the public domain.

Although the overall progression of public land use has moved toward incorporating an environmental ethic into their general management, there have been a few crucial junctures in which the overall trajectory of this policy shift has been interrupted by a backlash of regressive efforts at development. After introducing the overall phases of public land administration, this essay will trace the general chronology of legal restrictions on public land policy, explore how the harvest of specific resources has been effected by these laws, and, finally, shed light on a few of these interruptions in the overall evolution of public lands policy.

First Phase: Settlement

This phase of land use grew out of the nation's emphasis on settling or reclaiming nearly 1.8 billion acres of original public domain. Hundreds of laws were passed between 1785 and

1878 that were designed to encourage settlement and development of the timber, mineral, and forage supplies. Typically, such policies said little about neither the native peoples who were being displaced by this disposal of land nor about the manner in which the lands were to be developed.

This phase of land use had little to do with energy development, because it predominantly concerned an era with little use of fossil fuels. Growing out of this approach to land use, however, Congress created the General Mining Law of 1872, which remains in effect to govern mining on public lands. The General Mining Law grants free access to individuals and corporations to prospect for minerals in public domain lands. Once they discover a location, they can stake a claim on the mineral deposit. This claim provides the holder with the right to develop the minerals and may be "patented" to convey full title to the claimant.

In this stage of land use, the federal government clearly prioritized the stimulation of new use over any other consideration.

Second Phase: Resource Conservation

Growing out of the American conservation movement of the late 1800s, this era in public lands use emphasized Progressive ideals, particularly those expressed by President Theodore Roosevelt (1901–1909) and his chief advisor on land management, Gifford Pinchot. This era marked the beginning of the nation's first period of public land stewardship. President William Henry Harrison began the forest reserve system with land appropriations before solidifying it with the Organic Act of 1897.

Roosevelt and Pinchot emphasized forest management, expanding the reserves to 148 million acres. Energy resources, other than wood, attracted the attention of Progressives with the Coal Lands withdrawal in 1906 that allowed the government to reserve mineral rights on sixty-six million acres. Initially, this reservation included oil and gas, although it would be further expanded to include many other minerals. Roosevelt also withdrew other public lands from settlement, using them to create national parks, monuments, wildlife refuges, and military bases. Within such sites, this era saw an increased control over the use of the resources on public lands, including legislation such as the Mineral Leasing Act of 1920 and Taylor Grazing Act of 1934.

Pinchot's conservation ethic was reflected in laws that promoted sustainable use of natural resources. These ideas were also of interest to FDR, who used the CCC during the 1930s to carry out many of these ideas on public lands.

Third Phase: Post-World War II

During World War II and for most of the two decades following it, the federal management of public lands went along with nationwide initiatives to stimulate national growth and international security. In 1946, the Grazing Service merged with the General Land Office to form the Bureau of Land Management (BLM), the entity that oversees the majority of the public lands in the western United States. The pressure to harvest resources from these lands intensified with the need for materials for war and the expansion of American consumption with the massive growth of its middle class.

With the beginnings of environmentalism in the late 1950s, many Americans grew more concerned that public lands were not subject to environmental assessment before use. This shift in overall ethics is generally accepted to have grown from the effort to prevent the damming of the Green River in Dinosaur National Park in 1954. Whereas the environmental movement influenced land use in national parks to a much greater degree, the public lands shifted more toward a compromise philosophy of "multiple use."

Fourth Phase: Environmental Protection

While this era proceeds with significant interruptions, the overall pattern from 1960 to the present has been toward a more environmentally sustainable pattern of use on the public lands. The Multiple Use Sustained Yield Act in 1960 was followed by the public outcry related to Rachel Carson's publication of *Silent Spring* in 1962. The ensuing flurry of environmental legislation culminated in 1969 with the National Environmental Protection Act (NEPA), which created the Environmental Protection Agency (EPA) and required that any development on federally owned lands needed to be preceded by an environmental impact statement (EIS).

Ecological concepts, such as ecosystem and watershed, provided administrators with a new way of viewing the public lands. No longer simply a storehouse of potential resources, each of these sites could now be seen as part of larger, complex natural systems. This awareness crystallized during a serious period of contestation over the use of the public lands in the late twentieth century.

Establishing the Legal Framework for Energy Mining

The basic tool for defining each of these eras in public land use has been laws or regulations. Throughout the history of public lands, the primary rationale for their administration was their usefulness. If settlers found reasons to build communities in the area, it is likely that the site would not have become public land in the fist place. Vast tracts of the territories that became states in the late 1800s were simply unwanted by settlers and developers. Similar to the Homestead Act, the General Mining Law of 1872 provided incentives to settlement and development.

Typical of legislation of this era, the General Mining Law emphasizes individual rights with minimal government management. The law prioritizes free and open access to the land and relatively unlimited opportunity to explore and develop resources. Developers were offered access to permanent, exclusive resource rights and exempted from paying royalties to the federal government. Clearly, the law is based on the simple motive to spur development.

The success of this General Mining Law's effort to stimulate western development relied on allowing open exploration. During the late nineteenth century, little was done to stop exploration from growing rapidly into mine development and full-blown extraction. In these early years, miners were not required to share production, create joint ventures, or pay taxes or royalties. Many mining expeditions proceeded largely unnoticed on public lands. Today, the Mining Law continues to provide the structure for much of the western mineral development on public domain lands.

The unfettered basis of use and ownership established by the 1872 Mining Law continues to fuel the debate over how to harvest energy and mineral resources from public lands. The basis for such law was the assumption that there was no alternative criteria for land's value other than the resources that it held. The basis for much of the debate at the end of the twentieth century revolved around a simple question: what would happen when, for a variety of reasons, this land became desirable?

These debates in the twentieth century slowed the unlimited development of the earlier years. Many of the public lands had acquired additional importance, independent of energy and mineral resources. Additionally, such development now rarely proceeded unnoticed. Some Americans began calling for energy companies to share revenues with the land's owner through royalties. Other observers reflected a growing call for more stringent oversight of the extraction process.

This call resulted in the 1920 Minerals Leasing Act. For the first time, this act closed some federal lands to open access. Most often, such lands were now available for leasing. The 1920 act established a broad framework for leasing and for the payment of royalties. Lands were to be designated as "known mineral areas" and then administered by one of two methods: Preference Right Leasing, which provided exploration permits that would expire if the search was unsuccessful, or Competitive Bidding for Designated Tracts, which used lotteries, oral auctions, sealed bids, or other formats to disperse the rights to known mineral areas and transfer them into known leasing areas.

Eventually, the Competitive Bidding Act would be made even more specific. For coal, the law differentiated between new or maintenance tracts but still required bids for existing operators to continue mining an existing site. Oil and gas development used reservoir studies to determine the pattern of leasing and lottery systems. In 1987, lottery systems were discontinued for oil and gas and development and replaced by a familiar pattern in which industry representatives nominated tracts, which were then assessed for their environmental acceptability.

Neither the 1872 Mining Law nor the 1920 Mineral Leasing Act contained any direct environmental controls, but mining claims are subject to all general environmental laws as a precondition for development. These restrictions and requirements grew out of additional legislation, particularly the seminal environmental policies of the 1960s and 1970s, including the Multiple Use Sustained Yield Act, Wilderness Act, National Forest Management Act, NEPA, and Federal Land Policy Management Act (FLPMA). To varying degrees, each of these policies addressed environmental protection, multiple use, and management of federal land generally. By imposing new requirements on agency actions and by withdrawing some federal lands from development, these acts have significantly influenced mineral development on public lands.

By far, the most critical policies to the administration related to the formation and structure of the BLM. When it was formed by merging two portions of the U.S. Department of the Interior, the General Land Office and the Grazing Service in 1946, the BLM became entrusted primarily with the management of the remaining public domain and railroad/wagon road grant lands. The BLM also has the primary responsibility for the management oversight of the mineral domain that underlies the public lands. As with the sister agencies, the BLM is heavily involved in forest and range land planning, as well as keeping the official land status records of all public federal lands.

Over time, the BLM followed the lead of the USDA Forest Service and adopted policies of multiple-use and, eventually, of ecosystem management. In many cases, BLM lands are adjacent to or intermingled with other public lands, particularly national forest land. The BLM essentially operated under these previously existing acts until congressional discussion over forest management in the Forest Service came to a head in the mid-1970s. This debate resulted in the passage of the FLPMA of 1976, which provided the BLM with its "organic act," albeit three decades after its founding.

FLPMA brought the greatest change to the administration of public lands in American history. FLPMA brought the BLM unequivocal statutory basis for its public land management policy; however, debate continued over what exactly this meant. The 1934 Taylor Act provided for management of natural resources on public land "pending its final disposal." To some observers, the phrase spoke to the old policy of transfer of public lands into private ownership; to others, it provided an argument against federal investment in land administration or rehabilitation. FLPMA, however, clearly prescribed retention of public lands in federal ownership as the rule and made transfer a carefully circumscribed exception. FLPMA proclaimed multiple use, sustained yield, and environmental protection as guiding principles for public land management.

FLPMA established or amended many land and resource management authorities, including provisions on federal land withdrawals, land acquisitions and exchanges, rights-of-way, advisory groups, range management, and the general organization and administration of the BLM and the public lands. FLPMA also called for public lands and their resources to be periodically and systematically inventoried. The BLM was empowered to use these data to create a plan for each site's present and future. Additionally, after many years of debate, FLPMA also specified that the United States must receive fair market value for the use of the public lands and their resources.

Finally, the BLM was required to administer public lands in a manner that protected the quality of scientific, scenic, historical, ecological, environmental, air and atmospheric, water resource, and archeological values. When appropriate, the BLM was also empowered to preserve and protect certain public lands in their natural condition, particularly when such preservation would provide food and habitat for fish and wildlife. Of course, the BLM also remained responsible for providing outdoor recreation and human occupancy and use of such sites.

In short, FLPMA proclaimed multiple use, sustained yield, and environmental protection as the guiding principles for public land management. Thanks to FLPMA, the BLM took on the responsibility of administering public lands so that they are used in the combination that will best meet the present and future needs of the American people for renewable and nonrenewable natural resources. Although this remains the role of the BLM today, there have been many exceptions to this jurisdiction.

Making Legal Exception: Native American Communities

There remain areas of federal lands that do not cohere to these legal regulations. For instance, it is reported that 80 percent of the nation's uranium reserves, 33 percent of the nation's low-sulfur coal, and 3–10 percent of the recoverable oil and gas are located on Native American reservation land. Problems of jurisdiction and ownership have highlighted the history of resource extraction on native lands. More important, however, are the impacts that have been noted where such development has been allowed to take place.

Outside sources approached reservations for energy resources as early as 1920. For instance, the Navajo's prolonged involvement with oil development began with the influence of New Mexico's own Secretary of the Interior Albert B. Fall (who will be discussed later for his pivotal role in the management of energy resources on federal land). With one of interior's charges being to administer Native American reservations, Fall soon looked toward reservation land as a vast, untapped (and unprotected) resource. Native peoples, Fall wrote, were "not qualified to make the most of their natural resources" (Stratton 1998). Using the General Leasing Act of 1920, Fall attempted to clarify the status of treaty lands for leasing so that the government could do it for them.

Whereas Fall's perspective grew from a desire to battle waste, public sentiment after World War I held that the United States would run out of oil within ten years. Fall saw to it that oil companies won leases on reservation land, particularly that of the Navajos in New Mexico. In addition, he conspired with oil companies to provide native occupants with ridiculously low royalty rates. Historian Kathleen Chamberlain notes that oil intensified factions within the native community. In summary, she wrote, "oil also expedited a shift from subsistence and barter to a wage economy" (Chamberlain 2000). Although oil generated some jobs for the community, it never neared the promised levels.

Throughout the history of energy development on native lands, the development has been found to undermine tribal and family values and identity. For instance, in Hopi and Navajo experiences, wealth from royalties was spread unevenly and the large influx of wealth disturbed traditional hierarchies within the culture. Additionally, the non-Indian workers who were necessary for reservation development of energy resources represented a persistent threat to tribal sovereignty.

Possibly the most invasive change that energy development has wrought on native groups is on the traditional political process. In nearly every case, the debate over leasing lands for extraction has bitterly divided tribal leadership, ultimately resulting in long-term impacts on the political structure of the group. With such breakdowns, the reservations become what scholars have called a "colonial reservation economy."

Some of the most troubling instances involved negotiations with the U.S. Federal Government. During World War II and through the Cold War, uranium was mined from Hopi and Navajo reservations in Arizona. The mine remnants as well as the remaining waste have contributed to significant rises in cancer and other health problems. Additionally, many native men worked in the mines and have therefore contracted a variety of ailments, particularly lung cancer.

Since the 1970s, native groups have attempted to use the courts to acquire fair royalty rates and recompense for health hazards. This has been an uphill battle, with various successes starting in the late 1980s. The native community's lessons about the community impact of energy development blossomed into an organization in the Council for Energy Tribes, which helps each tribe to complete independent management of its energy resources. This has helped many tribes to acquire fair royalty rates as well as to initiate extensive development of alternative sources on western reservation land.

Getting Energy from the Public Lands

The laws and ideas relating to energy development on public lands have changed because of the larger ethics of American society, as well as changes in the type of resource being extracted. Primarily, the pressure to mine energy resources grew as resources such as petroleum became more and more critical to American life and national security. The most significant change followed the increased use of petroleum for transportation after 1900. For instance, by World War I, the energy reserves on public land had been transformed from an attraction for potential settlement to a matter of national security. In World War I, fossil fuels became directly linked to the nation's ability to protect itself and maintain global security.

Petroleum, however, is only one example. This section of the essay will consider three specific energy resources with the time period that they most influenced.

The Beginnings with Coal

The first energy resource to be granted unique status on public lands was coal. The coal act of 1864 gave the president authority to sell coal beds or fields for no less than $20 per acre. Lands not sold in this manner could be available to the general public at the minimum price under the general land statutes. In 1873, Congress authorized citizens to take vacant coal lands of not more than 160 acres for a minimum of $10 per acre. This law governed the disposal of coal lands until 1920. Precious metals were governed much more freely. In all probability, the rationale behind this distinction lay in the relative ease of locating coal compared with that of most precious metals.

The importance of the coal supply to national development would make it a major political issue in the early twentieth century. In rapid succession, the Roosevelt administration proposed controlling the federal grazing lands, increasing the national forests, and withdrawing the coal, oil, and potash lands from sale. In this same period, the administration also reserved potential waterpower sites and set out to reclaim the arid lands of the West.

In 1906, Roosevelt faced up to the rumors of random exploitation of coal supplies on federal lands by withdrawing coal-bearing lands from public availability. He responded to the industry's complaint that the acreage limitations in the 1873 statute impeded the development of certain types of coal. As a result, many coal and railroad companies evaded these limits by using dummy entries or entries processed under the agricultural land statutes.

To keep certain companies from acquiring a monopoly and to conserve mineral supplies, the withdrawals began in November 1906. Located in Colorado, North Dakota, Montana, Oregon, Washington, Utah, Wyoming, and the Territories of New Mexico, sixty-six million acres of land with supplies of "workable coal" were withdrawn. Roosevelt explained his logic as follows:

> The present coal law limiting the individual to 160 acres puts a premium on fraud by making it impossible to develop certain types of fields.... It is a scandal to maintain laws which sound well, but which make fraud the key without which great natural resources must remain closed. The law should give individuals and corporations under proper government regulation and control ... the right to work bodies of coal large enough for profitable development. My own belief is that there should be provision for leasing coal, oil and gas rights under proper restrictions. (Robbins 1976)

Many western congressmen reacted angrily to Roosevelt's withdrawal. They claimed that Congress and not the chief executive possessed the right to dispose of the public domain in such a manner. Some referred to Roosevelt's use of conservation as "state socialism."

Roosevelt revised his argument slightly in 1907 to include mineral fuels with forests and navigable streams as "public utilities." Between 1906 and 1909, public coal lands sold for between $75 and $100 per acre. During these same years, more than four million acres were segregated though many would be returned to the public domain.

Contested Development: Petroleum and Teapot Dome

After the 1859 strike of the first oil well in Pennsylvania, development in the United States moved to the Midwest. As oil exploration moved westward, it became evident that the public lands held great stores of "black gold." By the 1870s, oil fever had swept specifically into the public domain of California. In the early 1890s, oil was discovered near Coalinga, California, and, in 1892, E. L. Doheny drilled the first successful well in Los Angeles. The Placer Act of 1870 included "all forms of deposit" except veins of quartz. Although the Department of the Interior debated whether or not petroleum qualified, Congress passed an 1897 Placer Act that upheld that oil lands were administered by placer location. In 1909, President Taft temporarily withdrew three million acres of potential oil lands in the public domain.

Similar to Roosevelt's coal withdrawals, Taft's act garnered a great amount of debate, particularly in the West. Congress then passed the Pickett Act in 1910 to allow presidents to withdraw public lands for examination and reclassification for various "public purposes." During the 1910s, this act enabled the president to withdraw nearly all of the known oil lands. To conserve an adequate supply of oil for the Navy, two naval reserves were established by Executive Order in 1912: in Elk Hills and Buena Vista Hills, California. Similar to a forest reserve, the oil supply, administrators believed, was best preserved if left in the ground. These withdrawals were then followed in 1915 by reserves at Teapot Dome, Wyoming, and in 1916 by the naval oil shale reserves in Colorado and Utah. In 1923, a similar reserve was created in Alaska.

Although most of the public debate centered on the right of the executive branch to make such withdrawals, other issues emerged as well, including whether the land should be sold or leased to oil companies, whether or not land previously patented to railroad companies could be withdrawn by the federal government as well, and whether or not oil companies had any standing if the federal government wanted to withdraw land that they had already leased. These issues, of course, took on greater importance as the United States observed and then participated in World War I at the end of the decade. Petroleum had many military uses, including powering many of the ships of the new Navy.

Historian John Ise was one of the loudest critics of such government policy. His argument, however, differed from most of the critics'. He wrote in 1926, "During this time there was always overproduction of oil from privately owned lands and there was never any need or justification for opening any public lands." Although this argument shifted to the fore by the end of the twentieth century, there was little public support for the idea of conservation. Western lawmakers focused on ensuring federal supplies of petroleum on public lands when they passed the Mineral Leasing Act of 1920, otherwise known as the Smoot Bill. Congress intended this act to bring relief to some of these issues and to offer more equal treatment to interests of the western United States. In the naval reserves, leases were given only on producing wells, unless the president chose to lease the remainder.

Within six months after the passage of the Mineral Leasing Act, the secretary of the Navy was granted extensive jurisdiction over the naval reserves. In 1921, Albert B. Fall, a senator from New Mexico, was appointed secretary of the interior by Warren G. Harding.

The Teapot Dome episode of the 1920s was a very public scandal because it involved Secretary of the Interior Albert Fall, who became the highest-ranking official to be sent to prison. Although the public cared little about the grander implications behind the scandal, Teapot Dome set the stage for a century of battling over how energy resources existing on public lands should be treated and who should benefit from their harvest. Library of Congress.

In short order, Fall convinced Secretary of the Navy Edwin Denby to relinquish control of the reserves and had Harding sign a secret presidential order to this effect. Within a few months, word leaked to the public that Fall had leased reserves in California and Wyoming to his friends in the oil business.

The episode that transpired in full view of the public became the primary test of jurisdiction and law regarding energy development and the West. In the scandal known as Teapot Dome, a congressional committee found that Fall had acted secretly and without competitive bidding to lease the Teapot Dome to Harry Sinclair and the California tract to E. L. Doheny. Allowing a private company to develop naval oil reserves, Fall went outside the boundaries of existing law. As a result, he was one of the first members of a presidential cabinet to serve a jail term, convicted of bribery and ordered to serve a year in prison, as well as to pay a $100,000 fine.

Fall represented the western interests who felt betrayed by the overzealous control of the federal government over the development of public lands. Before Warren Harding appointed him as secretary of the interior, Fall had earned his spurs in farming and ranching. In New Mexico, he owned a ranch that spanned more than fifty-five miles long and thirty-five miles wide. He had been a judge and a sheriff. Throughout his public life in New Mexico, Fall referred to the territory as "corporation country." He hated government interference in business, and he did not have to struggle with torn loyalties when he became a federal employee.

"I have long believed," Fall said in 1918, "that the laws should protect ... capital and should be so enforced as to offer an inducement for its investment." In giving business the advantage of governmental favors, Fall had, however, acted as a man of a certain faith. The core of his faith was simple: because nature's resources were unending, there was no need for humanity to restrain its ambitions and desires. "All natural resources should be made as easy of access as possible to the present generation," Fall declared. "Man cannot exhaust the resources of nature and never will." This, of course, included those resources falling within the public domain.

When Americans began to hear about Teapot Dome in news reports in the 1920s, much of the public participated in a growing appreciation of nature. Fall's feelings were well known to many observers. On his appointment in 1921, one newspaper wrote, "To us Fall smells of petroleum and we know what that means among ... politicians." Another paper, also before the Teapot Dome scandal, described the new administrator of federal lands as a man who found "wise conservation of natural resources ... hateful to his every instinct."

Although Fall would be held accountable for his abuse of power on the federal lands, his basic idea of unfettered development and unending abundance remains on the table. This has been a major portion of the debate over petroleum development on the federal lands. Although Congress would act in 1928 to give the secretary of the Navy exclusive jurisdiction of the petroleum reserves, the matter would continue to be debated throughout the twentieth century.

Energy and National Security: Uranium

Similar debate followed the post–World War II search for uranium to be used in nuclear weapons and eventually in nuclear power generation. The demand for fissionable source materials after the war and the urgency of a government monopoly in them prompted the enactment of the Atomic Energy Act of 1946, which reserved all such ores in public domain land to the United States and required that they also be reserved in any mineral patent.

The uranium boom reached its height around 1954, when prospectors flooded western lands in a similar manner to the California gold rush. Millions of dollars was made by private prospectors leasing and locating uranium supplies through the end of the 1950s.

Applicable to both uranium and oil and gas deposits, the Multiple Mineral Development Act of 1954 clarified how lessees were to deal with dormant mining claims that may cloud their own title. The new act provided a procedure in which title examiners may be used. After the lessee filed in the proper county office, he or she then waited for ninety days before filing with the BLM.

Debating Ownership: The Sagebrush Rebellion

Picking up on the ideas of Albert Fall, a group of eleven western states rallied in the 1970s to gain control of the development of the resources on their public lands. In what has become known as the "Sagebrush Rebellion" and the "Wise Use Movement," these groups have gained national attention and often have been able to bypass FLPMA and other government regulations.

In the late 1970s, the movement flared up with the support of Adolph Coors and other westerners. Prominent in this legal effort, the Rocky Mountain Defense Fund used talented lawyers James Watt and Ann Gorsuch. Before his election, Ronald Reagan espoused support for the Sagebrush Rebellion. When he was elected in 1980, Reagan wasted no time in appointing Watt the secretary of the interior and Gorsuch to head the EPA. Watt pressed his ideas forward and was able to open thousands of acres of public land to coal mining. However, with his forced resignation in 1983, the movement lost a great deal of its support from the White House. Reagan and his successor, George H. W. Bush, each demanded that the EPA become not an advocate agency but a "neutral broker," which might actually foster development. Personnel in the EPA dropped by twenty-five percent, and its budget was sliced by more than half. Policies and limitations were put in place to limit the number of lawsuits that could be brought, and most other EPA cases were mired in inactivity (Opie 1998, 448–49).

These ideas, however, have found new life in the administration of George W. Bush. Guided by leaders such as the Assistant Secretary of the Interior J. Steven Griles, who served under Watt and then spent years as a consultant to the energy industry, the Bush administration pushed to open up vast new tracts of federal land to oil, gas, and coal development.

Conclusion: Battlelines in the Current Debate

Thanks to FLPMA, most new energy extraction requires an EIS, which is a detailed study of the proposed plan and how it will affect wildlife, plant life, and the quality of the water, air, and soil. The EIS also examines how the resource extraction will affect other uses of the public lands, such as recreation, cultural resource conservation, or grazing. If the impacts of the proposed well, dam, or mine seem too great, then alternatives are examined. These range from scrapping the plan entirely to using advanced technology to extract the resource, such as directional drilling, in which the well is sited in a less vulnerable area but the drill still reaches the underground reserves.

The debate over ANWR crystallizes many of the issues that now make up the larger debate about how energy development should proceed on federal lands. During the administration of William Clinton, oil and gas leasing increased considerably over previous administrations but with new environmental restrictions on industries that sought to develop energy resources on public lands. Clinton also designated twenty-one western national monument areas and fifty-six million acres of roadless area. Such changes, however, were susceptible to any changes in national politics. That change came in the election of 2000.

"Environmental issues today," instructed Bush's Interior Secretary Gale Norton, "are more complex and subtle than the ones we faced in the 1960s and 70s. With the growth of our population and expansion of our economy, pressures have increased correspondingly on our undeveloped land, water resources, and wildlife." Norton intended to put "collaboration ahead of polarization, Markets before mandates … and to transcend political boundaries." To do so, she and George Bush coined "New Environmentalism" to meet the United States' need "for a vibrant economy and energy security—while at the same time protecting the environment."

Thus far, environmental organizations have been disappointed by Bush and Norton's "New Environmentalism." "We're dismayed that Pres. Bush wants to turn the oil and gas industry loose on virtually all the lands in our national forests, national wildlife refuges, and other public systems," said Wilderness Society representative Bill Meadows. "We're beginning to wonder if James Watt is whispering in his ear," he said.

In addition to ANWR, federal lands along the Rocky Mountain Front in northwestern Montana to southern Wyoming's Red Desert and Colorado's Vermillion Basin are being explored for energy development. The BLM collected data on oil and gas deposits and their availability throughout the public lands to support the pro-development stance of the Bush administration. Many local landowners and environmentalists have fought these efforts. Norton has requested new financing to streamline permits and study sites to drill on federal lands. The Bush administration has also shown some interest in increasing the development of alternative energy sources on federal lands as well. The future of the public lands is certain to contain open debate over the domain's role in America's energy future.

Sources and Further Reading: American Frontiers: A Public Lands Journey. "Energy from Public Lands," http://americanfrontiers.net/energy/Energy2.php; Carstensen, *The Public Lands*; *Studies in the History of the Public Domain*; Cody, CRS Report for Congress, "Major Federal Land Management Agencies: Management of Our Nation's Lands and Resources," http://www.ncseonline.org/ NLE/CRSreports/Natural/nrgen-3.cfm?&CFID=8734533&CFTOKEN=91528013; Gerard, *1872 Mining Law: Digging A Little Deeper*; Robbins, *Our Landed Heritage: The Public Domain, 1776– 1970*; Stratton. *Tempest over Teapot*; For Bush Energy plan, see whitehouse.gov/energy/; For Energy Policy and Conservation Act, see http://www.doi.gov/epca.

GRAZING RIGHTS IN THE WEST AND ON PUBLIC LANDS

Time Period: 1890 to the present
In This Corner: Ranchers, western land owners
In the Other Corner: Federal administrators, farmers
Other Interested Parties: Taxpayers in other states
General Environmental Issue(s): Ranching, grazing; land administration, public lands

Conservation efforts also involved land resources, particularly in the western United States. Similar to river management, nineteenth-century land conservation was often organized around expansion, development, and profit. Different eras can be discerned in the formation of the federal lands in the West by the laws and regulations that administered them (Hays 1999, 66–72).

Once the railroad opened up remote sections of the West in the 1860s, vast sections of the region became available for development. In areas such as the Little Colorado River Basin, for instance, land ownership became very confusing between federal government, corporate organizations, and private settlers. Often, the common denominator was land use. Grazing quickly became the most popular way of deriving a profit from the West's vast open spaces. A very intensive land use in arid regions, grazing remained largely unregulated until well into the twentieth century. Although critics argued against the benign nature of this land use, they elicited little response until the Taylor Grazing Act in 1934.

Grazing in the West

Putting the West's open spaces to work actually took very little upfront capital. When the livestock industry came into prominence in the decades after the Civil War, a typical ranch included a ranch headquarters, a few cowboys, and a number of horses. Often, early ranchers had little more than a dugout for shelter and a corral for their horses. When the surrounding grass was eaten off, they simply moved their herds and headquarters to a new location.

For most of their existence on the ranch, the animals were left to fend for themselves until it was time for either branding and marketing. Although grazing could take place on privately owned land, many ranchers also used the vast stretches of land still owned by the federal government. Most cattlemen moved their herds between summer and winter ranges. Often in then midwestern United States, ranchers ranged their cattle on the federal lands during the summer and, before winter, moved their herds close to the home ranch where they could be fed hay.

The idea of making profit from, ostensibly, letting nature run its course (allow cattle to eat grass) eventually attracted investors and speculators from throughout the United States. Western livestock herds grew rapidly on the public range lands, and they were severely overcrowded and depleted by the late 1800s. It is estimated that, in 1870, seventeen western states held 4.1 million beef cattle and 4.8 million sheep. In the same states by 1900, there were 19.6 million beef cattle and 25.1 million sheep.

Case Study: Little Colorado River Basin

Similar to other sections of the West, railroad construction burdened companies with significant debt. In the late 1860s, the Atlantic and Pacific Railroad Company attempted to sell 5,424,800 acres of land granted to it by Congress in 1866. More than one million of these acres was acquired at a cost of 50¢ per acre by the Aztec Land and Cattle Company, a consortium of eastern businessmen and Texas ranching interests. By controlling the region's limited water resources, the Aztec Company, estimates historian William Abruzzi, monopolized more than two million acres of range land and, in effect, removed a substantial resource from local use.

Throughout most of the late nineteenth century, the Aztec Land and Cattle Company was among the largest ranches in North America. By the late 1890s, the Aztec Company imported between 33,000 and 40,000 head of cattle into Arizona. Overstocking this arid landscape contributed significantly to the deterioration of the range lands.

The open range of the American West largely came to an end in the 1880s with the introduction of barbed wire and windmills, together with the passage of legislation favorable to small farmers, which spelled the end of the open range. To recoup their investment before the encroachment of farmers resulted in the complete enclosure of grasslands, ranchers in western Texas dangerously overstocked their ranges.

Unable to sell their livestock at profitable prices, ranchers searched for a new range where cattle could be maintained until the market rebounded. The lush grasslands of eastern Arizona, combined with the financial difficulties of the Atlantic and Pacific Railroad, made the formation of the Aztec Land and Cattle Company and the relocation of thousands of starving Texas cattle on this cheaply acquired land an attractive investment opportunity.

In the end, the cumulative effect of drought, range deterioration, falling prices, and heavy losses of cattle from starvation and rustling forced the Aztec Company to declare bankruptcy in 1900. After only sixteen years of operation, the company had to liquidate its extensive holdings in the basin, thus ending the speculative cattle ranching era in this region.

Despite its brief life, Aztec, by Abruzzi's analysis, "had a devastating impact on local ranges and, therefore, a decidedly negative effect on the peoples and communities that depended on these ranges for their survival."

By excluding all competitors from more than two million acres of range land, the Aztec Company adversely effected many other sheep herders and cattle ranchers. In addition, the ecological implications of intensive ranching in the region also had lasting impacts. Cattle grazing led to widespread grassland deterioration throughout the region. Abruzzi wrote the following:

> Local authorities presently estimate that between seven and ten animal units could have been supported on one section of local rangeland prior to its deterioration during the 1890s. The Aztec Company clearly exceeded these figures. Grazing 60,000 head of cattle on two million acres, the company maintained animal densities of nearly 20 animal units per section, that is, between two and three times what the land could support. (Abruzzi 1995)

This overgrazing was accentuated, then, by disastrous droughts of the 1890s, which had a swift and devastating impact on the grassland community. Today, these impacts result in between 55 and 65 percent of the surface area within the grassland community being devoid of vegetation. The lack of the grasses' dense root system has also had a detrimental effect on surface water in the basin. Abruzzi wrote the following:

> The ultimate cause of regional grassland deterioration lay in the speculative nature of the nineteenth-century range cattle industry and in the effect that livestock speculation had on local range management policies. Because range-stocking decisions were based on national market considerations rather than on local environmental conditions, the number and density of cattle that were maintained on local ranges was both excessive and unresponsive to the marked variability which characterized local climatic conditions. (Abruzzi 1995)

Creating Federal Policies on Grazing

Throughout the history of public lands, the primary rationale for their administration was their usefulness. If settlers had found reasons to build communities in the area, it is likely that the site would not have become public land in the first place. Vast tracts of the territories that became states in the late 1800s were simply unwanted by settlers and developers.

Critics of grazing came from different directions, including scientists, federal administrators, and farmers. Serious consideration of the regional ecological impacts of grazing ultimately resulted in efforts by the federal government to regulate and limit this land use. Attempts by Congress during the first quarter of the new century to legislate some sort of "control" of the western federal lands failed. Drought and depression in the early 1930s set the stage for a renewed attempt at legislative intervention and the Taylor Grazing Act was enacted.

The Taylor Grazing Act was created after the creation of the Homestead Act of 1862, which granted early homesteaders 160 acres on which to farm and ranch. In the western half of the United States, 160 acres was too little property to produce a steady, marketable harvest of crops traditionally grown (Kelly 1999). The settlers also realized that the soil seemed resilient to the grazing of sheep and cattle (Kelly 1999). Ranchers took advantage of this observation and began to build fences and violate the open range policy on the western public lands. This course of action continued until the early 1900s when the public began to state concern with overgrazing. This concern, along with the creation of the National Forest Service in 1905, resulted in more extensive regulation of public range land. With more extensive regulation of public range land came a new debate focusing on the question of what constituted overgrazing. This debate was an issue for all people involved with public lands, but it was most intense between cattle ranchers and sheep ranchers. The conflict became so intense that, by 1915, bills were being introduced to regulate grazing, in part to minimize conflict among ranchers. However, even with tensions rising, a bill to regulate grazing was not passed until 1934 when the Taylor Grazing Act was adopted. This passage also coupled with the creation of the modern BLM.

The Taylor Grazing Act (43 USC 315) was the first federal act to regulate grazing on federal lands. The overarching goal of the act was to "stop injury to the public lands [excluding Alaska] by preventing overgrazing and soil deterioration; to provide for their orderly use, improvement, and development; [and] to stabilize the livestock industry dependent upon the public range" (Taylor Grazing Act 1934, §315). Signed by President Roosevelt and adopted in 1934, the Taylor Grazing Act established permits to manage grazing in newly created grazing districts.

In 1934, there were approximately eighty million acres of land available to be placed into grazing districts. These grazing districts were to be created in "vacant, unappropriated and unreserved" areas within public lands that were not national forest, parks, monuments, Native American reservations, or railroad or Coos Bay Wagon road land grants (Taylor Grazing Act 1934). The first grazing district was established in Wyoming in March 1935. By June 1935, the grazing districts included more than sixty-five million acres of land. All the grazing districts established in 1935 are still in effect today.

According to the Taylor Grazing Act, grazing districts are created by the secretary of the interior. These grazing districts were originally to be managed by the Grazing Service, an agency created by the Taylor Grazing Act, which today is the BLM. Before grazing districts are created, there must be public notice and the location of the grazing district needs to be considered convenient for state officials, settlers, residents, and livestock owners in the area. There must also be a hearing held by the state where the land is located.

Once grazing districts are established, adjacent landowners may apply for rights-of-way over these public lands, and stock owners may apply for permits to graze livestock within the district. Permits are granted for a period of not more than ten years, after which they may be renewed. The secretary of the interior specifies the number of stock allowed on each grazing district and the seasons within which use of each grazing district is allowed. The secretary of the interior must provide protection, administration, regulation, and improvement for the grazing districts, regulate occupancy and use, and preserve the land and resources from destruction or unnecessary injury (Taylor Grazing Act 1934). The act restricts who can graze livestock on the land, but it does not alter or restrict the right to hunt or fish or mine deposits within the designated grazing district.

The actual text of Title 43, Chapter 8A, Subchapter I, §315 of the Taylor Grazing Act describes all of these issues:

To promote the highest use of the public lands pending its final disposal, the Secretary of the Interior is authorized, in his discretion, by order to establish grazing districts or additions thereto and/or to modify the boundaries thereof, of vacant, unappropriated, and unreserved lands from any part of the public domain of the United States (exclusive of Alaska), which are not in national forests, national parks and monuments, Indian reservations, revested Oregon and California Railroad grant lands, or revested Coos Bay Wagon Road grant lands, and which in his opinion are chiefly valuable for grazing and raising forage crops: Provided, That no lands withdrawn or reserved for any other purpose shall be included in any such district except with the approval of the head of the department having jurisdiction thereof. Nothing in this subchapter shall be construed in any way to diminish, restrict, or impair any right which has been heretofore or may be hereafter initiated under existing law validly affecting the public lands, and which is maintained pursuant to such law except as otherwise expressly provided in this subchapter nor to affect any land heretofore or hereafter surveyed which, except for the provisions of this subchapter, would be a part of any grant to any State, nor as limiting or restricting the power or authority of any State as to matters within its jurisdiction. Whenever any grazing district is established pursuant to this subchapter, the Secretary shall grant to owners of land adjacent to such district, upon application of any such owner, such rights-of-way over the lands included in such district for stock-driving purposes as may be necessary for the convenient access by any such owner to marketing facilities or to lands not within such district owned by such person or upon which such person has stock-grazing rights. Neither this subchapter nor the Act of December 29, 1916 (39 Stat. 862; U.S.C., title 43, secs. 291 and following), commonly known as the "Stock Raising Homestead Act", shall be construed as limiting the authority or policy of Congress or the President to include in national forests public lands of the character described in section 471 of title 16, for the purposes set forth in section 475 of title 16, or such other purposes as Congress may specify. Before grazing districts are created in any State as herein provided, a hearing shall be held in the State, after public notice thereof shall have been given, at such location convenient for the attendance of State officials, and the settlers, residents, and livestock owners of the vicinity, as may be determined by the Secretary of the Interior. No such district shall be established until the expiration of ninety days after such notice shall have been given, nor until twenty days after such hearing shall be held: Provided, however, That the publication of such notice shall have the effect of withdrawing all public lands within the exterior boundary of such proposed grazing districts from all forms of entry of settlement. Nothing in this subchapter shall be construed as in any way altering or restricting the right to hunt or fish within a grazing district in accordance with the laws of the United States or of any State, or as vesting in any permittee any right whatsoever to interfere with hunting or fishing within a grazing district. (Taylor Grazing Act 1934)

The Taylor Grazing Act was largely supported by local, state, and the national stockmen's association. Its passage not only eased the conflict between cattle ranchers and sheep

Gullies and bunch grass are signs of overgrazing in this pasture in the mountains of Bernalillo County, New Mexico, around 1940. Although overgrazing presented westerners with short-term gains, it often denuded and damaged an already fragile landscape. Library of Congress.

ranchers, but it also solved some major public lands problems (Kelly 1999). Additionally, the Taylor Grazing Act created a system of fees, in which the monies involved go to both the federal government and the states to help maintain the grazing systems. More specifically, fees for permits go to the U.S. Treasury, except for 12.5 percent of the fee, which is paid to the state in which the grazing district is located (50 percent of the money collected from so-called isolated grazing districts is paid to the state) (Taylor Grazing Act 1934, §315m).

With the passage of the Taylor Grazing Act and establishment of permit requirements, much of the free-wheeling, frontier method of operating and administering western ranches was eliminated. The Taylor Grazing Act placed controls on public land grazing and established specific grazing allotments or areas of use. In some cases, this policy forced operators to make more use of private lands through purchasing or leasing private pasture.

Conclusion: Contemporary Developments

Although the Taylor Grazing Act was preempted by the FLPMA of 1976, today ranchers, among others in the West, believe the Taylor Grazing Act established a way of life in the West. FLPMA was partly initiated by changing social values with respect to environmental protection and conservation of natural resources. The main emphasis, however, was the interest of western states to generate tax revenue from the massive blocks of land in their states that was reserved by the federal government for various purposes.

Furthermore, a case can be made that the Taylor Grazing Act also continues to affect the landscape in new and previously unimagined ways. Today, groups, particularly environmental groups, are attempting to use elements of the Taylor Grazing Act to resolve what they view as new public lands issues: environmental and ecosystem concerns.

To protect the health of the ecosystems within the grazing districts, particularly during periods of severe drought or other situations that deplete the range, the secretary of the interior may "remit, reduce, refund in whole or part, or postpone payment of grazing fees for the time the emergency exists" (Taylor Grazing Act 1934, §315b). Environmentalists appeal to this section of the act when they believe range land is being depleted. The debate focuses on the fact the often ranchers and environmentalists determine what conditions constitute "depletion" differently. Furthermore, environmental groups are now competing with ranchers to purchase the grazing permits within the grazing districts created by the Taylor Grazing Act. To support this, the U.S. Supreme Court upheld regulations that eliminated the limitation that only stock owners "engaged in the livestock business" could receive grazing permits (Stern and Long 2000). Over the years, the uses of the Taylor Grazing Act have changed, but the importance of the act remains.

Today, livestock grazing, as a legitimate use of public lands, is increasingly competing with other legitimate uses of public lands, such as recreation, wildlife habitat, riparian management, endangered species management, mining, hunting, cultural resource protection, wilderness, and a wide variety of other uses. There are increased expectations from the public to reverse unacceptable livestock impacts on public lands. Overall, changing social values and competition for land use in the United States in general have required that public-land management decisions achieve greater balance among sometimes conflicting resource uses. In recent years, this has made many ranchers claim that their way of life is under attack.

Sources and Further Reading Abruzzi, *The Social and Ecological Consequences of Early Cattle Ranching in the Little Colorado River Basin*; Donahue, *The Western Range Revisited: Removing Livestock from Public Lands to Conserve Native Biodiversity*; Federal Land and Policy Management Act, http://www.blm.gov/flpma and http://www.blm.gov/flpma/snapshot.htm; Kelly, "Grazing Act Still at Work to Protect Grasslands"; McGrory Klyza, *Who Controls Public Lands?: Mining, Forestry, and Grazing Policies, 1870–1990*; Opie, *Nature's Nation*; Stern and Long, "U.S. Supreme Court Upholds 1995 Department of the Interior Grazing Regulations," http://www.modrall.com/articles/article_68.html#; Taylor Grazing Act 43 U.S.C. §§315-316o, June 28, 1934 as amended 1936, 1938, 1939, 1942, 1947, 1948, 1954, and 1976, http://ipl.unm.edu/cwl/fedbook/taylorgr.html and http://www4.law.cornell.edu/uscode/html/uscode43/usc_sup_01_43_10_8A_20_I.html; *Waste of the West*, http://www.wasteofthewest.com/Chapter1.html; White, *It's Your Misfortune and None of My Own*; Worster, *Dust Bowl: The Southern Plains in the 1930s*.

MISSION 66 AND THE ERA OF TOURISM DEVELOPMENT IN THE NATIONAL PARKS

Time Period: 1950–1970
In This Corner: National park administrators, tourist developers
In the Other Corner: Preservationists, environmentalists
Other Interested Parties: Political leaders, park administrators elsewhere
General Environmental Issue(s): National parks, preservation, management

Scenes from *Yogi Bear* and Jellystone National Park informed a generation of Americans about the role of the national parks. Similar to much of popular culture, there was at least a modicum of reality to Yogi's ongoing story. As the NPS matured after World War II, administrators and politicians reconsidered many of its efforts and priorities. If national parks were truly the property of the American people, went one line of thought, then the chief responsibility of the NPS was to make the facilities as useful and useable as possible. Operating much as a tourist bureau from 1950 forward, the NPS sought out new ways to spur visitation to its units.

But how useful did it need to be? Did preservation really include making a site more able to be used? Or was it simply concerned with setting aside and leaving a site alone? As the NPS internally debated its responsibilities to preservation, Americans, especially those interested in the history of a specific site, asked these sort of questions about NPS activities. In fact, they continue to do so today.

Accessing Wild America

Railroads were the key to early tourist development in national parks. Discussed in another essay, Western parks were often defined by railroad companies. In the eastern United States, however, preservationists were beginning to define the ethics that they would use everywhere. One of the key concepts was that of wilderness.

The intellectual formation of ideas of wilderness at the Adirondacks in the late 1800s was one of the first expressions of the public's growing interest in nature preservation. This same sentiment, however, would soon attract interested Americans westward to existing national parks. Railroads seized this potential and began to develop lines that could specifically take advantage of the American passion for wilderness at the end of the 1800s. Often, the railroad constructed lodges in the parks that were modeled after the Adirondacks' rustic camps. By the 1910s, park areas had begun to be accessed on an individual basis thanks to a new technology made possible by the petroleum discovery at Spindletop: the automobile. Ironically, the access to wild areas grew from a technology that would ultimately prove injurious to the environment.

The interest in the outdoors, wrote historian Paul Sutter, stemmed from a variety of cultural forces, including "fears of Anglo-Saxon racial degenerations," feminization caused by urban living, and a culture of authenticity to combat the mass productions of modern technology (Sutter 2002, 22). Individuals became specifically interested in spending leisure time outdoors and, potentially, in wild areas between 1880 and 1920. This era coincides with the popularization of the automobile in the United States.

By the 1920s, the *New York Times* estimated that at least five million autos per year were being used for auto-camping (particularly surprising given that there were a total of approximately ten million cars on the road) (Sutter 2002, 30). Before roads had been improved, driving long distances was a certain version of a survival experience in its own right. Referred to as auto-camping or "gypsying," drivers often stopped to sleep in open fields or along the side of the road. By the 1930s, pull-along campers (including the well-known Airstream) had gone into production and helped to make traveling by car even simpler. The development of this impulse included national conferences organized by President Calvin Coolidge in 1924 and 1926, which were each dubbed the National Conference on Outdoor Recreation (Sutter 2002, 39–41).

Some travelers sought the outdoors only to escape the "civilized" world. Others sought out activities such as hunting and fishing. Each of these constituents played an important role in defining the role of nature in everyday American life.

The Hess Camp cottages were one of the many retreats or resorts available for visitors in New York's Adirondacks. This photo was taken around 1903 when state legislation was put in place prioritizing wilderness preservation in the park. Library of Congress.

Many of these impulses congealed into complicated forms that were related to the "rustic" and "wilderness" characteristics that were part of the cultural interest in locales such as the Adirondacks. The use of natural forms in design and home layout was a well-established portion of traditions such as English gardens and urban parks, which were advocated by A. J. Downing and others during the mid-1800s. The blending of this current with modern sensibilities, however, produced a uniquely American rustic aesthetic that had never before been applied with such intensity to buildings.

For instance, as tourists became increasingly interested in visiting national parks, the architectural example used by the NPS often derived almost directly from the Adirondack camps. The use of log poles closely associated with Adirondack camps was copied elsewhere in rustic resorts and recreational architecture, appearing in signs, gateways, bridges, and cabins from the White Mountains to Camp Curry in Yellowstone by the turn of the century. A number of the early hotels in national parks, such as those of Glacier National Park and Yellowstone National Park's Old Faithful Inn, were influenced by the architecture as well as the decorative arts characteristic of the Adirondack camps (Carr 1988, 189–90).

What began the century as a token interest of the wealthy had grown into a federal mandate for the conservation of America's natural resources by 1920. It was not a perfect ethic: priorities still included park tourists over ecology or any natural inhabitants. However, the stage was set for American society to begin to consider the human role in nature in a more complicated manner than it had ever been capable of previously.

Case Study: Mission 66 at Gettysburg

On the selected site, the NPS contracted with architect Richard Neutra to design the new structures. The Neutra structures became part of a sweeping new era of design and planning in the entire national park system. Referred to as the "Mission 66 era," which included the decades leading up to 1966 when nationwide the NPS wanted to be prepared for a swell of eighty million visitors, each park endured significant internal improvement and change. Mission 66 prioritized tourism, however, and its impact was most clearly seen on landscapes intended for visitation. Weeks wrote that "Neutra's design and the building's location must be viewed within the context of the Cold War" (Weeks 2003). Although its mix of medias and compromising of the sacred may have distorted the image of the battle somewhat, Weeks wrote, "Its juxtaposition with the real battlefield brought to life a memory that monuments no longer could evoke. The painting still offered both sensationalism and edification, but now for those who hoped the American past might provide fun and togetherness during an annual family vacation." Very few Americans thought to question the decisions and action of the NPS at Gettysburg. Most often, tourists appreciated the user-friendly landscape that Mission 66 planning brought to the Gettysburg National Military Park (GNMP).

Tilberg believed that by the mid-1950s, tourists coming to GNMP had changed considerably. He wrote the following:

> During the first half century after the battle, the veteran came to the field to review the ground over which he had fought.... [In recent years, however,] the interest in the battlefield now centers more in the story of the battle, famous landmarks,... and how each of these well-known places fits into the course of the fighting. Although a large proportion of people who come to Gettysburg have only a curiosity to see the battlefield and to learn something of the overall story of the battle, many have made a study of it and have become tacticians in their own way, hoping to find an answer to the failure of certain actual battle maneuvers. (Unrau 1991)

In particular, Gettysburg fit well into the post-war and Cold War nationalism that swept up the growing middle class. Consider these figures: a respectable 659,000 visitors in 1950 rose to 1,300,000 in 1960 and then, by 1963, it topped two million (Unrau 1991). The results of Mission 66 planning and the rise in tourism resulted in a new landscape at GNMP, including resurfacing park roads, construction of a new visitor center–cyclorama complex with space for park offices, new field exhibits, pull-outs from auto tours, and a new high-water walking tour.

The sensibility to decrease surrounding blight also planted the seeds of the NPS's interest in authentic views. Agricultural fields that would be seen by auto tourists became an emphasis for NPS planners in 1958 when they established the policy that "the agricultural use of some 1100 acres of the park, leased for such use as a means of maintaining character of the land as it was at the time of battle, will be continued. In some instances, woodlands will be declared to open up vistas, and in a few cases the growth of trees and orchards will be encouraged to enhance the authenticity of the historic scene."

Such statements were ripe with irony when made by a public agency that was simultaneously opening the new visitor center and cyclorama (opened in 1962) in the middle of some of the battlefield's most significant terrain. Although some criticism of this location came from employees of the NPS, the service's director, Conrad L. Wirth, made the final decision on the location of the buildings and he concurred with Tilberg's analysis. Ironically, the NPS formed a new master plan that was released in 1963:

> GNMP's most *significant resources* are the land on which the battle of Gettysburg was fought.... A less significant but important feature of the Park is its monumentation.... Some of Gettysburg's monuments are genuine works of art, others merely commemorative stones, but each in itself is an interpretive device and all are reminders of the greatness of the American heritage. A primary mission of GNMP is to assist the visitor to convert the Park's resources, the battlefield and its monuments and the National Cemetery, into meaningful concepts.

By the 1960s, the culture of preservation stressed democratic access and use of the landscape. Few comments were registered about authenticity and, instead, NPS planning stressed modernist designs, such as that created by Neutra. Decisions such as the placement of the visitor center and cyclorama reflected a fairly typical modernist rejection of symbol and even of the sacred for the sake of rational planning.

Conclusion: Mission 66 Alters the Role of National Parks

National Parks were not run as tourist meccas before Mission 66. Although critics argue that emphasizing tourism belies the NPS's primary mission to preserve and conserve natural resources, new facilities allowed more Americans than ever before to access the nation's natural wonders.

Sources and Further Reading: Black, *Contesting Gettysburg: Preserving an American Shrine*; Carr, *Wilderness by Design*; Nash, *Wilderness and the American Mind*; Runte, *National Parks: The American Experience*; Sellars, *Preserving Nature in the National Parks: A History*; Sutter, *Driven Wild: How the Fight Against Automobiles Launched the Modern Wilderness Movement*; Unrau, *Gettysburg Administrative History*.

ENVIRONMENTAL CONSEQUENCES AND LESSONS OF AUTOMOBILE MANUFACTURING

Time Period: 1920 to the present
In This Corner: Auto manufacturers
In the Other Corner: Unions and workers at times, foreign competitors
Other Interested Parties: Automobile consumers
General Environmental Issue(s): Automobiles, pollution, green design

Throughout the United States, industry and new technology continued to lead the American economy forward in the early twentieth century while the federal government attempted to establish a balance that would allow it some control for the sake of issues such as human health. By the 1920s, the greatest example of this rapid industrial growth could be seen around Detroit, Michigan, the hub of the nation's emerging automobile industry. Although the growth of the auto industry provided jobs for generations, it has also created problems in its wake.

Manufacturing Perfection

The scale of auto manufacturing was unlike any manufacturing undertaking that the nation had ever seen before. Ford's plants around Detroit, Michigan, created the model of the single-factory site that could combine the assorted materials and tasks necessary to turn out mass-produced automobiles. The most famous of these manufacturing sites was the River Rouge Plant. To create the ideal environment for his ideas of mass production, Ford created the Rouge Plant in the late 1920s and early 1930s. It was the world's largest industrial complex, containing cutting-edge technology of many types. By concentrating the entire manufacturing process at one site, Ford continued his effort to rationalize and simplify industrial processes. Efficiency was his goal (Brinkley 2003, 200).

One aspect of Ford's pursuit of efficiency was independence. He wanted his manufacturing of autos to be as free as possible from the constraints of other businesses and regions. Therefore, the Rouge Plant was built as a nearly self-sufficient and self-contained industrial city. After a decade of construction, the Rouge Plant released the new Model A Ford in September 1927. At that time, the facility contained ninety-three structures, ninety miles of railroad tracks, twenty-seven miles of conveyors, and 53,000 machine tools. The Rouge Plant alone had 75,000 employees. By the early 1940s, the Rouge Plant had produced more than fifteen million cars.

To realize Ford's dreams of efficiency, the Rouge Plant incorporated all phases of auto production, including steel forging and stamping operations, manufacturing of parts, and assembling automobiles. In addition, however, the Rouge Plant included its own power plant, glass plant, cement plant, and a byproducts plant, which produced petroleum products, such as paints, fertilizers, and charcoal. To feed the factory's fires, Ford used its own iron ore mines in northern Michigan and Minnesota and coal mines in Kentucky and West

The green roof of the Dearborn Truck Plant, which opened in 2004. The "living roof" helps to heat and cool the truck assembly plant at the former site where Henry Ford turned raw iron ore into Model A cars. AP Photo/Carlos Osorio.

Virginia. To transport these raw materials, Ford used his own railroad lines or ships. Over the years, Ford even established a lumber operation in northern Michigan and a rubber plantation in Brazil for tire production.

Remarkably, the Rouge Plant grew even larger during the next decade. In the 1940s when it employed more than 100,000 workers, the factory grew to include thirty miles of internal roads and 120 miles of conveyors. During the war years, each day the Rouge Plant consumed 5,500 tons of coal and 538 million gallons of water. With this input, the plant could smelt more than 6,000 tons of iron and roll 13.4 miles of glass each day. At its most efficient, Ford estimated that the Rouge Plant produced a new car every 49 seconds (Brinkley 2003, 154)!

Concentrating the production also concentrated the toxic byproducts of manufacturing. Heavy manufacturing created toxic byproducts, soot, and smoke wherever it took place. In the case of the Rouge Plant, the concentration of production ability corresponded with concentrated levels of pollution in the area. When other manufacturers concentrated their production facilities in Detroit, Michigan, as well, this region was promised decades of environmental problems (Hurley 1995, 2–14).

Green Manufacturing Meets the Automobile

Given this early history of heavy industry, it was particularly striking in recent years when Bill Ford, the great-grandson of Henry, set out to blaze a new trail for American automobile manufacturing. Whereas a generation earlier, the Rouge Plant modeled a new century of mass production concentrated in one site, Ford has now redesigned the plant to bring automobile manufacturing more in line with the environmental era.

While refitting the factory, Ford also has spent more than $2 billion to transform the gritty piece of American history into a modern model of environmental responsibility and manufacturing flexibility. The result is a marriage of natural and industrial systems in what Ford hopes will be both a productive and profitable environment.

To rethink the Rouge Plant, Ford hired Virginia architect William McDonough, known for his environmentally sensitive designs. Similar to the Gap offices that McDonough designed in California, the new truck factory is constructed under a 10.4-acre "living roof," which is designed to keep the plant warmer in winter and cooler in summer. The roof accomplishes at least two benefits: sedum plants absorb and filter water from rain and snow, absorb carbon dioxide, and give off oxygen. Additionally, on the roof of the new visitor's center, photovoltaic panels turn sunlight into electricity to supplement the building's power supply.

On the ground, solar collectors heat water for the same building. The spaces around the factory are considered to be important now for more than storage. Green space around the center includes natural water-treatment wetlands and wildlife habitat. Approximately 85,000 flowering perennials were planted along a new greenbelt parkway, along with 20,000 shrubs and hundreds of new trees. McDonough's plan also uses wind patterns to pump outside air into buildings as well as install skylights to add natural light. Plants, trees, and shrubs are added around buildings (hundreds of varieties were planted at the Rouge Plant) to boost oxygen and wildlife counts.

When McDonough began the project, more than eighty years of industrial pollution from steel operations, glass production, and vehicle assembly at the site had left its mark on the surrounding landscape.. There were few fish, birds, mammals, reptiles, or amphibians that could be found on the site. Trees, flowers, and shrubs were virtually nonexistent. This caused related problems with stormwater runoff, which intensified pollution problems from the site. By adding several acres of landscaped ditches, or swales, along with wetlands and native plants to clean the soil, McDonough estimates the team saved Ford $5 million over a system that uses chemicals to treat stormwater. Other benefits emerged as well. Since the wetlands were installed last summer, area wildlife experts report rising levels of non-employee residents, including birds, fish, mammals, and reptiles.

Conclusion: Can Car Manufacturing Help the Environment?

Preliminary data since 2002 showed that dissolved oxygen in the Rouge Plant is on the rise, the result of Ford's new wetlands and a host of other cleanup enhancements up and down the waterway. These rising oxygen levels help support fish and other aquatic life, which show that Bill Ford's ideas might be paying dividends of many sorts. The new Rouge Plant could eventually help to make automobile production less problematic for the environment.

Sources and Further Reading: Brinkley, *Wheels for the World: Henry Ford, His Company and a Century of Progress*; Flink, *The Automobile Age*; Gorman, *Redefining Efficiency: Pollution Concerns, Regulatory Mechanisms, and Technological Change in the U.S. Petroleum Industry*; Jackson, *Crabgrass Frontier: The Suburbanization of the United States*; Jacobs, *The Death and Life of Great American Cities*; Kay, *Asphalt Nation*; McShane, *Down the Asphalt Path*; Motavalli, *Forward Drive: The Race to Build "Clean" Cars for the Future*.

CONCEIVING OF HUMAN EVOLUTION

Time Period: 1900s to the present
In This Corner: Scientific community, some educators
In the Other Corner: Religious representatives, some educators, some members of scientific communities
Other Interested Parties: Students, educators
General Environmental Issue(s): Human species, evolution, scientific thought

In terms of the essential ecological question of how humans fit into the world around them, one of the essential quests has been to understand basic human origins. Archaeology, of course, has provided essential theoretical explanations about the movement of humans from their point of origination in eastern Africa. This explanation, however, does not cohere with the origin myths of many societies, including the Judeo-Christian tradition that prevails in the United States. Therefore, efforts to teach these ideas in schools, even if the theories are accepted by the scientific community, have proven to be a flash point of contention in what by the end of the 1900s had become known as the "culture wars." Similar ideological conflicts, however, can be found much earlier in American history.

Creation-Evolution Controversy

In the late 1700s, geological discoveries began to suggest that the biblical story accepted by many Europeans and Americans was not viable. Considered blasphemous, such ideas threatened the fixed social order of many societies and religions. When the Church of England reacted vigorously against these ideas, some other groups such as Unitarians, Quakers, and Baptists endorsed them.

The 1859 publication of Charles Darwin's *On the Origin of Species by Means of Natural Selection* brought scientific findings that seemed to endorse the geological findings. More than settling the issue, however, Darwin's theories enlarged the controversy surrounding the idea of evolution.

In the United States of America, there was no official resistance to evolution by mainline churches. In fact, early in the twentieth century, most high school and college biology classes taught scientific evolution. Revival of Christian fundamentalism in the 1910s, however, brought the issue of teaching evolution into public debate. In 1925, the State of Tennessee passed the Butler Act to prohibit the teaching of any theory of the origins of humans that contradicted the teachings of the Bible.

Scopes Trial and the Theory of Evolution

The inconsistent reach of scientific knowledge and modern ways of thinking about the human's place in the world was never more obvious than in the Tennessee trial of 1925 that became known as the "Scope's Monkey Trial." The school board brought suit against a Tennessee biology teacher who insisted on teaching evolution. In the early 1920s, the older Victorians worried that the traditions that they valued were ending, whereas younger modernists sought to apply the new intellectual revelations that took shape. Intellectual experimentation flourished. Americans danced to the sound of the Jazz Age, showed their contempt for the prohibition of alcohol, debated abstract art and Freudian theories. In a response to the new social patterns set in motion by modernism, a wave of revivalism developed, becoming especially strong in the American South. The nature of the human species became one of the primary battlegrounds for these groups of Americans.

In the Dayton, Tennessee, courtroom during the summer of 1925, John Scopes defended the teaching of a nature-based theory of evolution against the teachings of the Judeo-Christian Bible. His opponent was represented by William Jennings Bryan, three-time Democratic candidate for president and a well-known populist. With his oratorical skills, Bryan led a Fundamentalist crusade to banish Darwin's theory of evolution from American classrooms. In the courtroom, Bryan squared off against the well-known attorney Clarence Darrow, who defended Scopes.

The Scopes trial revealed deep divisions in the nation's interest in and abilities to conceive of scientific findings that might challenge the religious convictions of many citizens. Viewed in this light, social Darwinism and even segregation become part of a complicated cultural reaction to modern ideas.

The Butler Law was upheld by the Tennessee Supreme Court and remained on the books until it was repealed in 1967, by which point most public schools had dropped the subject of evolution from their textbooks. In 1968, however, the U.S. Supreme Court ruled in *Epperson v. Arkansas 393 U.S. 97* that such bans contravened the Establishment Clause because their primary purpose was religious.

Conclusion: Revisiting the Evolution Debate

The emergence of the politically powerful religious right in the late twentieth century ensured that evolution would be an important portion of what became known as the "culture wars."

In today's debate, the mainstream scientific consensus on the origins and evolution of life is challenged by creationist organizations and religious groups who desire to uphold some form of creationism as an alternative. In 2005–2006, the movement centered around the need to teach evolution as a theory. Therefore, argued many critics, other theories should be taught as well.

In particular, Christian groups organized around a theory titled "intelligent design" in which one being is believed to have been the essential originating point of all life. In the 2005 court case *Kitzmiller v. Dover Area School District, PA*, U.S. District Judge John E. Jones III ruled that intelligent design is not science but is grounded in theology and "cannot uncouple itself from its creationist, and thus religious, antecedents." Also in 2005, the main instigator of the intelligent design movement, the Discovery Institute, arranged to conduct state hearings in Kansas to review the teaching of evolution. Although the state school board initially voted to adopt changes to the state curriculum, four of the committee members supporting the referendum were voted out in 2006. The new board returned to previously held standards.

Will this be the end of the evolutionary controversy? In May 2007, the Gallup Poll found that 43 percent of Americans felt that God created man in present form, 38 percent that man developed under God's guidance, and 14 percent that God had no part in man's development.

Sources and Further Reading: Darwin, *Origin of Species*; Miller, *An Evolving Dialogue: Theological and Scientific Perspectives on Evolution*; NPR, *Teaching Evolution: A State-by-State Debate*, http://www.npr.org/templates/story/story.php?storyId=4630737; Numbers, *The Creationists: The Evolution of Scientific Creationism*.

LIVING WITH LEAD

Time Period: 1930s to 1970s
In This Corner: Lead producers, auto manufacturers
In the Other Corner: Scientists, consumers, public health advocates
Other Interested Parties: Federal, state, and local government officials
General Environmental Issue(s): Pollution, autos, lead

Human relationships with chemicals do not remain static, they change over time. In the case of lead, a mined metal has at times appeared to offer wondrous opportunities, but at other times, science has taught us some of lead's dangerous attributes. This has led to a most problematic relationship because many building practices and mechanical processes have brought lead into close proximity with humans, including into our homes and into the water we drink and the air that we breath.

Putting the Lead In

Lead was one of the earliest metals discovered by humans, estimated to have been in use by 3000 BC. In ancient Rome, lead was used for water pipes and for lining baths. During the Medieval Era, lead was used for the construction of cisterns, roofs, coffins, tanks, and gutters.

The dull gray lead is ideal for such uses because it is not at all easily corroded, or eaten away. Unlike iron and steel, it does not need protection by painting. Even so, lead is quite soft and pliable. It is this softness that makes it easy to squeeze or roll lead into different shapes. For instance, when used as a covering for electric cables, because of its pliability and its resistance to corrosion, the inner wires are protected without making the cable too stiff to bend.

Similar to other minerals, lead is mined from an ore called galena. Lead is obtained by crushing the galena and then roasting it to drive off the sulfur. The roasted galena is mixed with coke and limestone and put into a furnace. Air is blown into the lower part of the furnace to make a draft for burning the coke, and the molten lead is drawn off from the bottom.

Given this process, lead is obtained in small quantities, similar to gold, silver, copper, and other metals. After further purification, the lead is cast into lumps called pigs. (Additional supplies of lead are obtained from scrap from old batteries and pipes.)

Domestic Uses in the United States

The United States, more than possibly any other human civilization, has integrated lead into our lives. As Americans have slowly accepted the toxic nature of the metal, it has led to serious health consequences and difficult remediation processes.

The pipes and lines supporting the infrastructure for the U.S. water system were one of the first uses of lead in the American domestic environment. In addition to some use in the piping run through American homes, the larger service lines that brought water from reservoirs and connected to individual homes and apartment buildings to street mains were also made of lead. Once again, it was the malleability and durability that made lead desirable to some engineers over plain iron or steel. Malleability reduced labor costs by making it easier to bend the service main around existing infrastructure and obstructions, and, compared with iron, lead was a soft and pliable metal. As for durability, the life of the typical lead service pipe was thirty-five years, which is approximately twice that of iron or steel pipe.

One of the most troublesome uses of lead, however, followed the motto that "Lead Makes Your Engine Sing." By the early 1900s, urban reformers, including Alice Hamilton, had begun complaining about the health implications of lead. However, many Americans, very likely the vast majority, were just getting used to the conveniences of the modern era. The idea of questioning their worth or safety was lessened by the immediate gratification of the possible. One example of this is the use of lead in gasoline used in automobiles powered by the internal-combustion engine.

Charles "Boss" Kettering had made a name for himself by inventing the self-starting ignition system that eliminated the hand crank used to start early autos. Whereas reformers such as Hamilton focused on elements of human health, Kettering set about to solve a technological ill that plagued the new automobile culture: engine knock, which he referred to as "the noisy bugbear." During the 1920s, the automobile industry desperately sought a solution to the engine knocking and power lapses that plagued almost every internal-combustion engine. Iodine, aniline, selenium, and other substances had all fallen by the wayside in the frantic search for a fuel additive that would improve engine performance and reduce engine knock (Gorman 2001). As vice president of research at General Motors (GM), Kettering guided the scientific team that in December 1921 unveiled a gasoline additive that diminished the knocking sound that came from poorly running engines. GM teamed with DuPont and Standard Oil to market ethyl, their brand name for the first leaded gasoline on the market.

As tests were run on ethyl at a Standard Oil facility in 1924, several workers died from a form of sudden lead poisoning in which they became delirious and violent. Reports surfaced that DuPont workers had suffered similar deaths. These incidents gave credence to complaints already being voiced by public health reformers. In actuality, the use of tetraethyl lead created almost as many problems as it solved. A total of fifteen workers died from poisoning. Although a movement began to remove the additive from gasoline, scientists were slow to bring data to bear against the powerful industry. Such action clearly required a wholesale change in the ability of the government to regulate industries such as automobile manufacturers (Gorman 2001). Public outcry followed when it was learned that Standard Oil had already put ethyl on the market based only on the results of its own tests with the substance. In some communities, boards of health blocked any sale of ethyl until it had been further tested.

Unfortunately, there was no federal entity that could investigate the new fuel without bias. Examining the impacts of ethyl's use fell to the U.S. Surgeon General's office. At this time, the surgeon general was overseen by the Treasury Department. In May 1925, Surgeon General Hugh Cumming formed a panel of experts to consider the new fuel compound, ethyl.

Also indicative of the era, the hearings were dominated by the corporate entities who had been responsible for developing and testing the new gasoline. The hearing heard experts report that the workers who had died from manufacturing ethyl breathed much greater amounts of lead than would the general public from the burning of such fuel. Although reformers pointed out the clearly proven dangers of lead poisoning, citing significant health risks from just small amounts of lead, the corporate effort to prove ethyl's safety was overwhelming. One Standard Oil spokesman likened ethyl to a "gift of God." Leaded gasoline, they explained, would significantly increase the utility of American automobiles. Ultimately, the panel agreed to lift the ban on the sale of leaded gasoline (Gorman 2001, 139–40).

Although the effort to control lead failed, the attempt to place expectations relating to health on large American industries proved a harbinger of things to come. Unfortunately, in the 1920s, the federal government often could not muster the influence to fully control powerful corporations, such as the petroleum and automobile manufacturers. There were, however, some important areas in which the regulative authority of the federal government could have great sway.

Changing Our Minds about Lead

Although engine performance won out over public health, in other sectors, attitudes changed faster. For instance, the movement against the use of lead in paint began as early as the 1930s, as seen in this account from 1930:

Lead-free Paint on Furniture and Toys to Protect Children

Lead poisoning as a result of chewing paint from toys, cradles and woodwork is now regarded as a more frequent occurrence among children than formerly, and all children's hospitals, realizing the extent of the dangers from this source, are coming to use a lead-free paint on their beds, toys, furnishings and interior decorations, it was stated orally Nov. 19 on behalf of the Public Health Service.

Children are very susceptible to lead, it was stated, and show a higher fatality rate than adults. Frequently, it was said, small amounts of lead which may cause only chronic lead poisoning in an older person may be of sufficient quantity to cause acute poisoning, leading to death, in an infant.

The following information was also furnished by Dr. Russell:

The most common sources of lead poisoning in children are paint on various objects within the reach of a child and lead pipes which are used to convey drinking water. Various manufacturing companies, however, are now beginning to make paints for indoor purposes which are lead-free and lead is being replaced in pipes by other metals.

Lead poisoning in infants is not so often heard of because the condition is frequently unrecognized by physicians. The poisoning creates certain disturbances which are common to various diseases which occur during infancy.

Acute lead poisoning in children is very painful, one of the symptoms being severe cramps in the stomach. The poisoned child becomes intensely irritable and has convulsions and tremors. Chronic lead poisoning leads to a gradual deterioration of numerous parts of the body. The nervous system in general is affected and the result may be nervousness, insomnia and neuritis. The kidneys and blood vessels are also affected. In general, lead poisoning is apt to lead to chronic invalidism.

Children who have been exposed to lead should have a diet rich in calcium and vitamins. Fruits are very desirable and sunshine aids greatly.

(The United States Daily, November 20, 1930 Presenting the Official News of the Legislative, Executive and Judicial Branches of the Federal Government and of Each of the Governments of the Forty-Eight States)

Unfortunately, paint companies were not only slow in responding; in most cases, these scientific findings were disregarded, and companies even went as far as to exploit the relationship between lead paint and children. National Lead Company, for instance, used the famous logo of the Dutch Boy. Ironically, their ads alerted Americans "Don't Forget the Children." The reason for not forgetting them, however, was that one of the unique attributes of paint using white lead was that it would be less likely to be marked, particularly if touched by children (lead advertisements).

National Lead produced several paint books for children, including *The Dutch Boy's Lead Party* and *Dutch Boy Conquers Old Man Gloom* in which the illustrations show the Dutch Boy mixing white lead with colors and painting walls and furniture. Another ad shows an infant crawling on the floor and touching a painted wall. The caption states "There is no worry when fingerprint smudges or dirt spots appear on a wall painted with Dutch Boy white-lead" (lead advertisements).

Ad campaigns by the Lead Industries Association (LIA) in the late 1930s sought to dispel any customers' apprehension about using leaded paint in their homes. Because lead paint was less expensive, LIA ads specifically targeted low-income housing. This same logic led the LIA to expand its campaign by 1940 to include municipal, state, and county institutions. In this capacity, lead paint was almost always used in public schools and health facilities (Markowitz 2002).

This culture of lead began to change with a cover story in *Time* magazine in December 1943 that proclaimed childhood lead poisoning to be a national issue. Blaming a lack of information for American parents, the article demonstrated the significant decline caused in IQ scores in schools with lead paint. The LIA refuted this report, and, by 1945, it had begun another set of ad campaigns to combat the negative image of lead. Continuing into 1952, the LIA promoted the usefulness of white lead for both interior and exterior covering, while ignoring the mounting evidence of its harmful qualities.

Public sentiment continued to grow against LIA's statements during the 1950s. In 1952 alone, there were 500 newspaper reports detailing lead poisoning. Finally in July 1956, *PARADE* magazine and CBS television ran an article titled "Don't Let YOUR Child Get Lead Poisoning" (lead advertising).

Making Policy to Control the Use of Lead

The era of environmental policies began during the 1960s, and this proved to be an important new era for attitudes toward lead. By this point, the amount of lead added to a gallon of gasoline hovered in the vicinity of 2.4 grams. The Department of Health, Education, and Welfare, which was home to the surgeon general starting with the Kennedy administration, had authority over lead emissions under the Clean Air Act of 1963. The criteria mandated by this statute were still in the draft stage when the act was reauthorized in 1970 and a new agency called the EPA came into existence. The days of lead's use in American gas tanks and on American walls was clearly on the wane.

In terms of the use of lead paint, federal legislation prohibited it in 1970. The Consumer Products Safety Commission also passed a ban on the use of all lead paint after February 1978. Of course, this legislation was most concerned with preventing further use of the paint, not on remediating the lead paint already on the walls of homes, schools, and offices. Even today, the National Safety Council estimates that thirty-eight million homes still contain lead paint and 25 percent of homes contain some type of a lead hazard. The Center for Disease Control and Prevention estimates that there are currently 434,000 children between the ages of one and five with elevated levels of lead in their blood.

For leaded gasoline, the end was near as well. In January 1971, the EPA's first administrator, William D. Ruckelshaus, declared that "an extensive body of information exists which indicates that the addition of alkyl lead to gasoline ... results in lead particles that pose a threat to public health." The resulting EPA study released on November 28, 1973 confirmed that lead from automobile exhaust posed a direct threat to public health. As a result, the EPA issued regulations calling for a gradual reduction in the lead content of the nation's total gasoline supply, which includes all grades of gasoline.

Starting with the 1975 model year, U.S. automakers responded to the EPA's lead phase-down timetable by equipping new cars with pollution-reducing catalytic converters designed to run only on unleaded fuel (Gorman 2001).

Conclusion: Lead All Around

Today, the image of lead has changed considerably. Scientific findings have demonstrated conclusively that lead and the compounds that contain lead are poisonous. Although it has been used in numerous consumer products, lead is a toxic metal now known to be harmful to human health if inhaled or ingested. Lead poison collects in the body over time. If a person is exposed to lead over a long period, poison builds up and causes damage to the brain and nervous system. This effect can cause weakness and loss of coordination and mental powers. Smaller bodies, such as children and even fetuses, seem to be even more susceptible to lead.

It is estimated that lead in drinking water contributes between 10 and 20 percent of total lead exposure in young children. In the last few years, federal controls on lead in gasoline have significantly reduced people's exposure to lead (Davis 2002). The single biggest source of lead poisoning, however, remains lead-based paint. Houses built before 1979 that are poorly maintained or being renovated are especially hazardous.

Sources and Further Reading: *Child Lead Poisoning and the Lead Industry*, http://www.sueleadindustry.homestead.com; Cincinnati Children's Hospital Medical Center. *History of Lead Advertising*, http://www.cincinnatichildrens.org/research/project/enviro/hazard/lead/leadadvertising/industryrole.htm; Davis, *When Smoke Ran Like Water*; *Gallery of Lead Pollution Promotions*, http://www.uwsp.edu/geo/courses/geog100/lead-ads.htm; Kinder, *Preventing Lead Poisoning in Children*, http://www.yale.edu/ynhti/curriculum/units/1993/5/93.05.06.x.html#b; Markowitz, *Deceit and Denial*; Opie, *Nature's Nation*.

DONORA SMOG PROVIDES A LESSON ABOUT INDUSTRIAL POLLUTANTS

Time Period: 1948
In This Corner: Residents of industrial communities, health professionals
In the Other Corner: Industrial interests, government officials
Other Interested Parties: Working class
General Environmental Issue(s): Pollution, industrial waste

The perception of environmental hazards in the 1920s and 1930s was very crude. Individual reformers such as Alice Hamilton and other early critics of urban pollution concentrated on the observable and easily traceable impacts that industry had on urban populations. Still, the immediacy of such health concerns was difficult to prove. An important event in this progression occurred on October 30–31, 1948.

Up to this point, air pollution was suspected to cause significant health problems around many factories; however, it had proven to be difficult to specifically trace. On these days in 1948, atmospheric conditions in the vicinity of Donora, Pennsylvania, helped to concentrate toxic emissions and contributed to the deaths of nineteen people within a twenty-four-hour period. Of the fatalities, two had active pulmonary tuberculosis. The other seventeen were known to have had chronic heart disease or asthma. All were between 52 and 85 years of age. In addition, approximately 500 residents of the area became ill, reporting symptoms of respiratory problems. No doubt, countless others suffered in silence (Opie 1998, 454).

Around 1900, Donora had become a prototype of many other industrial cities of the modern era. Incorporated in 1901, its name combined that of Nora Mellon, wife of R. B. Mellon, and W. H. Donner, the purchasers of the land along the river on which their Union Steel Company constructed a rod mill that later became the American Steel and Wire

Works. In 1902, the Carnegie Steel Company completed a facility that consisted of two blast furnaces, twelve open hearth furnaces, and a forty-foot blooming mill furnace. At the same time, the Matthew Woven Wire Fence Company erected a facility. A third rod mill was constructed in 1916. A year earlier, the Donora Zinc Works began production. Such industrial expansion required more effective transportation facilities than the river barges and short-line railroads could provide. The Pennsylvania Railroad bought what had been the Monongahela Valley Company and expanded rail service. By 1908, the Donora station had the largest volume of freight in the "Mon Valley." Of course, these industries needed workers, and job seekers flocked to the area, especially recently arrived immigrants. In 1948, 14,000 people resided in Donora, and additional thousands lived in towns in the immediate vicinity.

For years before 1948, residents complained about the industrial pollutants. An investigation by the state government's Bureau of Industrial Hygiene revealed an extraordinarily high level of sulfur dioxide, soluble sulphants, and fluorides in the air. According to the agency's report and complaints by residents, such contamination of the atmosphere was caused by different portions of the steel-making process, including the zinc smelting plant, steel mills' open hearth furnaces, a sulphuric acid plant, slag dumps, coal-burning steam locomotives, and river boats. Typically, the surrounding hills kept such pollutants from dissipating very quickly, but in addition, in October 1948, an unusually dense fog combined with these other factors kept the pollutants close to the earth's surface where the residents inhaled them, killing nineteen.

The Donora Smog Disaster attracted attention to the potential problems related to air pollution from industrial plants. Donora's experience had been an odd confluence of factors; however, similar environmental contaminants could be found around many American communities. As a result of the Donora events, Pennsylvania established the Division of Air Pollution Control to study the matter in 1949.

The mechanisms for creating laws and agencies to make significant changes on the federal level, however, were decades away. Eventually, members of Pennsylvania's General Assembly felt the pressure to cleanse Pennsylvania's atmosphere of harmful substances. Consequently, the legislature passed the Clean Streams Law in 1965 and began to enact statewide clean air regulations in 1966.

Sources and Further Reading: Davis, *When Smoke Ran Like Water*; Opie, *Nature's Nation*.

THE CONTROVERSIAL EMERGENCE OF ECOLOGY

Time Period: 1900–1950
In This Corner: Trained biologists
In the Other Corner: Older guard of scientific community, defenders of status quo
Other Interested Parties: American public, government officials
General Environmental Issue(s): Ecology, evolution of science

While conservation was taking active form during the early 1900s, an intellectual revolution was changing the scientific ideas of humans' relationship to the natural world. Ultimately, these ideas would converge; however, during this era, conservation remained largely focused on managed development, and the science of ecology remained limited to the biology community. In fact, the public at large heard very little about new ideas of the ecological complexity of the world around humans. This would begin to change in the 1930s (Worster 1994, 221–24). The intellectual shift, however, had roots in the late 1800s.

Centered around the midwestern United States, a group of scientists participated in and eventually led the development of the field of ecology. Henry Chandler Cowles, a plant ecologist, helped to lead this group of scientists. When he began graduate studies in 1895, his faculty at the University of Chicago introduced him to the ideas of a Danish scientist named Eugenius Warming. Cowles supplemented his study of botany with the science of physiography. This combination helped him to better appreciate the importance of landforms as a factor in the shaping patterns of plant life. He incorporated these combined approaches to form a theory of dynamic "vegetational succession" that he first expressed in his 1898 PhD thesis, "The Ecological Relations of the Vegetation on the Sand Dunes of Lake Michigan" (Worster 1994, 206–9).

In his thesis, Cowles used the southern shore of Lake Michigan, which is an area of beaches, sand dunes, bogs, and woods, to demonstrate that the natural succession of plant forms over time could be traced in physical space as one moved inland from the open lake beach across ancient shorelines through the shifting dunes to the interior forest. Along this route, scrubby beach grass would give way to flowers, more substantial woody plants, cottonwoods, and pines would be seen yielding to oaks and hickories, and one would finally encounter the climax forest of beeches and maples. If left alone, Cowles argued, nature had a systematic structure of growth and development all of its own (Cowles, *Ecology and the American Environment*). Of course, this made it clear that humans were a disturbing agent in the natural world. Cowles's thesis had an immediate and far-reaching impact. Published serially in 1899 in the *Botanical Gazette*, "The Ecological Relations of the Vegetation on the Sand Dunes of Lake Michigan" became one of the most influential works in American plant science and quickly established Cowles's reputation as a pioneering American ecologist.

Cowles went on to apply the theory of ecology that he had developed in the Indiana Dunes to the entire range of plant communities found throughout the midwest. He demonstrated that the natural processes of succession and climax were not confined to the isolated dunes. This demonstrated that plant life in any setting had a great deal in common. Therefore, the patterns of change in plant communities could be more effectively tied to climatic or regional variables.

Henry C. Cowles and his theories attracted students from all over the world. As these former students became active scientists, the influence of Cowles's ideas grew larger. One study of scientific influences by Douglas Sprugel in 1980 concluded that, of the seventy-seven recognized American scientists dominant in the field of ecology from 1900 to the early 1950s, no fewer than forty-six were students of Cowles or were directly influenced by professional mentors who had been students of Cowles (Cowles, *Ecology and the American Environment*).

These new concepts possessed an intrinsically new way of viewing nature outside of human existence. Although it would not immediately impact American life, the intellectual principles clearly constructed a worldview that relegated humans to simply being one part of a larger natural story.

The International Phytogeographic Excursion of 1913

The United States acted as an incubator for some of these new scientific understandings that eventually moved abroad. One catalyst for the spread of these ideas occurred from July 1913 to September 1913, when American ecologists led by Cowles hosted the International Phytogeographic Excursion in America. The excursion was a scientific tour of significant natural environments in the United States by a visiting party of the leading European botanical experts of the time (Worster 1994, 209).

Members of the party traveled between cities by rail and made tours of local environments and plant communities. The route of the excursion was east to west, beginning in New York City on July 27 (Tansley, *Ecology and the American Environment*). English botanist and ecologist Arthur Tansley reported on the International Phytogeographic Excursion and wrote, "Certainly no member of the international party will ever forget the overwhelming impressions we received of American landscapes and vegetation, designed truly on the grand scale." In particular, he praised the work of American ecologists when he wrote the following:

> In the vast field of ecology America has secured a commanding position and from the energy and spirit with which the subject is being pursued by very numerous workers and in its most varied aspects, there can be little doubt that her present pre-eminence in this branch of biology—one of the most promising of all modern developments— will be maintained. (Tansley, *Ecology and the American Environment*)

Ultimately, ecology spread the concept of ecosystems throughout the biological sciences. This term is credited to Tansley, who in the 1940s argued that nature occurred in self-sufficient (except for solar energy) ecological systems. He would go on to add that such systems could overlap and interrelate. The existence of such systems, of course, began to suggest that the human agent existed as the interloper in any system. Although the spread of ecological understanding among scientists was a significant change in humans' relationship with nature, the new ideas of ecology now needed to find their way to the general public.

Clements's Idea of Plant Succession

Cowles was not alone in recognizing the significance of dynamic succession for the study of plant communities. Independently, Frederic E. Clements (1874–1945) of the University of Nebraska and the University of Minnesota developed the principles of ecology and based them on his studies of the grasslands and sand hills of Nebraska. Simultaneously, Cowles's former student Victor E. Shelford extended ecological theory into the realm of animal communities.

Clements began his ecological work shortly after Cowles published his influential work. In his own work, Clements argued that vegetation must be understood as a complex organism. Beginning in 1913, Clements and his wife established a laboratory on Pikes Peak in Colorado where they began conducting systematic studies of plant succession (or self-replacement) in the surrounding mountains. Clements documented the environmental influences (temperature, sunlight, evaporation, etc.) on specific plants so that any shift could be explained. Their work demonstrated the complex interrelationship that mountain plants have with surrounding insects and animals. Ultimately, his work established verifiable natural patterns for succession within plant species.

In Shelford's work, he applied Cowles's theories of ecology to the animal world. Specifically, Shelford analyzed the impact of ecological variables on the life histories and habits of tiger beetles. His later studies applied these ideas to species including fishes, moths, antelope, lemmings, owls, and termites. Ultimately, Clements and Shelford combined their ideas of plant and animal ecology in *Bio-Ecology* (1939). Their essential argument was that neither plants nor animals exist in a vacuum; instead, they must be understood within a complex set of factors and variables. This was a crucial step toward creating the ecosystem concept that would organize ecological thought. Ultimately, these ecological ideas would form the foundation of all of environmental thought (Worster 1994, 209–18).

Aldo Leopold and the Land Ethic

The ideal of wilderness received scientific definition through the growing science of ecology and the related development of the concept of ecosystems. One important figure who carried forward the ideas of Clements, Cowles, and others was Aldo Leopold. Eventually, Leopold would be one of the earliest voices to bring these new scientific principles to the public.

After completing a degree in forestry at Yale in 1909 (from the school of forestry begun by Pinchot), Leopold worked for the U.S. Forest Service for nineteen years. Primarily, Leopold worked in the Southwest (New Mexico and Arizona) until he was transferred in 1924 to the Forest Products Laboratory in Madison, Wisconsin. It was while working in the Gila National Forest that Leopold came to a new understanding about the role of the U.S. Forest Service in managing nature. In *A Sand County Almanac, and Sketches Here and There*, he wrote the following:

> We were eating lunch on a high rimrock, at the foot of which a turbulent river elbowed its way. We saw what we thought was a doe fording the torrent, her breast awash in white water. When she climbed the bank toward us and shook out her tail, we realized our error: it was a wolf. A half-dozen others, evidently grown pups, sprang from the willows and all joined in a welcoming melee of wagging tails and playful maulings. What was literally a pile of wolves writhed and tumbled in the center of an open flat at the foot of our rimrock.

> In those days we had never heard of passing up a chance to kill a wolf. In a second we were pumping lead into the pack, but with more excitement than accuracy; how to aim a steep downhill shot is always confusing. When our rifles were empty, the old wolf was down, and a pup was dragging a leg into impassable side-rocks.

> We reached the old wolf in time to watch a fierce green fire dying in her eyes. I realized then, and have known ever since, that there was something new to me in those eyes—something known only to her and to the mountain. I was young then, and full of trigger-itch; I thought that because fewer wolves meant more deer, that no wolves would mean hunters' paradise. But after seeing the green fire die, I sensed that neither the wolf nor the mountain agreed with such a view.

> Since then I have lived to see state after state extirpate its wolves. I have watched the face of many a newly wolfless mountain, and seen the south-facing slopes wrinkle with

a maze of new deer trails. I have seen every edible bush and seedling browsed, first to anaemic desuetude, and then to death. I have seen every edible tree defoliated to the height of a saddlehorn. Such a mountain looks as if someone had given God a new pruning shears, and forbidden Him all other exercise. In the end the starved bones of the hoped-for deer herd, dead of its own too-much, bleach with the bones of the dead sage, or molder under the high-lined junipers. (Leopold 1948, 129–32)

Aldo Leopold, a leading voice for ecology and wilderness, examining tamarack, presumably at his Sand County, Wisconsin, retreat in 1947. Library of Congress.

In 1928, Leopold quit the Forest Service, and, in 1933, he was appointed professor of game management in the agricultural economics department at the University of Wisconsin, Madison. Leopold taught at the University of Wisconsin until his death in 1948. While in Wisconsin, he purchased a worn-out farm and began to experiment with ways of reinvigorating the soils and of managing the site as a cohesive ecosystem. His efforts may have been the first such ecological preservation effort in the United States (Worster 1994, 271–74).

While working in Wisconsin, Aldo Leopold wrote his best known work, *A Sand County Almanac, and Sketches Here and There*, which was published in 1949. A volume of nature sketches and philosophical essays, the *Almanac* is now recognized as one of the enduring expressions of an ecological attitude toward people and the land. Within its pages, Leopold penned a concept known as the "land ethic," which was rooted in his perception of the human's need to exert itself as one component in a larger environment. Ultimately, the land ethic simply enlarges the boundaries of the human community to include soils, waters, plants, and animals, or collectively, the land.

Conclusion: Inspiring Modern Environmentalism

Contemporary environmentalists use this ethic as a way of measuring the impact and implications of human activity on the environment. They ask: can we say that we are giving equal standing to soils, bugs, and the air and water? Can we give standing to an element of nature, such as a species of snail or a fish, when it slows or limits human opportunity and development? The land ethic strives to give more equal standing to these other elements of nature. Contemporary environmentalists who are able to fully commit to this ethic often refer to themselves as "deep ecologists." Throughout the late twentieth century, however, some variation of Leopold's land ethic could be found in much of the environmental policies that became a normal part of American life.

Although Leopold's writing would construct a new environmental ethic for generations to come, federal programs had begun applying some of the lessons of the new ecology by the 1930s. However, these early efforts rarely reached far enough to meet with approval from Leopold and other idealists.

Sources and Further Reading: Clements, *Ecology and the American Environment*; Cowles, *Ecology and the American Environment*; Leopold, *A Sand County Almanac, and Sketches Here and There*; Worster, *Nature's Economy: A History of Ecological Ideas*.

FRANKLIN D. ROOSEVELT IMPLEMENTS CONSERVATION POLICIES

Time Period: 1930s
In This Corner: Roosevelt, scientific progressives
In the Other Corner: Existing government representatives
Other Interested Parties: American public
General Environmental Issue(s): Policy, conservation, New Deal

Ecology only emerged as a bona fide field in the 1930s; however, it quickly became the basis of portions of the massive works projects of the New Deal. The policies of the New Deal were intended to be about new ideas; however, the pace of change and the use of the federal government to administer significant portions of everyday American life left many Americans concerned. The policies of the New Deal, therefore, did not arrive without debate and criticism.

A New Deal of Conservation

When Franklin Delano Roosevelt (FDR) took office in 1933, he sought the advice of modern-thinking experts in many fields. Looking to colleges and universities, FDR inserted intellectuals immediately into the emergency of the Great Depression. Both he and the American people expected results. With a long-term interest in the science of forestry and resource management, FDR was particularly struck by the waste of American natural resources at a time of great need. In his inaugural address, FDR stated, "Nature still offers her bounty and human efforts have multiplied it. Plenty is at our doorstep, but a generous use of it languishes in the very sight of the supply." His initiatives sought to intelligently use these resources while creating jobs for out-of-work Americans (Henderson and Woolner 2004, 35).

These policies incorporated the emerging ecology with federal policies to manage watersheds, maintain forests, teach agriculture, and hold fast the flying soils of the southern plains. The main impetus for federal action derived from a national surge in joblessness. The

CCC enrollees doing soil conservation work. The workers, usually white men, were employed on similar projects throughout the United States. Library of Congress.

economic collapse of 1929 left millions of Americans incapable of making a living. Nowhere was this more evident than on the American southern plains (Worster 1979, 182–85).

Terrible drought combined with economic difficulty to make many farmers in the rural midwestern United States incapable of farming. Residents of Oklahoma (nicknamed Okies) fled westward to California, creating resettlement problems as well. In the southern plains, the loose topsoil was lifted by heavy winds, creating dust storms of epic proportions. Press coverage of the Dust Bowl of the 1930s presented a natural disaster caused by drought and bad luck. Through government-made documentary films such as *The Plow That Broke the Plains*, the New Deal infused a bit of ecological background to explain desertification and agricultural practices that can be used to combat it. In the process of a natural disaster, the American public learned a great deal about its role within the natural environment. Proper land use could be taught, and the federal government installed extension agents to do so.

This was also apparent in New Deal river projects, particularly the Tennessee Valley Authority (TVA). Finally, FDR's pet project, the CCC, merged the previous Roosevelt's trust in the importance of work in the outdoors for the development of young Americans with scientific understandings of agriculture and watershed management. CCC projects often grew from lessons of ecology—for instance, the need to construct "shelter belts" of trees to help block the wind on the plains and to keep topsoil in place—but their most important priority was creating employment opportunities for young men.

Overall, the emergence of ecology had brought a new utility for science in the everyday life of Americans. Scientific knowledge, however, was still largely controlled by experts, often working for the federal government. During World War II, science would be placed more in the public eye than ever before (Worster 1994, 339–41).

Tennessee Valley Authority Creates a Model of Planning

Moving beyond City Beautiful, aesthetic design merged with science and sociology during the 1930s into a bona fide scholarly field known as "planning." Although many CCC efforts fell under the rubric of planning, the greatest example of New Deal planning was the TVA. The entire watershed of the Tennessee River contributed to flooding problems on its banks and along the Mississippi River before 1933. In this year, FDR created the TVA to manage the entire watershed through a system of dams and other structures. The land management system, based in ecology, would restore lost topsoils, prevent floods, stabilize transportation possibilities, and create the opportunity for recreation (Henderson and Woolner, 182–84).

TVA was an idea based in idealism, and the New Deal offered the forum through which utopian schemes might be put into practice. The TVA Act of 1933 called for the U.S. government to finance, plan, and carry out the revitalization of a depleted region by constructing a series of dams along the Tennessee River to harness the river's potential for generating power while also tempering its flow to prevent flooding. The first TVA project, Norris Dam, cost $34 million to erect. During the 1930s, the United States invested $300 million in TVA projects, creating eight dams along the Tennessee. By 1945, TVA would double the number of dams and in the process put thousands of people to work. These projects, of course, also created controversy by pushing hundreds of residents from their homes.

From the TVA perspective, untrained humans could no longer be entrusted with this vulnerable ecological legacy. Such planning was based on the conservation ethic of efficiency and functionality but especially on limiting waste. In his inaugural address, FDR revealed the forces that would drive many of his New Deal policies when he said, "Nature still offers her bounty and human efforts have multiplied it. Plenty is at our doorstep, but a generous use of it languishes in the very sight of the supply." The wealth was available, he suggested, if the management of such resources were conducted with more care.

TVA began with similar aspirations in the minds of Senator George Norris of Nebraska and Arthur Morgan, the founding director of TVA. The regional planning movement grew out of the modern field of human engineering that Norris and others believed could resurrect a downtrodden portion of society. Morgan arrived as a visionary engineer with extensive experience in river management. For his chief planner, Morgan, the utopian dam builder, resisted pressure by Lewis Mumford, the modernist thinker and designer, and others to hire a regional planning enthusiast; instead, Morgan chose Earle Draper, a landscape architect.

Unlike many previous products of landscape architecture, however, the TVA landscape was intended as a multiuse area based in conservation. Morgan and Draper's TVA represented a national trend. As journals for planning were founded and forty-two state planning boards were established under the guidance of one national planning board in the 1920s, planners and landscape architects found themselves in great demand. The magical hand of federal authority gave designers carte blanche to make this valley an ideal, no matter if the plan necessitated moving families, towns, or forests to more convenient locations.

Taming the river led to many ancillary duties, including relocation, town planning, power distribution, and recreational administration. The Electric Home and Farm Authority, for instance, facilitated the purchase of low-cost appliances that could be powered by TVA electricity. Before TVA, 97 percent of those living in the area had no electricity. TVA's control of the river had other accomplishments beyond flood control, including the reduction of cases of malaria, which was thought to be endemic to the region. With these additional duties in mind, TVA conceived of its purpose in a much more far-sighted manner than simply constructing dams. Once the river was controlled and soil runoff eased, TVA planners set out to make sure that the land would not be depleted again. This was to be accomplished through education.

For 1930s Americans, the practices and products of TVA acted as concrete symbols or archetypes to help the public understand conservation as a concept. The public's education represented the clearest way to solidify support of TVA and conservation. The visual aesthetic of well-managed land, specifically landscape architecture, knitted TVA's system of conservation into a cohesive form and impressed Americans with its pleasing appearance. Landscape architects carried out this vision as they shaped the scenery and functionality of the 41,000 square miles of the Tennessee Valley.

Undoubtedly, TVA designers construed their dams much differently than had the Corps of Engineers or Bureau of Reclamation. The act of stopping a river becomes so momentous that observers equate all such structures with this awesome intention. However, there is a great deal of structural diversity within dam design, and many of these variations suggest different intentions motivating construction. Dams of TVA were neither planned as landmarks nor intended to dominate the natural environment; instead, they were a portion of an integrated system of managed nature that would conserve the resources of the entire river valley. They were a cog in the machine of conservation.

In addition to controlling the floods and erosion of the Tennessee, TVA dams helped to solidify the effectiveness of scientifically based planning and land use in the popular consciousness. One 1940 observer commented that TVA had perfected "the architecture of public relations." For the first time, conservation was able to be envisioned clearly as a distinct scientific act of planning followed by action that led to improved natural resources.

These systems of recovery, development, and conservation proved revolutionary in American land use. However, the idea for a national system of regional planning authorities dissolved in the shadow of TVA experience. The legal fights that befell TVA in the late 1930s crystallized the American discomfort with the blurring of the line between government and private economic development (these cases essentially involved the right of the federal agency to seize private property to build facilities such as dams and power plants). Indeed, such wrangling seems a precursor of current angst focused against the EPA and general environmental regulation.

After legal decisions favoring TVA, it may have been possible for regional planning to regain the support seen in the early 1930s; however, World War II forever changed TVA and also, by association, took with it the dreams of utopian planning. In essence, the war burst the original aspirations of planning and quickly changed the job and image of TVA to that of a wartime agency. The outbreak of World War II brought more power needs to TVA, including that needed by the Oak Ridge facility to produce atomic weapons from uranium 235. TVA's power capacity more than doubled to 2.5 million kilowatts in 1946. TVA also diversified its modes of energy production to include coal mining, particularly large-scale strip mining. The emphasis on energy production made TVA's original mandate fade, permanently divorcing the agency from its utopian roots in regional planning.

"The usefulness of the entire Tennessee River," FDR had instructed in 1933, "transcends mere power development; it enters the wide fields of flood control, soil erosion, reforestation, elimination from agricultural use of marginal lands, and distribution and diversification of industry." By 1945, TVA by most assessments had accomplished these goals and more. These early accomplishments place TVA as a watershed in American land-use planning, even suggesting an important role for TVA in the development of modern environmentalism. However, as TVA's mandate shifted to emphasize power generation after World War II, it rightfully earned the ire of many environmentalists, causing most observers to forget its crucial early role in redefining the federal government's role in planning the use of the natural environment.

TVA's role in the complicated twentieth-century movement toward federal environmental regulation suggests the limits of engineering and planning's role in social or ethical change in American society. The ethical role of the TVA utopia only functioned while it was supported by New Deal rhetoric and propaganda. These ideas linger in American society but find new applications and vehicles. TVA's planned landscape, instead of symbolizing effective conservation, has become an accepted, functioning portion of the natural landscape. TVA's basic legacy has become a defining role for the federal government in regulating and orchestrating the American relationship with the natural environment (Henderson and Woolner, 148).

Conclusion: The Depression and Modern Land Planning

TVA marked a linkage between ecological regions and community development. However, planning was also being applied to residential communities that would ultimately become an important step in the growth of the American suburb after World War II. In the 1910s, philanthropists and investors launched the first two experiments in Letchworth and Welwyn,

England, and then in the 1920s came Radburn, New Jersey. These projects marked the beginning of a new genre of community planning. The emergency of the Great Depression and the Dust Bowl brought the need for extreme measures in planning and development. Under the auspices of Roosevelt's New Deal, Rexford Tugwell initiated the Farm Resettlement Administration, which planned the construction of four greenbelt towns that prioritized new ideas for community design. The first of these would be named Greenbelt, Maryland, which proved to be the largest New Deal project in the Washington, DC area.

The economic difficulties of the 1930s resulted in a significant displacement of rural Americans. The primary goal of the Resettlement Administration was to establish, maintain, and operate communities for the resettlement of destitute or low-income families. For Greenbelt, the Resettlement Administration acquired nineteen square miles near Berwyn, Maryland. In 1937 and 1938, approximately 9,000 workers constructed the new community at a price of approximately $13 million.

Serving as the New Deal's primary model community, Greenbelt marked the most ambitious housing project in American history to that point. Because such centralized planning required authoritative action, many critics called the undertaking socialism. Anti–New Dealists dubbed Greenbelt "the first communist town in America" (Callcott 1985, 73). Enemies of the project protested the excessive power the government would gain over the development of the community. In 1952, Congress forced the government to sell the experiment. Before this, however, a community had been formed and informed by the innovative planning and design elements incorporated into the fabric of Greenbelt.

After World War II, American housing became dominated by prefabricated suburbs; however, in the late twentieth century, architects and designers searched for a better model. The greenbelt towns provided an important example that resembled the ecosystem concept. Green planners sought to create a human living environment that possessed the least possible impact on the surrounding environment. Eventually, these planned communities inspired architects to design more sustainable communities, a project described as "New Urbanism."

In housing and other areas, the inter-war years brought ideas of conservation to action. Combined with strands of the ecological ideas reforming the biological sciences during these years, human planning and resource management took on a scientific quality not seen previously. Most often, the conservation impulse required the involvement of the federal government as an authoritative or regulative influence. Although this model met with some success in the 1920s and 1930s, it also linked environmental planning to politics in a way that would remain volatile. Throughout the twentieth century, the government's ability to integrate and administer new scientific understanding would vary with social, political, economic, and foreign affairs factors.

Sources and Further Reading: Callcott, *Maryland and America*; Henderson and Woolner, *FDR and the Environment*; Maher, "Neil Maher on Shooting the Moon"; Worster, *Nature's Economy: A History of Ecological Ideas*.

NONGOVERNMENTAL ORGANIZATIONS EMERGE AS AN ESSENTIAL MECHANISM FOR ENVIRONMENTAL ACTION

Time Period: 1970 to the present
In This Corner: Environmental organizations, concerned citizens

In the Other Corner: Government leaders, lobbyists
Other Interested Parties: Government leaders, industry, consumers, politicians
General Environmental Issue(s): Policy, environmental action, NGOs

The 1960s counterculture contributed to the development of institutions that would change basic relationships in American life. The American relationship with nature was one of the most prominent shifts. Much of what became known as the modern environmental movement was organized around groups and organizations that prospered with the influence of 1960s radicalism; however, the real impact of these organizations came during the later 1960s and 1970s when their membership skyrocketed with members of the concerned middle class, such as Lois Gibbs and members of the Love Canal Homeowners Association (LCHA).

A prime catalyst for the growth in grassroots environmental efforts was Earth Day 1970. Events all around the United States included acts of civil disobedience, as well as educational lectures and teach-ins. In many communities, Earth Day spurred the establishment of ecology centers and programs to implement environmentally friendly practices such as organic farming and recycling. In many locales, these efforts became institutionalized on a permanent basis. Nationally, these ideas were dispersed through other means.

During the late 1900s, many of these environmental special-interest groups would evolve into major political players through lobbying. Nongovernmental organizations (NGOs) broadened the grassroots influence of environmental thought; however, they also created a niche for more radical environmentalists. The broad appeal as well as the number of special-interest portions of environmental thought stood in stark contrast to nineteenth-century environmentalism. Whereas early conservationists were almost entirely members of the upper economic classes of American society, the new environmentalists came most from the middle class that grew rapidly after World War II (Opie 1998, 418–25).

During the 1970s and 1980s, these NGOs helped to bring environmental concern into mainstream American culture. Some critics argue that American living patterns changed little; however, the awareness and concern over human society's impact on nature had reached an all-time high in American history. These organizations often initiated the call for specific policies and then lobbied members of Congress to create legislation. By the 1980s, NGOs had created a new political battlefield as each side of environmental arguments lobbied lawmakers.

Often, the American public financially supported organizations that argued for their particular perspectives. Even traditional environmental organizations such as the Sierra Club (1892), National Audubon Society (1905), National Parks and Conservation Society (1919), Wilderness Society (1935), National Wildlife Federation (1936), and Nature Conservancy (1951) took much more active roles in policy making. The interest of such organizations in appealing to mainstream, middle-class Americans helped to broaden the base of environmental activists. However, it also contributed to the formation of more radical-thinking environmental NGOs that disliked the mainstream interests of the larger organizations. In fact, many devout environmentalists argued that some of these NGOs were part of the "establishment" that they wanted to fight.

Sources and Further Reading: Gottleib, *Forcing the Spring: The Transformation of the American Environmental Movement*; Nash, *Wilderness and the American Mind*; Opie, *Nature's Nation*.

PLASTICS FORM THE CORE OF AMERICA'S DISPOSABLE SOCIETY

Time Period: 1930s to the present
In This Corner: Plastics manufacturers, petroleum suppliers, homemakers and others bene-
   fiting from labor savings
In the Other Corner: Waste disposers/managers, environmentalists
Other Interested Parties: Policy makers
General Environmental Issue(s): Wastes, trash, plastics

When is the price of convenience too high? As the consumerism of the late twentieth century fed American enchantment with a "disposable" society, plastics—created from petroleum—became a mainstay of our culture. Since the 1950s, plastics have grown into a major
industry that affects all of our lives, from providing more effective packaging of our food or
other store-bought items to enabling wondrous new gadgets such as laptop computers, televi-
sions, and cell phones. Plastics now even allow doctors to replace worn-out body parts. In
fact, since 1976, plastic has been the most used material in the world, and its evolution has
been called one of the top 100 news events of the century (Meikle 1997, 12).

Plastics started well before Dustin Hoffman's character in *The Graduate* referred to them
in the 1960s. The creation of synthetic materials that are related to plastics began in 1907
when a New York chemist named Leo Baekeland developed a liquid material that when
cooled hardened into a replica of whatever form one chose. He called this resin material Bakelite. This new material was the first thermoset plastic, which meant that it would not lose
the shape that it had taken. In fact, Bakelite would not burn, boil, melt, or dissolve in any
commonly available acid or solvent!

In this same general product genre, inventors in the early 1900s developed products
such as rayon and cellophane. Very often, large chemical companies such as DuPont had
researchers constantly working in laboratories to develop any synthetic material that might
prove to be useful. DuPont developed nylon during the 1930s but did not make the first
pair of stockings until 1939. Many similar innovations also occurred in the 1930s, including polyvinyl chloride, vinyl Saran wrap, Teflon, and polyethylene. Although each of these
items possessed well-known domestic uses, most of them were first used in other substances. For instance, during World War II, polyethylene was used first as an underwater cable
coating and then as a critical material insulating radar units. By decreasing the weight of
radar units, this material made the technology more portable so that it could be placed on
planes.

After World War II, many of these innovative substances found roles in the new American culture of conspicuous consumption. For instance, polyethylene became the first plastic
in the United States to sell more than a billion pounds per year. It is currently used for most
soda bottles, milk jugs, and grocery and dry-cleaning bags. Of course, the downside of the
durability of many plastics is that now they are a major component of the solid waste being
thrown away by American consumers.

Today, five resins account for nearly 60 percent of all plastics used by consumers: (1)
low-density polyethylene, used in garbage bags, (2) polyvinyl chloride, used in cooking oil
bottles, (3) high-density polyethylene, used in milk jugs, (4) polypropylene, used in car battery cases, and (5) polystyrene, used in disposable food containers.

Each year more than 93 billion plastic drink containers end up in U.S. landfills. Recycled plastic bottles have had a second use in commercial products such as plastic chairs, kayaks, jewelry and even in clothing. AP Photo/Rich Pedroncelli.

One last resin, polyethylene terephthalate, is produced in much smaller quantities. It is used as a plastic in soft drink bottles. In 2002, about 107 billion pounds of plastic were produced in North America. Of all of this plastic, one-third is estimated to be packaging for other products. Of course, this packaging plastic goes directly into the waste stream once the product has been purchased. Nondegradable plastic packaging is blamed for filling commercial landfills, increasing their operational expense, contaminating the environment, and posing a threat to animal and marine life.

Together, this plastic waste accounts for about one-quarter of all municipal solid waste. This is a particularly big problem because of plastic's remarkable durability. In addition, there are significantly toxic wastes involved in manufacturing plastics. Many companies have recently been experimenting with making polymers that degrade more rapidly (such as with lactic acid) and could be used for packaging and other products that do not require the durability of some others.

Sources and Further Reading: Meikle, *American Plastic: A Cultural History*; Society of Plastics Industry, *History of Plastics*, http://www.plasticsindustry.org/industry/history.htm.

NATIVE RIGHTS TO WHALE, FISH, AND HAVE WATER RIGHTS IN CONTEMPORARY AMERICA

Time Period: 1900s
In This Corner: Native American tribes
In the Other Corner: Legal precedent, non-native fisherman
Other Interested Parties: Political officials
General Environmental Issue(s): Native Americans, fishing, whales, coastal areas

The rights of native groups have been debated since the settlement era of the United States. Certain topics, however, have kept this issue alive into the present day. Most often, these issues involve the use of common resources, such as water, petroleum, or animals. Debate over treaty and property rights has often focused on issues such as the taking of fish and whales, because very often native groups claim that such activities possess cultural significance. In such instances, native residents might claim that state or local laws have no jurisdiction over them.

In the case of fishing rights, the claims of native groups derive at least partly from the unique cosmology of many Native American people. Traditionally Native Americans have had an immediate and reciprocal relationship with their natural environments. When Europeans arrived around 1500, most native groups defined themselves by the land and regional characteristics of their location. Having relied on the resources of the natural world for many years, each people formed explanations and ways of perceiving the cycles of the world around them. These cultural ideas shaped each people's identity—its view of how it fit into the larger world. Today, scholars refer to this idea as a philosophy or cosmology.

For most native communities, their cosmologies connected them in mutual relationships with all animate and inanimate beings. In most cases, their cosmology allowed natives to acknowledge the power of Mother Earth and the mutual obligation between hunter and hunted as coequals. Using rituals, many native groups made preparing and killing animals or growing crops in fields a way of connecting themselves into the larger natural cycles around them. Often, men and women used song and ritual speech to modify their world, whereas their labor functioned to physically alter the landscape in methods that were productive for them. This is not to suggest that they passively adapted to their surroundings; instead, they continuously adjusted their surrounding environments to meet their cultural as well as material desires.

The impact of European settlement on these traditional ways of life was significant. Although these changes will be discussed in other essays, the designation of reservations often displaced peoples from their traditional regions. Many reservations were located on land that was largely unwanted or remote. In many of these cases, the lands were without economic value and almost certainly different from previous sites of native residence. The problem became even more complicated when a great deal of this reservation land was lost with the General Allotment Act of 1887, which divided reservations into individual holdings. By the beginning of the twentieth century, Native Americans controlled mere remnants of their former estates, most in the trans-Mississippi West.

Collective Rights and Environmental Problems

One of the specific purposes of the Allotment Act was to undermine or even destroy opportunity for collective action by native peoples. In addition to multiple factors that threatened their cultural traditions, native people throughout the United States (but particularly in the American West) lost much of their ability to politically call for proper treatment. In the case of issues relating to the environment and natural resource use, this development was particularly problematic.

Whereas environmental problems are shared by other minority groups in the United States, Native Americans add the unique consideration that, in addition to being U.S. citizens, many of them are also members of federally recognized sovereign nations that, in theory, have the authority to manage their environmental problems independently. This issue is referred to as sovereignty.

As sovereign nations, tribes have a number of opportunities not open to most Americans, including making laws governing the conduct of Indians in "Indian country" (an all-encompassing term that refers to all existing American Indian tribes, governments, people, and territory), establishing tribal police and court systems, levying taxes, and regulating hunting, fishing, land use, and environmental pollution. Thus, tribal nations may have it both ways: they can try to take advantage of their sovereignty from laws and regulations or, similar to individual states, they can apply for and assume enforcement responsibility for federal environmental programs.

Given this confusing status, negotiating environmental and natural resource issues has been very complex over the last century. One of the most problematic issues is the very existence of tribes as sovereign nations (561 federally recognized American Indian tribes are supposed to negotiate with the United States on a government-to-government basis) and the existence of a variety of treaties and stipulations, many more than two centuries old. The most recent era and this legal interpretation is referred to as the "Self-determination Era," in which the United States shifted as much regulatory authority to the tribes as possible. This policy has been reinforced by the Clinton administration, which in 1994 issued a presidential memorandum entitled "Government to Government Relations with Native American Tribal Governments." In this memorandum, the United States encouraged federal agencies to negotiate with tribal authority over natural resource issues in a "knowledgeable, sensitive manner respectful of Tribal sovereignty."

Negotiating the policies is just one portion of the problem, however. If federal agencies approve tribal regulatory programs, enforcement responsibility on the reservations generally still falls to the EPA or similar agencies. For instance, there are approximately 850 industrial facilities located on tribal lands that must be tracked and monitored. Limited resources have contributed to such federal agencies consistently neglecting their managerial and enforcement responsibilities.

In the twentieth century, however, a series of important developments influenced the small amount of reservation land: first, in some cases, the valueless land proved to possess valuable resources unappreciated by previous generations; second, the Indian movement grew a political voice capable of voicing protest; and, last, contemporary Native Americans became increasingly intrigued by reconnecting with cultural traditions, including the adherence to and appreciation of their people's cosmology. Hunting and fishing was one of the traditional activities that most appealed to native peoples wanting to reconnect with their cultural roots.

Tribal policies vary with political preferences, of course, and also with court rulings. The rest of this article will look at a series of case studies of recent cases related to issues of Native American resource use.

Case Study: Native American Fishing Rights

Although there are fishing issues related to native peoples throughout the Great Lakes region, Washington State has come to represent an important focus point for litigating the legal frame-work of native rights. By most estimations, many regional tribes had been ensured the right to fish at "usual and accustomed grounds and stations" by federal treaties signed in 1854 and 1855. As settlement increased through 1900, however, Euro-American immigrants displaced native peoples and usurped the rights that had been guaranteed by a previous generation.

In the early twentieth century, under pressure from commercial and sports fishermen, state and federal officials limited Indian off-reservation hunting and fishing. These regula-tions hit native fishermen in the Northwest particularly hard. They were competing with a growing number of commercial operations and losing native fishing sites to dams. As the fight for Native American rights gathered steam in the 1960s, regional tribes organized the first "fish-ins" on the Puyallup River in 1964. These organized protests sought to openly defy Washington State's attempts to regulate their fishing.

As suits and countersuits flew between native and government groups, in 1970, sixty per-sons (Native Americans and their supporters) were arrested in Tacoma for failing to disperse during a fish-in on the Puyallup River. The trial that followed in 1973 turned into a watershed ruling, normally referred to as the Boldt ruling (named for the presiding Judge George Boldt).

The "Boldt Decision" reaffirmed the rights of Washington's Indian tribes to fish in accus-tomed places. In addition, the ruling allocated 50 percent of the annual catch to treaty tribes. This allocation, of course, enraged non-native fisherman, who felt that they caught well more than half of the annual amount. In the same ruling, however, Boldt denied any federal recog-nition to tribes such as the Samish, Snoqualmie, Steilacoom, and Duwamish who owned no land. The ruling specifically stated that the government's promise to secure the fisheries for the tribes was central to the treaty-making process and that the tribes had an original right to the fish, which they extended to white settlers. It was not up to the state to tell the tribes how to manage something that had always belonged to them. Based on this logic, Judge Boldt actually ordered the state to take action to limit fishing by non-Indians. The Boldt Decision led to violent clashes between tribal and nontribal fishermen and regulators. Once the Boldt ruling was upheld by the U.S. Supreme Court in 1979, its logic has been applied to other resources, including shellfish.

In other areas of the United States, similar treaty issues have been litigated. In Wis-consin, whites and Chippewa fishermen disputed over the natives' right to practice off-reservation hunting and spearfishing. In this case, native groups qualified for 50 percent of the annual harvest, which was well beyond what they achieved as fishermen.

In other parts of the United States, white sportsmen and environmentalists have joined together to protest native rights to kill bald eagles, bowhead whales, Florida panthers, or other endangered species guaranteed under the NEPA of 1969. The debate has emphasized which federal law is dominant, the ESA or the American Indian Religious Freedom Act of 1978. The acts have safeguarded and allowed Indian access to sacred nonreservation areas and resources and injected a level of legal tolerance to native religious practices that revolve around resource use. Together, they have created a confusing and ongoing debate over native rights over fishing.

Case Study: Native American Whaling

The issue of whaling has not involved many tribes; however, one example in particular has attracted international attention. In 1999 and 2000, after a hiatus of seven decades, Makah Indian whalers again hunted gray whales from their ancestral lands around Cape Flattery on the Olympic Peninsula of Washington. The Makah, whose whaling tradition dates back thousands of years, are the only tribe in the United States with a treaty guaranteeing the right to hunt whales.

The Makah had last taken a whale in the 1920s. They argued that their younger generation needed to find a link to their traditional heritage. When the gray whale was removed from the Endangered Species List in 1994, they announced that they would resume their traditional hunt.

In an editorial, the Makah chief responded as follows:

Have I lost my culture? No. I come from a whaling family. My great grandfather, Andrew Johnson, was a whaler. He landed his last whale in 1907. My grandfather, Sam Johnson, was present when the whale was landed and told me he played on the whale's tail. I lived with my grandfather for 16 years and heard his stories about our whaling tradition and the stories of family whaling told by my father Percy and my uncle Clifford. When I was a teenager I was initiated into Makah whaling rituals by my uncle Clifford. While I cannot divulge the details of these rituals, which are sacred, they involve isolation, bathing in icy waters and other forms of ritual cleansing. These rituals are still practiced today and I have been undergoing rituals to prepare me for the whaling which is to come this year. Other families are using their own rituals.

I can tell you that all of the Makah whalers are deeply stirred by the prospect of whaling. We are undergoing a process of mental and physical toughening now. I feel the cultural connection to whaling in my blood. I feel it is honoring my blood to go whaling. We are committed to this because it is our connection to our Tribal culture and because it is a treaty right—not because we see the prospect of money. We are willing to risk our lives for no money at all. The only reward we will receive will be the spiritual satisfaction of hunting and dispatching the whale and bringing it back to our people to be distributed as food and exercising our treaty right.

He continued that, in 1855, the treaty was negotiated based on the stated understanding that "The Great Father knows what whalers you are—how you go far to sea to take whale."

With this fact in mind, the U.S. territorial governor promised U.S. assistance in promoting the Makah whaling commerce, including the only treaty known to contain a guarantee of the Makah right to take whales. The chief continued, "The Treaty was ratified by the Congress in 1855 and has since been upheld by all the Courts and the Supreme Court. To us it is as powerful and meaningful a document as the U.S. Constitution is to you, because it is what our forefathers bequeathed to us. In fact, one of our whalers has said that when he is in the canoe whaling, he will be reaching back in time and holding hands with his great grandfathers, who wanted us to be able to whale."

The contemporary decision by the Makah became an international flash point for native rights legal cases. Some animal rights activists bitterly denounced the Makah, but other groups, from advocates for indigenous rights to the United States government, supported the tribe's right to hunt. The Makah whaling crew continued to prepare and practice, paddling the canoe *Hummingbird* out to sea almost daily. After legal battles and physical confrontations with protestors, Makah whalers landed their first whale in more than seventy years in May 1999.

Even during this historic hunt, as the *Hummingbird* approached the whale, speedboats and Zodiacs from the protest group Sea Defense Alliance tried to stop them. Protestors threw things at the canoe and fired fire extinguishers. Twice, harpooner Theron Parker threw the harpoon at a whale but missed. When the Makah finally harpooned a thirty-foot gray whale they subsequently shot and killed it with the rifle.

Although the tribe argues that such a hunt is a guaranteed treaty right of the Makah, they specifically believe the hunt represents an important way for them to perpetuate age-old cultural traditions.

Case Study: Dams and Water Use

One last dimension of water and fish rights for native people occurs with the interest of state and local governments to construct dams on rivers used by Indians. As federal efforts to develop western rivers took off after the Newlands Act, many native users found that they had very little power to stand up to American irrigation interests. In the case of Truckee River in Washington, for instance, the dam helped to cause the extinction of the Lahontan trout, which was the primary source of subsistence for certain native groups, including the Pyramid Paiutes.

The most serious issue, however, has evolved since the dams built on the Columbia River beginning in the 1930s have been found to impede the migration of salmon and other anadromous species. Indian fisheries such as Celilo Falls in Washington have been permanently altered, both in terms of their sacredness to local cultures and their ability to act as spawning grounds. In a similar situation on the Missouri River, the Pick-Sloan Plan for damming and flood control destroyed agricultural lands and sacred areas of the Indians of the Standing Rock, Cheyenne River, Crow Creek, and Fort Berthold reservations. In another case, reservations along the Grand Canyon have battled over aquatic species preservation and Indian fishing rights and the ownership and sale of water.

Federal Attempts to Manage Native Water Quality

Tribes today have a variety of legal mechanisms available to regulate and manage reservation water quality. The Clean Water Act provides most of this legal opportunity; however, this act

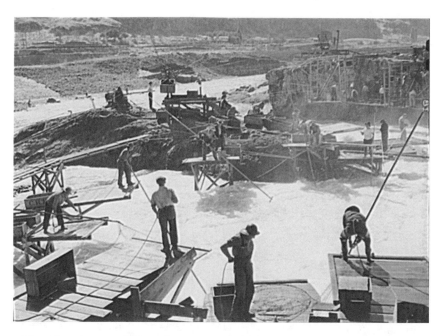

Indians fishing for salmon at Celilo Falls, Oregon, one of the sites of debate over treaty rights for fishing. Library of Congress.

does not provide protection for reservations against water quality degradation from upstream water diversions and uses. In such cases, tribes must file suits that assert their right not only to a specific amount of water but also for that water's quality to be sufficient for the tribal needs.

The legal expectation for water quality is a requirement of the federal government when it reserves public land as an Indian reservation (or also as a military reservation, national park, forest, or monument). In the 1908 U.S. Supreme Court ruling *Winters v. United States*, the government was required to reserve sufficient water to satisfy the purposes for which the reservation was created. In this case, the U.S. Supreme Court actually found that an Indian reservation (the Fort Belknap Indian Reservation) might reserve future water for an amount necessary to fulfill the identified purpose of the reservation. The Winters Doctrine, as the ruling is known, established that, when the federal government created a reservation, it also was duty bound to consider its future water supplies for agriculture or whatever was identified as the typical activity of the tribe.

Even before the Reclamation Act, the Winters Doctrine deviated from the established convention in that water law was purely a state matter. Although it continues to control such cases today, there have been legal adjustments. In 1952, Congress passed the McCarren Amendment and returned substantial power to the states with respect to the management of water. Essentially, the McCarren Amendment requires that the federal government waive its sovereign immunity regarding water issues and at least participate in the state's water allocation system.

Federal court decisions since the McCarren Amendment have placed additional limitations on the federal government.

The Rio Grande River, in the Bosque del Apache National Wildlife Refuge, New Mexico. In 2004, more than three dozen endangered Rio Grande silvery minnows were found dead and more than 4,200 of the tiny fish were rescued as a stretch of the Rio Grande went dry due to heavy use of the water upstream. AP Photo/*Albuquerque Journal*, Marla Brose.

Rulings were especially stringent on limiting tribes to primary purposes of the reservation. Less and less consideration was allowed for secondary uses of reservation land. Today, these rulings have narrowed the scope of the Winters Doctrine until the federally reserved water rights may only apply to the minimum requirements for the primary purpose.

Conclusion: A Legacy of Conquest

Through all of these issues, the effort of native peoples to exert their rights is complicated by the fact that there is no Indian consensus, no tribal consensus on how to proceed. These issues have been adjudicated on a case-by-case basis as tribes address use and preservation issues specific to their people. In many cases, tribal decisions and preferences have not been environmentally benign. There are many cases in which Indian individuals have not always acted in the best long-term interests of the environment. However, similar to American society at large, native environmentalists have challenged the politics of their own tribal development and faced down government agencies and multinational corporations.

Some of these native organizations include the Indigenous Environmental Network, Native Americans for a Cleaner Environment, Kaibab Earthkeepers, and Dinè CARE. For tribal groups, their fight for rights over resource use begins with maintaining their sovereignty. This autonomy, then, allows them to file suits against the U.S. government concerning the future of their land and resource use.

Sources and Further Reading: Ambler, *Breaking the Bonds: Indian Control of Energy Development*; Colorado Plateau Land Use History of North America, *Native Americans and the Environment: A Survey of Twentieth Century Issues with Particular Reference to Peoples of the Colorado Plateau and Southwest*, http://cpluhna.nau.edu/Research/native_americans4.htm; Eichstaedt, *If You Poison Us: Uranium and Native Americans*; Fixico, *The Invasion of Indian Country in the Twentieth Century: American Capitalism & Tribal Natural Resources*; Krech, *The Ecological Indians: Myth and History*; Lewis, *Neither Wolf nor Dog: American Indians, Environment, and Agrarian Change*; Limerick, *A Legacy of Conquest*; National Council for Science and the Environment, *Chippewa Treaty Rights: History and Management in Minnesota and Wisconsin*, http://ncseonline.org/nae/docs/chippewa.html and http://ncseonline.org/NAE/fishing.html; Royster, "Water Quality and the Winters Doctrine"; White, *It's Your Misfortune and None of My Own*.

INTERNATIONAL WHALING COMMISSION EMERGES FROM GLOBAL ENVIRONMENTAL CONCERN

Time Period: 1930 to the present
In This Corner: International whalers
In the Other Corner: Environmentalists, industrial leaders
Other Interested Parties: Scientific community
General Environmental Issue(s): Species management

While the commons idea fueled new ideas about global policies in the late twentieth century, the sea commons had already been responsible for bringing nations together for the sake of one of its most well-known beings: the whale. By the early 1900s, the international whaling fleet had become one of the world's most efficient industries. With whale products used in many manufactured goods (including explosives), whalers from Britain, Norway, and Japan perfected the location and processing of whales in the international waters of both polar areas. By the 1920s, the League of Nations urged the international community to begin monitoring whale populations before they had dropped too far to repair.

In 1930, the Bureau of International Whaling Statistics was created to monitor the number of whales taken. This led to the first international regulatory agreement signed by twenty-two nations in 1931. Without signatures from Germany and Japan, however, whalers still killed 43,000 whales in 1931. In 1948, the International Convention for the Regulation of Whaling recognized that many whale species neared extinction. As a result, the International Whaling Commission (IWC) was established by fourteen member nations. Today, sixty-one nations are members of the IWC, and whales are only hunted for purposes of scientific or cultural tradition.

Although these early initiatives were largely carried out by the whaling industry, the IWC took advantage of the interest in the environment after the 1960s. Similar to rain forests, whales became one of the first venues through which many people came to know ideas such as extinction and sustainable management. Even with the policies in place, however, stopping illegal practices being carried out by whaling pirates often fell to well-trained activists.

Paul Watson is credited with the first example of "direct action" on behalf of whales. In 1975, Watson and other activists spun their inflatable boat into the path of a Soviet whaler trying to harpoon a sperm whale. Organizing a group known first as the Sea Shepherds and later as Greenpeace, Watson and others used larger ships and actions to block illegal whaling. They also became quite adept at staging publicity stunts (such as using a small submersible) that would attract the global press and draw attention to the pirates. When Greenpeace ships rammed whaling ships, however, some onlookers thought they had gone too far.

Thanks in part to the activists' actions, the IWC's global moratorium on commercial whaling came into law in 1986. This provision, although controversial, was based on scientific advice and thus has managed to remain in place. That said, the ban is kept under review and is periodically challenged by pro-whaling countries. Furthermore, as more and more scientific evidence suggests that some whale populations are rebounding, the attempts to overturn the ban have become more intense. Despite the controversy and the regular challenges to the ban, as of 2007, it still stands.

Whaling Ban Fuels Global Cooperation

The anti-whaling movement began in the 1970s and was substantially strengthened in 1972 at the U.N. Conference on the Human Environment, held in Stockholm, Sweden. At this conference, the anti-whaling movement proposed a ten-year moratorium on commercial whaling. Adoption of this resolution at the conference was the first major victory for the anti-whaling movement. Additional support for the ban came in 1977 and 1981 with the reports of the Convention on International Trade in Endangered Species. These reports identified many species of whales in danger of extinction. Finally, in 1982, the whaling moratorium was voted through by the IWC, the body that mattered most for stopping whaling. The moratorium stated that all commercial whaling would end in 1986.

The original IWC moratorium on whaling was accepted on July 23, 1982 by a three-quarters vote by the commission: 25 to 7, with five members abstaining from the vote. At the time of the vote, the IWC was changing its primary focus from one of promotion and maintenance of whale fishery stocks to one for the protection of all species of whales. Four voting countries, Norway, Japan, the Soviet Union, and Peru, were not in support of the change in focus or the moratorium on whaling. These countries lodged official objections to the ban rooted in the fact that the moratorium, although based on scientific advice, was not based on advice from the IWC Scientific Committee. Because of pressure from the United States, Japan and Peru later withdrew their formal objections to the ban (Black 2007).

That same year, the U.N. Convention on the Law of the Sea directly addressed the issue of whales and stated that nations shall work through the appropriate international organizations for the conservation and management of whales. The anti-whaling movement thought they had won, but whaling continues today. What went wrong?

Since the late 1980s, many of the countries opposed to the whaling ban have questioned the protective role of the IWC. Even before the ban, the most depleted species were already protected by the commission, and there were ever-declining annual quotas for most of the other species (Black 2007). The IWC maintained its call for protection of all whales despite the fact that populations of some species of minke whale were arguably sufficient to allow limited hunts. This, along with general disagreement on the management of this natural resource, created controversy between the pro-whaling countries and those against whaling. It has led to numerous heated debates and votes concerning the possibility of allowing whaling. Ultimately, the outcome of these debates, continued restrictions on whaling, has led the pro-whaling nations to question the legitimacy of the IWC's decisions.

Despite suggestions that the IWC is a protectionist agency, it does grant two exemptions to the whaling moratorium: scientific whaling and aboriginal whaling. Both pro-whaling and anti-whaling countries use these exemptions. The United States allows first nations to whale under the aboriginal whaling exemption, and Japan began whaling under scientific research permits in 1986. Controversy surrounding these exemptions has most recently focused on suggestions that Japan's scientific whaling is really a front for the Japanese fondness for whale meat. Japan's response is to point out the similarities between U.S. aboriginal subsistence whaling and various Japanese fishing communities that traditionally hunted whales.

Beyond the exemptions, in 1992, Norway lodged a protest to the moratorium removing them from the restrictions; they began commercially whaling in 1994. Iceland put forth a reservation to the moratorium in 2002, although this reservation is not viewed as legal by many of the IWC's member countries. Based on this reservation, Iceland began hunting whales commercially in 2006.

In June 2005, there was again a push led by Japan, Norway, and Iceland to end the moratorium on commercial whaling. These nations view eating whale as no different than eating beef and suggest that it is necessary to control whale numbers to protect fish stocks. Anti-whaling countries countered these arguments with observations that, to know how many animals to take, managers must have an idea of population sizes. This is no easy task with whales; population surveys are generally done by surface observation, a method with a large margin of error. Without an end to the moratorium, Japan suggested that they would simply increase their self-determined quota for scientific hunts, a number that does not have to be justified to the IWC. After this debate, at the IWC meeting in 2006, a pro-whaling majority passed a statement suggesting that the moratorium was no longer required. However, in May 2007, the IWC reauthorized the moratorium on commercial whaling. A majority of anti-whaling countries (thirty-seven in total) voted to adopt a resolution stating that the whaling ban remains valid. This vote to overturn the 2006 statement indicated a renewed strength of the anti-whaling nations. The pro-whaling countries abstained.

Sources and Further Reading: BBC News, "Iceland Violates Ban on Whaling," http://news.bbc.co.uk/2/hi/europe/6074230.stm; Black, "Did Greens Help Kill the Whale?" http://news.bbc.co.uk/2/hi/science/nature/6659401.stm; Clapham and Baker, "*Modern Whaling*"; Dorsey, *The Dawn of Conservation Diplomacy*; Kirby, "Extinction Nears for Whales and Dolphins," http://news.bbc.co.uk/2/hi/science/nature/3024785.stm; Melville, *Moby Dick*; Mulvaney, *The Whaling Season: An Inside Account of the Struggle to Stop Commercial Whaling*; Schweder, "Protecting Whales by Distorting Uncertainty: Non-precautionary Mismanagement?"

DOWNWINDERS AND ATOMIC TESTING AS A QUESTION OF NATIONAL SECURITY

Time Period: 1945–1980
In This Corner: Western residents, residents of other testing areas
In the Other Corner: Scientists, military, federal government
Other Interested Parties: Soviet Union
General Environmental Issue(s): Human health, the West, nuclear weapons

When the first atomic weapons exploded over Japan in 1945, observers from all over the world knew that human life had changed in an instant. In the years since, proponents of nuclear technology have struggled to overcome its dangerous potential and to have it be identified as a public good. Although technology rapidly became a tool for environmental action, it also presented a downside that inadvertently helped to propel environmental concern. Nothing else embodies Americans' ambivalence about the development of new technologies like atomic technology, including the bombs developed to end World War II. However, this era of nuclear successes created an exuberance for many other technologies as well. Together, the technological developments of the post–World War II era brought the United States a standard of living unrivaled in the world. Atomic technology serves as a symbol of the emergence of the United States as a world power based on its ingenuity and technological development. Although these strengths translated into economic progress, they also fueled military dominance as well.

Although nuclear technology began under the control of the military, its implications influence every American. The dangerous implications of nuclear reactions have made it an important part of domestic politics since the 1960s. All of this together illustrates that twentieth-century life has obviously been significantly influenced by "the bomb," although it has been used sparingly, nearly not at all. A broader legacy of atomic technology can be seen on the landscape, from Chernobyl to the Bikini Atoll or from Hiroshima to Hanford, Washington. With this broader legacy in mind, nuclear technology may have impacted everyday nature more than any other innovation. In fact, between its explosive ability and its toxic products, atomic technology changed the order of the natural world, clearly placing with humans the ability to destroy everything.

During the Cold War era, Americans focused on stabilizing everyday life and improving the quality of living for the growing middle class. Nuclear weapons were an important part of this stability. Atomic bombs, which were being tested in the American West, were not known to be a threat to the environment. Information about radiation was not released to the American public.

Nuclear Testing

To keep the Cold War from heating up, the technology needed to be demonstrated for Soviets and others to see and tested to perpetuate the race to one-up the American enemy. The focus of American testing was unpopulated areas of the West, particularly Nevada.

The Nevada Test Site (NTS) is located approximately sixty-five miles northwest of the city of Las Vegas. Formerly known as the Nevada Proving Ground, the 1,350-square-mile site was established on January 11, 1951 for the testing of nuclear weapons. Between 1951 and 1992, there were a total of 925 announced nuclear tests at the NTS. Eight hundred twenty-five of them were underground. Other sites used ocean-based testing as early as 1946. Most of these sites were in the Marshall Islands in the Pacific Ocean, including Bikini. These sites will be dealt with at length below.

In terms of the western United States, in the opening excerpt from *Refuge*, Terry Tempest-Williams describes the plight of many westerners who became "downwinders." Although it is difficult or impossible to prove that their illness and cancer derives from radioactive fallout from nuclear testing, many residents of the region share Williams's perspective that they were unwittingly part of the cost of the technological effort to win the Cold War. The following is an excerpt from Terry Tempest-Williams, *Refuge*:

The Clan of One-Breasted Women: Epilogue

I belong to a Clan of One-Breasted Women. My mother, my grandmothers, and six aunts have all had mastectomies. Seven are dead. The two who survive have just completed rounds of chemotherapy and radiation.

I've had my own problems: two biopsies for breast cancer and a small tumor removed between my ribs diagnosed as "a borderline malignancy."

This is my family history.

Most statistics tell us breast cancer is genetic, hereditary, with rising percentages attached to fatty diets, childlessness, or becoming pregnant after 30. What they do not say is living in Utah may be the greatest hazard of all.

Over dessert, I shared a recurring dream of mine. I told my father that for years, as long as I could remember, I saw this flash of light in the night in the desert. That this image had so permeated my being, I could not venture south without seeing it again, on the horizon, illuminating buttes and mesas.

"You did see it," he said.

"Saw what?" I asked, a bit tentative.

"The bomb. The cloud. We were driving home from Riverside, California. You were sitting on your mother's lap. She was pregnant. In fact, I remember the date, September 7, 1957. We had just gotten out of the Service. We were driving north, past Las Vegas. It was an hour or so before dawn, when this explosion went off. We not only heard it, but felt it. I thought the oil tanker in front of us had blown up. We pulled over and suddenly, rising from the desert floor, we saw it clearly, this golden-stemmed cloud, the mushroom. The sky seemed to vibrate with an eerie pink glow. Within a few minutes, a light ash was raining on the car."

I stared at my father. This was new information to me.

"I thought you knew that," my father said. "It was a common occurrence in the fifties."

It was at that moment I realized the deceit I had been living under. Children growing up in the American Southwest, drinking contaminated milk from contaminated cows, even from the contaminated breasts of their mothers, my mother—members, years later, of the Clan of One-Breasted Women.

It is a well-known story in the Desert West, "The Day We Bombed Utah," or perhaps, "The Years We Bombed Utah." Aboveground atomic testing in Nevada took place from January 27, 1951, through July 11, 1962. The winds were blowing north, covering "low use segments of the population" in Utah with fallout and leaving sheep dead in their tracks, and the climate was right. The United States of the 1950s was red, white, and blue. The Korean War was raging. McCarthyism was rampant. Ike was it and the Cold War was hot. If you were against nuclear testing, you were for a Communist regime. (Tempest-Williams 1992)

Health Impacts for Downwinders

What has been the effect of these tests on the human population? In a 1997 report by the National Cancer Institute, it was determined that, between 1952 and 1957, ninety atmospheric tests at the NTS deposited high levels of radioactive iodine-131 across a large portion of the contiguous United States. The report further states that the doses of iodine-131 were large enough to produce 10,000 to 75,000 cases of thyroid cancer.

Although it is problematic for any citizen to be infected without his or her knowledge, native peoples held little legal or political standing with which to file their complaints. Because of the long-term use of the NTS, which is located on traditional Shoshone land, the Western Shoshone nation has become known as "the most bombed nation on earth." Groups bringing suit against the military state evidence that, since 1951, 928 American and nineteen British nuclear explosions in this area have been classified by the Western Shoshone National Council as bombs rather than "tests" (*Downwinders*). The complainants choose their terms carefully because "the purpose of a bomb is to destroy while the idea of a test is to introduce something new."

By the estimation of their complaints, 1,350 square miles of the Shoshone total territory of about 43,000 square miles has been destroyed with craters and tunnels, including unregulated underground nuclear waste dumps. No permission was sought for these activities, because the work was considered to be in the U.S. national interests. The Shoshone point of view reads as follows:

Environmental monitoring reports for the NTS from the 1950s until 1991 document substantial low level releases of radioactive iodine, strontium, cesium, plutonium and noble gases that have contaminated lands in Nevada and Utah. The Western Shoshone reservations, Duckwater and Ely, are within a fifty-mile radius of the NTS and were more heavily contaminated. Residents reported unusual animal deaths, hair loss and gardens turning black. The health of the population still remains at high risk from cancers and birth defects. (*Downwinders*)

Ironically, this is the general location of the area known as Yucca Mountain, which the U.S. government has designated to become the final repository for the high-level nuclear waste from the U.S. nuclear industry.

Compensating Downwinders and the Displaced

The effort to compensate or recognize the impact of nuclear testing on western residents, native and non-native, has been most complex. In 1990, Congress passed the Radiation Exposure Compensation Act to aid American civilians who had been subjected "… to radiation that is presumed to have generated an excess of cancers among those individuals." The Downwinders Act, as it was called by the public, offered $50,000 to residents of the testing areas who were suffering from certain illnesses that could most likely be traced to exposure to fallout from the testing. As of January 2006, more than 10,500 of such claims had been approved and around 3,000 have been denied, for a total amount of more than $525 million in compensation dispensed to "downwinders" (*Downwinders*).

The downwinder problem is further complicated for the Pacific Islands, which were also used for U.S. weapons testing. Although tribunals were established to figure out proper compensation for residents of locations such as the Marshall Islands, reliable information about the exposure level of individuals in other atolls was needed to establish the baseline. Such data, simply, did not exist. Therefore, most personal injury claims were abandoned. Eventually, Congress established a program compensating U.S. civilians in the affected area for diseases. The period for Nevada was normally considered to be January 1951 and October 1958 or during July 1962.

In this legislation, the affected area included at least fifteen counties covering more than 83,000 square miles in the states of Nevada, Utah, and Arizona. Some of these locations were almost 500 miles away from the NTS. In the Downwinders Act, the U.S. Defense Nuclear Agency states that eighty-seven atmospheric nuclear tests were conducted at the NTS during the period. The largest of those tests was 100 kilotons (0.1 megaton), and the total yield of all eighty-seven tests was approximately 1.1 megatons.

Case Study: The Marshall Islands

Located in the Pacific chain of islands, Marshall Islanders became aware of the 1990 Compensation Act after its Nuclear Claims Tribunal had already been established in 1988. The islands are estimated to have been subject to approximately ninety-nine times the total of the Nevada atmospheric tests; therefore, the tribunal expected similar treatment to that received by the downwinders in the United States. Similar to the U.S. program, the tribunal's program, which was begun in 1991, assumed two things of downwinders: first, all residents of the Marshall Islands were assumed to have received exposure to levels of ionizing radiation that caused the possible medical conditions; and second, the infecting agent was presumed to have been exposure to radiation as a result of the testing program.

The tribunal operated with a list of twenty-five specific diseases or conditions, which were drawn from those used in the Downwinders program, but also included additional assessments by the tribunal. For information on long-term health outcomes of radiation exposure, the tribunal considered the findings of the Life Span Study of atomic bomb survivors conducted by the Radiation Effects Research Foundation in Japan and other scientific findings.

The tribunal's list was extended in 1996 to add seven new conditions and extended again in 1998 and 2003, adding one new condition each time. Today, the tribunal's program covers thirty-six medical conditions. By 2003, the tribunal had awarded a total of $83 million to 1,865 individuals who suffered from one or more of those conditions. Thyroid disorders have been the basis for approximately two-thirds of the total number of awardees (*Downwinders*).

This process at the Marshall Islands has contributed significantly to the scientific understanding of the impact of radiation on humans. One basic concept that changed was the evolution of the concept of "Maximum Permissible Exposure Levels." In recent years, negotiators have located scientific reports that were previously unavailable. Data concerning the Islanders from immediately after the testing show that ten of the twenty-two populated atolls listed exceeded the National Council on Radiation Protection 1957 maximum limit of 500 mrem over the period of an entire year for the general public and an additional ten populated atolls exceeded the International Commission on Radiological Protection 1959 general public limit of 170 mrem for a whole year. These classified reports were not available to those negotiating the settlement on behalf of the Marshall Islands (Marshall Island).

Conclusion: A Question of National Security

Developing nuclear weapons required continuous testing during the Cold War, and it was determined to be in the best interest of the United States to steer the altercation with the Soviet Union toward an arms race. Thus, the United States used planned tests on its own soil as well as marine locations such as the Marshall Islands to develop weapons that would ensure its national security. It is a policy that, in hindsight, is ripe for criticism; however, in the panic of a Cold War, many Americans considered it critical to our own safety.

Sources and Further Reading: Downwinders: http://www.atsdr.cdc.gov/hanford; Eichstaedt, *If You Poison Us: Uranium and Native Americans*; Hanford Downwinders Litigation Information Resource, http://www.downwinders.com/index.html; Marshall Island: http://www.nuclearclaimstribunal.com/appro.htm.

NATIVE RIGHTS AND LIVING WITH MINING'S RESIDUE

Time Period: Post-1950
In This Corner: Some native residents, health officials
In the Other Corner: Other native residents, mining companies, litigators
Other Interested Parties: Political officials
General Environmental Issue(s): Mining, Native American rights

The confusion and debate over the status of resources in relation to native communities may have been most acute in relation to valuable mineral resources. The mineral wealth of some modern western Indian reservations has proved both a blessing and a curse. Although the discovery of valuable minerals such as gold—or even false reports of it—significantly influenced episodes of conflict between settlers and natives, this article is most concerned with conflict that arises during the twentieth century over issues of mining on reservation land for non-renewable energy resources.

Terraces cut into the hillside at the Santa Rita copper mine in Grant County, New Mexico, demonstrate methods used throughout the West. Once a significant portion of public lands, companies in Canada and seven other foreign countries have obtained hardrock mining rights. AP Photo/*Albuquerque Journal*, Richard Pipes, File.

General Mineral Extraction

Beginning as early as 1900 with the discovery of oil on Osage land in Oklahoma, nonrenewable resource development has unleashed some of the most significant environmental destruction on or near reservation land in the American West. Today, mine and drilling sites, roads and machinery, tailing piles, and settling ponds threaten tribal land, water, air, health, and lifestyles. These problems are exacerbated by a century of inequitable leases and mismanagement by federal, state, and tribal authorities.

To help this situation in the late twentieth century, many tribes banded together to form pan-Indian organizations, including the Council of Energy Resource Tribes, to balance use and protection of resources. However, much of the damage was already done by mining and oil and gas exploration over the twentieth century.

In the case of coal, British-owned Peabody Coal Company operates two huge strip mines on the remote and sacred Black Mesa, leased from the Hopi and Navajo tribes. The purpose of this extraction is to provide electricity to Americans in neighboring areas. The operation is massive, and the tribe has certainly reaped financial reward from their decision to lease the property to the coal company.

The leases, dating back to 1964 and renegotiated in 1987 because of a very low royalty rate, allows them to mine 670 million tons of high-grade, low-sulfur coal from a 64,858-acre site. Each year, seven million tons of coal from the Kayenta Mine are shipped by electric railroad seventy-eight miles to the Navajo Generating Station near Page, Arizona. Coal from the smaller Black Mesa Mine travels 273 miles in three days through an eighteen-inch-diameter coal slurry pipeline around the Grand Canyon to the Mohave Generating Station near Laughlin, Nevada. To move each ton of coal through this pipeline requires 270 gallons of water. For this purpose, the Peabody Coal Company pumps 3.9 million gallons of water from the Navajo Sandstone Aquifer, 3,000 feet below Black Mesa, each day. This totals more than 1.4 billion gallons of water each year to transport five million tons of coal.

Once the coal slurry reaches Mohave, the water is separated to cool the power plant's cooling towers. The coal, however, becomes electricity for more than two million residents of southern California and Nevada. That, of course is not the end of the story for native residents in the area. The Navajo and Mohave generating plants return a sulfur dioxide haze that hangs over the Grand Canyon and its surroundings.

Is this permanent pollution and aquifer depletion worth the benefits brought to the Hopi? The industry provides jobs for many members of the tribe, and millions of dollars in annual coal royalties make up approximately 70 percent of the tribe's budget. However, the Hopi claim has been that Peabody has not adhered to safety testing that it would have been required to carry out elsewhere. They have filed suit to ask the U.S. federal government for help.

Similar problems have occurred in other areas where pollutants from mining and processing plants migrate into reservation air and water. For instance, cyanide heap-leach mining in Montana is polluting water on the Fort Belknap Reservation. In Washington, radioactive pollution and toxic waste from the Hanford nuclear weapons plant threatens all tribes who depend on the Columbia River salmon. In the midwest, the Mdewakanton Sioux of Prairie Island, Minnesota, fear the health impacts of a nuclear power plant built on the edge of their small reservation, while Western Shoshones protest the use of their land as a nuclear test site. In such cases, tribes often fall into a classification referred to as environmental racism: their long-term personal and environmental health may be threatened, but they lack the political power to fight.

Industrial waste dumps surround the St. Regis Indian Reservation, fouling the St. Lawrence River. Poorly treated urban waste and agricultural effluent threatens nearby reservation environments. Today, groups such as the Standing Rock Sioux and Northern Cheyenne are beginning to enforce federal laws protecting their land, water, and air from such pollution. However, they must do much of the enforcement with little help from federal authorities.

Mining for Oil on Reservation Land

In a move that would ultimately hold bitter irony, the U.S. government in 1933 enlarged the Navajo reservation in southeast Utah. As part of this agreement, 37.5 percent of any future oil or gas royalties for Utah Navajos was to be placed in trust by the state. These funds would then be administered by the state. The 62.5 percent that remained would belong to the Navajo Nation and could be used as they saw fit. The irony of this minor portion of the agreement became clear in 1956 when significant reserves of oil were discovered on the land near the Aneth and Montezuma Creek.

During the rest of the twentieth century, oil companies drilled 577 wells and pumped an estimated 370.7 million barrels of oil and another 339,100 cubic feet of natural gas from the area. Since 1956, the oil royalties are estimated to have totaled at least $180 million for the Navajo, including $60 million to the Utah Navajo Trust Fund (Chamberlain 2000). Over the last few decades, however, critics have complained about how little of this money actually makes it to residents of Aneth, who live in overwhelming poverty. In fact, although they have millions in the bank, seventy-five percent of the 6,500 Utah Navajos in the region have no electricity or running water. To purchase water, many residents must travel one hundred miles (Chamberlain 2000).

Similar to other industrial undertakings on reservations, oil extraction is not without its impact on the surrounding environment. Often, water on the reservation can no longer be used because it has been contaminated. Oil companies have injected carbon dioxide and salt-water into oil wells to increase production. In 1990, for instance, there were ninety-nine spills of oil, saltwater, and chemicals in the Aneth fields, damaging 36,622 acres. There has been little effort to rectify these problems or to compensate the Navajos.

Uranium Mining

Uranium is at once the oddest and most problematic portion of the story of mining on Native American reservations. Uranium, of course, is a naturally occurring radioactive element. Dr. Richard Pierce reportedly mined it first in the western United States in 1871. He shipped 200 pounds to London from the Central City Mining District near Denver, Colorado. Its use at this time was largely experimental as a possible additive in the fabrication of steel alloys, chemical experimentation, and as pigments for dyes, inks, and stained glass.

In 1898, Pierre and Marie Curie and G. Bemont isolated the "miracle element" radium from the uranium pitchblende. That same year, uranium, vanadium, and radium were found to exist in carnotite. Records showed that this colorful red and yellow ore had been used as body paint by early Navajo and Ute Indians on the Colorado Plateau. This realization initiated the first prospecting boom in southeastern Utah, and radium mines opened as the West became a major source of ore for the Curies.

Utah continued to see a slow, steady search for uranium, radium, and vanadium through the early 1900s. The real boom, of course, was spurred by the federal government's decision that uranium would play a major role in national defense. Uranium, by this time, had become a waste product in the vanadium mines (this additive made paints glossier). During the early years of the Manhattan Project, almost 90 percent of the United States' uranium supply was imported from the Belgian Congo and Canada, but scanty amounts being filtered from abandoned radium and vanadium dumps on the Colorado Plateau gave promise of an untapped domestic source. The Manhattan Project of the U.S. Corps of Engineers, charged with development of an atom bomb to end the war, instituted a covert program to mine uranium from the vanadium dumps and sent geologists to scour the region in search of new lodes.

Of course, the need for these chemicals did not end with World War II. A federally sponsored boom began when the Atomic Energy Commission (AEC) promised $10,000 bonuses for new lodes of high-grade ore. They also guaranteed minimum prices and paid up to $50 per ton on 0.3 percent ore, constructed mills, helped with haulage expenses, and posted geologic data on promising areas tracked by federal geologists using airborne scinillometers and other sophisticated radiation detection instruments.

Prospectors with Geiger counters combed the entire "Four Corners Area," where Utah, Colorado, Arizona, and New Mexico meet. Most often, the surface rock guided them with the Geiger counter, which detects radiation. The prospectors then used diamond drills to core test holes that would show what minerals were present. Between 1946 and 1959, 309,380 claims were filed in four Utah counties. A center of activity, the once sleepy farming town of Moab became known as "The Uranium Capital of the World." By 1955, there were approximately 800 mines producing high-trade ore on the Colorado Plateau. Utah alone had produced approximately nine million tons of ore valued at $25 million by the end of 1962.

Of course, it was just a matter of time until it was discovered that significant stores of uranium could be found on reservation land, particularly that of the Navajo. With few other employment options, Navajo families were thankful when mining started on the reservation. They were not given information regarding the danger that was associated with uranium mining. These lessons had to be learned from their own experience (Brugge 2000). Medical literature dating as early as the sixteenth century documented cancer deaths among radium miners in the Erz Mountains of Germany and Czechoslovakia, but neither the AEC, state governments, nor the mining companies would take responsibility to regulate mine and mill ventilation and safety practices. After the mining had stopped, the U.S. Congress voted to provide financial assistance to those miners effected by the industry. For the Navajo, however, the danger remained part of their everyday lives.

In the uranium mines and mills, some Navajo workers were exposed to high levels of radioactivity. One 1959 report found radiation levels ninety times acceptable limits (Land Use History of North America 2002). Although very few statistics were kept, one startling record is very revealing: of the 150 Navajo uranium miners who worked at the uranium mine in Shiprock, New Mexico, until 1970, 133 died of lung cancer or various forms of fibrosis by 1980 (Ali 2003).

Working in the mines and mills, however, was not the only danger. Although mining companies left the area when mining ceased in the late 1970s, Navajo remained in their reservation land. Few of the companies completed the safety measures that were required, including sealing the tunnel openings, filling the gaping pits, sometimes hundreds of feet deep, or removing the piles of radioactive uranium ore and mine waste. It is estimated that more than 1,000 of these unsealed tunnels, unsealed pits, and radioactive waste piles remain

on the Navajo reservation today. Navajo families live among this residue and use the surrounding land to graze their livestock.

The complex legal decisions involved in assisting Navajos have dragged on for decades. In
the important ruling regarding Kerr-McGee Corporation, one sees the problem of jurisdiction and responsibility:

> Ultimately, this case requires an attempt to accommodate two independent and impor-
> tant congressional concerns: comity interests flowing from tribal sovereignty and nu-
> clear energy regulation. Though a close question, we cannot conclude that the 1988
> Amendments, or the PAA's scope generally, create an "express prohibition" to tribal
> court jurisdiction. Congress intended to control where and how nuclear incident litiga-
> tion is to take place, but did not take the next step of specifically divesting tribal courts
> of jurisdiction. While the precise scope of retained tribal jurisdiction in this case is
> subject to reasonable debate, tribal adjudicatory authority over this nuclear incident is
> not "patently violative of an express jurisdictional prohibition...."

> The third comity consideration, obtaining the benefit of tribal court expertise may be
> of value in this case. 42 USC §2014(hh) requires that the substantive rules of decision
> in such actions are to be derived from the "law of the State in which the nuclear inci-
> dent involved occurs, unless such law is inconsistent with the provisions of such sec-
> tion." At oral argument, counsel for the Tribal Claimants argued that the substantive
> rule of decision in this case should be derived from Navajo tribal law.

> Upon reviewing the comity factors, we conclude that abstention is appropriate. Kerr-
> McGee contends that it is inappropriate to analyze the tribal court jurisdiction ques-
> tion and the comity factors separately. Specifically, it argues that our comity analysis
> must consider the strong role afforded federal courts in resolving Price-Anderson suits.
> We disagree. The National Farmers framework does not accord countervailing federal
> concerns a place in the comity analysis. Consideration of federal jurisdictional concerns
> is only appropriate in the context of determining whether to engage in the comity
> analysis at all.

> It does not serve Congress's interest in promoting development of tribal courts, see
> Iowa Mutual, 480 U.S. at 19, to second-guess the jurisdictional determinations of the
> Navajo district court before the tribal appellate process has run its course. The district
> court's judgment is AFFIRMED. (Wise Uranium Project, *Uranium Mining and Indig-
> enous People*)

The district court's decision, however, still left issues to be considered. In 2005, "Post-71"
uranium miners demanded compensation for work that had occurred after the 1971 dead-
line. Antonio Sena and Margarito Martinez, former Kerr-McGee mine workers, argued for
these miners to be included. The 1990 Radiation Exposure Compensation Act legislation
established a cutoff of 1971 based on government liability related to the uranium procure-
ment program. The report's authors said that there was a perception among many in government and the uranium industry that the "4 Working Level Months" (WLM) standard
passed in 1971 would provide adequate protection for miners. In their complaints, "Post-71"
workers argued that this scheduling model did not obviate risk.

Scientific reports supported the "Post-71" workers. They also issued a word of caution in interpreting working level findings "because of validity and reliability issues." Findings in a 1980 National Institute for Occupational Safety and Health report, based on scientific study, showed that the Mine Safety Health Administration standard of 4 WLM "does not provide an adequate degree of protection for underground miners exposed to radiation when it is evaluated over their exposure lifetime" (*Gallup Independent* 2005).

Conclusion: Native Mineral Rights and the Law

Many members of the native communities have argued that the lack of oversight and regulation over reservation lands marks a bitter irony of American conquest. Others, however, rail against the inability of native communities to select and pursue their own paths toward economic development. As resources become more scarce, those still held on reservation property will increase in value and the problem will likely intensify.

Coal and uranium mining on the Navajo reservation has destroyed large areas of land, polluted water and air, and caused untold long-term health problems. In one of the most famous examples, the Alaska Native Claims Settlement Act (1971) and the subsequent North Slope energy boom with its drilling sites, pipelines, and access roads have transformed the landscape, threatening migratory mammals and waterfowl and contributing to changes in Native Alaskan land use and ownership patterns.

In each case, many observers claim that long-term health concerns of native communities have often been overlooked for short-term royalty profits.

Sources and Further Reading: Chamberlain, *Under Sacred Ground: A History of Navajo Oil, 1922–1982*; Eichstaedt, *If You Poison Us: Uranium and Native Americans*; Krech, *The Ecological Indian: Myth and History*; Ringholz, *Uranium Frenzy, Boom and Bust on the Colorado Plateau*; White, *It's Your Misfortune and None of My Own*; Wise Uranium Project, *Uranium Mining and Indigenous People*, http://www.wise-uranium.org/uip.html (see also http://www.wise-uranium.org/ulite.html#booksgen); Science Education Resource Center, *Impacts of Resource Development on Native American Lands*, http://serc.carleton.edu/research_education/nativelands/index.html.

NUCLEAR POWER AND ITS ROOTS

Time Period: 1945 to the present
In This Corner: Atomic engineers, military leaders
In the Other Corner: Power developers, political leaders
Other Interested Parties: Electricity consumers
General Environmental Issue(s): Nuclear power, energy

When the first atomic weapons exploded over Japan in 1945, observers from all over the world knew that human life had changed in an instant. In the years since, nuclear technology struggled to define itself as a public good when the public seemed much more inclined to view it as an evil. Its proponents argue that electricity made from nuclear reactors has the capability to power the world more cleanly than any other resource. Critics are less sure. As the debate rages, nuclear power has become an increasingly important part of the world energy picture.

Beginning as a Bomb

By the late 1930s, World War II threatened the globe. Leaders of every nation searched for any edge that would defeat the enemy forces. Scientists in America and Germany actively experimented with atomic reactions. In Germany, leaders felt such a technology might prove a decisive force in the war effort. In reaction, American scientists enlisted Albert Einstein to write a letter about their research to FDR. In this letter, he stressed the technology's potential, particularly if it were developed by the enemy. In October 1939, Roosevelt authorized government funding for atomic research.

Eventually, science and the military would be linked in a way never before seen. However, first scientists needed to demonstrate the viability of such a reaction. Of course, today the concept of force generated by separating atomic particles is fairly well known; however, in 1940, such ideas smacked of science fiction. In 1940, Enrico Fermi and Leo Szilard received a government contract to construct a reactor at Columbia University. Other reactor experiments took place in a laboratory under the west stands at the Stagg Field stadium of the University of Chicago. In December 1942, Fermi achieved what the scientists considered the first self-sustained nuclear reaction. It was time to take the reaction out-of-doors, and this process would greatly increase the scope and scale of the experiment.

Under the leadership of General Leslie Groves in February 1943, the U.S. military acquired 500,000 acres of land near Hanford, Washington. This served as one of three primary locations in "Project Trinity," which was assigned portions of the duty to produce useful atomic technology. The coordinated activity of these three sites under the auspices of the U.S. military became a path-breaking illustration of the planning and strategy that would define many modern corporations. Hanford used water power to separate plutonium and produce the grade necessary for weapons use. Oak Ridge in Tennessee coordinated the production of uranium. These production facilities then fueled the heart of the undertaking, contained in Los Alamos, New Mexico, under the direction of J. Robert Oppenheimer.

Oppenheimer, a physicist, supervised the team of nuclear theoreticians who would devise the formulas using atomic reactions within a weapon. Scientists from a variety of fields were involved in this highly complex theoretical mission. Once theories were in place and materials delivered, the project became assembling and testing the technology in the form of a bomb. All of this needed to take place on the vast Los Alamos compound under complete secrecy. However, the urgency of war revealed that this well-orchestrated, corporate-like enterprise remained the best bet to save thousands of American lives.

By 1944, World War II had wrought a terrible price on the world. The European theater would soon close with Germany's surrender. Although Germany's pursuit of atomic weapons technology had fueled the efforts of American scientists, the surrender did not end the project. The Pacific front remained active, and Japan did not accept offers to surrender. "Project Trinity" moved forward, and it would involve the Japanese cities Hiroshima and Nagasaki as the test laboratories of initial atomic bomb explosions. *Enola Gay* released a uranium bomb on the city of Hiroshima on August 6, and *Bock's Car* released a plutonium bomb on Nagasaki on August 9. Death tolls vary between 300,000 and 500,000, and most were Japanese civilians. The atomic age, and life with the bomb, had begun.

Envisioning the Atomic Future

Historian Paul Boyer wrote that "Along with the shock waves of fear, one also finds exalted prophecies of the bright promise of atomic energy" (Boyer 1994, 109–14). Many of the scientists involved believed that atomic technology required controls unlike any previous innovation. Shortly after the bombings, a movement began to establish a global board of scientists who would administer the technology with no political affiliation. Wresting this new tool for global influence from the American military proved impossible. The AEC, formed in 1946, would place the U.S. military and governmental authority in control of the weapons technology and other uses to which it might be put. With the "nuclear trump card," the United States catapulted to the top of global leadership.

In the 1950s, scientists turned their attention to taking the nuclear reaction and applying it to peaceful purposes, notably power generation. The reaction was a fairly simple process. Similar to fossil-fuel powered generators, nuclear plants use the heat of thermal energy to turn turbines that generate electricity. The thermal energy comes from nuclear fission, which is made when a neutron emitted by a uranium nucleus strikes another uranium nucleus, which emits more neutrons and heat as it breaks apart. If the new neutrons strike other nuclei, a chain reaction takes place. These chain reactions are the source of nuclear energy, which then heats water to power the turbines.

Soon, the AEC seized this concept as the foundation for plans for "domesticating the atom." It was quite a leap, however, to make the American public comfortable with the most destructive technology ever known. The AEC and others sponsored a barrage of popular articles concerning a future in which roads would be created through the use of atomic bombs and radiation used to cure cancer.

The atomic future in the media included images of atomic-powered agriculture and automobiles. In one book published during this wave of technological optimism, the writer speculated that, "No baseball game will be called off on account of rain in the Era of Atomic Energy." After continuing this litany of activities no longer to be influenced by climate or nature, the author summed up the argument: "For the first time in the history of the world man will have at his disposal energy in amounts sufficient to cope with the forces of Mother Nature" (Boyer 1994, 109–15). For many Americans, this new technology meant control of everyday life. For the Eisenhower administration, the technology meant expansion of our economic and commercial capabilities.

As the Cold War took shape around nuclear weapons, the Eisenhower administration looked for ways to define a domestic role for nuclear power even as Soviet missiles threatened each American. "Project Plowshares" grew out of the administration's effort to take the destructive weapon and make it a domestic power producer. The list of possible applications was awesome: laser-cut highways passing through mountains, nuclear-powered greenhouses built by federal funds in the midwest to enhance crop production, and irradiated soils to simplify weed and pest management. Although domestic power production, with massive federal subsidies, would be the long-term product of these actions, the atom could never fully escape its military capabilities. This was most clear when nuclear power plants experienced accidents.

Developing U.S. Nuclear Power

Begun under Project Plowshares, Eisenhower's Atoms for Peace program emphasized that it was impossible for a nuclear plant to behave as a bomb would; it could not explode. The

AEC, which took over the oversight of nuclear technology after World War II, worked tirelessly to encourage the commercial use of nuclear reactors for the generation of electricity. Lewis L. Strauss, chair of the AEC, proclaimed to the public that the production of nuclear power was "too cheap to meter." This was especially true if the federal government helped to finance the construction, operation, and insurance of atomic power plants (Opie 1998, 473–74).

In 1951, the first experimental reactor went on line near Idaho Falls, Idaho. Initially, it produced only enough power to light four 150-watt light bulbs. The lessons learned in Idaho, however, led to the AEC-sponsored pilot project with Duquesne Light Company in Shippingport, Pennsylvania. This sixty-megawatt breeder reactor opened in 1957 to serve as a model for future projects. In addition, however, the Westinghouse-designed plant served as a model for the Navy program to use nuclear power for submarine propulsion. Ultimately, the Shippingport reactor became the first licensed American commercial reactor.

During the ensuing decades, new reactors would be constructed throughout the United States. Although the AEC and the federal government offered assistance, these power plants were normally constructed by private utilities. The electricity that they generated was placed on the utility's grid and sold with power made from coal, hydroturbines, and oil.

Accidents Fuel Public Doubt: Three Mile Island

A number of accidents occurred before the late 1970s, but they went largely unnoticed by the American public. Nuclear power became increasingly popular, although critics continued to argue issues of safety. Then in 1979, the United States experienced its first nuclear accident in a residential area outside of Harrisburg, Pennsylvania.

Located within a working-class neighborhood outside a major population center, the Three Mile Island (TMI) nuclear power plant experienced a partial core meltdown in 1979. As pregnant women and children were evacuated from Pennsylvania's nearby capital, Harrisburg, the American public learned through the media about potential hazards of this technology as well as about many other sources. More than the potential danger specifically from nuclear power, however, TMI marked a watershed in the expression of "Not in My Backyard" (NIMBY) ideas in American life.

The accident at the TMI Unit 2 nuclear power plant is still recognized as the most serious accident at a U.S. commercial nuclear power plant. Although it led immediately to no deaths or injuries, the publicity that it stirred instigated sweeping changes in emergency response planning, reactor operator training, and almost every other area of nuclear power plant operations. It also caused the U.S. Nuclear Regulatory Commission (NRC) to tighten and heighten its regulatory oversight (Rothman 2003, 146–48).

Although the sequence of events at TMI led to a partial meltdown of the TMI Unit 2 reactor core, only very small off-site releases of radioactivity occurred. The accident, however, revealed a massive lack of knowledge about the technology of atomic power production. The lack of information about the impact of the accident led to a few fearful days and, ultimately, the demise of the industry.

The actual events of the accident began at about 4:00 A.M. on March 28, 1979, when the plant experienced a failure in the secondary, non-nuclear section of the plant. A mechanical failure stopped the flow of feedwater to the reactor, and therefore the reactor automatically shut down. This led to an increase in the pressure within the reactor (Opie 1998, 447).

To prevent that pressure from becoming excessive, the pilot-operated relief valve (a valve located at the top of the pressurizer) opened and released water and pressure. Unknown to the pilot, however, the valve stuck open, and cooling water poured out of the stuck-open valve and caused the core of the reactor to overheat. Plant operators then took a series of actions that made conditions worse by simply reducing the flow of coolant through the core. It was later found that about one-half of the core melted during the early stages of the accident.

In short, the accident caught federal and state authorities off-guard in a dramatic public moment. On March 30, a significant release of radiation from the plant's auxiliary building caused a great deal of confusion. In an atmosphere of growing uncertainty about the condition of the plant, the governor of Pennsylvania, Richard L. Thornburgh, consulted with the NRC about evacuating the population near the plant. Eventually, Thornburgh advised pregnant women and pre-school-age children within a five-mile radius of the plant to leave the area.

Soon, observers began discussing a large hydrogen bubble in the dome of the pressure vessel. Speculation suggested that the bubble might burn or even explode and rupture. Of course, the public very quickly undid all that Eisenhower and others had done to distance atomic power from atomic weapons. The bubble bursting became the equivalent in the popular imagination of an atomic explosion. On April 1, experts established that the bubble could not do any damage and the panic passed.

Although there has been considerable debate about the health impact of TMI, the NRC studies estimated that two million people in the area received about one millirem of radiation (chest X-rays expose patients to approximately six millirem). The real impact of TMI came through the public's lesson that technical answers were not always the best answers.

Events at TMI effectively squelched further development of nuclear reactors not yet planned in 1979. Although the number of nuclear units increased through 1990, most of the contracts had been drawn up many years before. By 1990, most Americans refused to help finance new reactors. Most communities lobbied to remove or shut down reactors near residential areas. The dangers of radiation were now more widely understood, and Americans also now had an expectation of physical safety that is unique in human history. In short, Americans expected to live in uncontaminated safety. Responsibility for ensuring this safety would fall to the federal government.

Accidents Fuel Public Doubt: Chernobyl

Of course, the international community noted TMI; however, it clearly did not present a grave threat to the world. When the Chernobyl meltdown occurred in 1986, however, residents of every developed nation took note. During a test in April, the fuel elements ruptured and resulted in an explosive force of steam that lifted off the cover plate of the reactor, which released fission products into the atmosphere. A second explosion released burning fuel from the core and created a massive explosion that burned for nine days. It is estimated that the accident released thirty to forty times the radioactivity of the atomic bombs dropped on Hiroshima and Nagasaki. Hundreds died in the months after the accident, and hundreds of thousands of Ukrainians and Russians had to abandon entire cities.

The implications of nuclear weapons and power had already been of great interest to environmental organizations before Chernobyl. Now, international organizations such as Greenpeace dubbed nuclear power a trans-border environmental disaster waiting to happen.

General view of empty houses and trees showing new growth in the town of Pripyat, with the closed Chernobyl nuclear power plant in background, May 2007. Two decades after an explosion and fire at the nearby Chernobyl nuclear power plant sent clouds of radioactive particles drifting over the fields and homes, natural wildlife is returning to the area as scientists watch to chart the radioactivity's impact. AP Photo/Efrem Lukatsky.

Interestingly, even within the environmental movement, nuclear power maintained significant support because of its cleanness. Whereas almost every other method for producing large amounts of electricity created smoke or other pollution, nuclear power created only water vapor, yet, at least in the public's mind, there remained the possibility of atomic explosions.

Growth in the International Market

Although accidents greatly slowed the domestic interest in nuclear power generation, the international community refused to be so quick to judge the technology. Since the early 1990s, nuclear power has become one of the fastest growing sources of electricity in the world. Today, 16 percent of the world's energy derives from nuclear power.

Although only eight nations possess nuclear weapons capability, fifty-six operate civil research reactors, and thirty-one nations have 440 commercial nuclear power reactors. As many as forty more reactors are scheduled to be built over the next few years. Nations that depend on nuclear power for at least a quarter of their electricity, include Belgium, Bulgaria, Hungary, Japan, Lithuania, Slovakia, South Korea, Sweden, Switzerland, Slovenia, and Ukraine.

Currently, fewer nuclear power plants are being built than during the 1970s and 1980s. However, the newer reactors are much more efficient and capable of producing significantly more power. Additionally, nuclear power is being used for needs other than public electricity. In addition to commercial nuclear power plants, there are more than 280 research reactors operating, in fifty-six countries, with more under construction. These have many uses, including research and the production of medical and industrial isotopes, as well as for training. Reactors are also used for marine propulsion, particularly in submarines. More than 150 ships are propelled by more than 200 nuclear reactors.

Conclusion: Nuclear's Nagging Waste Problem

Regardless of its use, nuclear energy continues to be plagued by its most nagging side effect: even if the reactor works perfectly for its service lifetime, the nuclear process generates dangerous waste. In fact, reactor wastes from spent fuel rods are believed to remain toxic for 50,000 years. At present, each nuclear nation makes its own arrangements for the waste. U.S. nuclear utilities now store radioactive waste at more than seventy locations while they await the fate of the effort to construct and open a nuclear waste repository inside Nevada's Yucca Mountain in 2002.

Internationally, the situation is not much clearer. Opponents in Germany have obstructed nuclear waste caravans, and shipments of plutonium-bearing waste to Japan for reprocessing are often placed under dispute. Some observers have voiced concern that less-developed nations will offer up their nations as waste dumps for the more developed nations. The financial income from such an arrangement may be too much to turn down for many nations.

In the energy industry, many observers continue to believe that atomic power remains the best hope to power the post-hydrocarbon age. The issues of safety and waste removal need to be dealt with. However, in nations with scarce supplies of energy resources, atomic power, even with its related concerns, remains the most affordable alternative.

During the 1950s and 1960s, nuclear technology symbolized the stable future that could be ensured and maintained through technological innovation. This confidence, however, shielded Americans from a more serious line of questioning about nuclear technology and other innovations. Technological progress, we would learn, was not an automatic good. Nuclear weapons and energy production, for instance, carried extreme health risks at each stage of their existence: production, storage, and after-life. As Americans began to consider the broader implications of nuclear testing, the arms race, and energy production, the popularity of nuclear technology would plummet over the next decades.

Sources and Further Reading: Boyer, *By The Bomb's Early Light*; Hampton, *Meltdown: A Race against Nuclear Disaster at Three Mile Island: A Reporter's Story*; Hughes, *American Genesis*; Hughes, *Networks of Power: Electrification in Western Society, 1880–1930*; Josephson, *Red Atom: Russia's Nuclear Power Program from Stalin to Today*; May, *American Cold War Strategy*; McNeill, *Something New Under the Sun: An Environmental History of the Twentieth-Century World*; Nye, *Electrifying America: Social Meanings of a New Technology*; Smil, *Energy in World History*; Weiner, *Models of Nature: Ecology, Conservation, and Cultural Revolution in Soviet Russia*.

DAVE FOREMAN, EARTH FIRST!, ECO-RADICALISM, AND THE GREAT DIVIDE IN ENVIRONMENTALISM

Time Period: 1970s to the present
In This Corner: Environmental extremists
In the Other Corner: Wise users, government officials, industry
Other Interested Parties: Environmentalists
General Environmental Issue(s): Environmentalism, extremism

Serious environmental concerns fueled grassroots movements, such as that in Love Canal, to demand governmental or industrial action. Social movements such as the one initiated by Lois Gibbs were often linked together under the NIMBY motto. At the root of this movement was a change in middle-class Americans' expectations of their own safety and health. This represented just one faction of environmentalism, however.

Mainstream environmental NGOs helped to clear the way for more extreme organizations that held ideas unpalatable to most middle-class Americans. Many critics claim that these extremists give environmentalism a bad name. The activities of environmental extremists, argue critics, make many Americans less likely to support environmental ideals.

One reason for this criticism is that in addition to holding extreme philosophical stands, many of these organizations also go about their activities in a much more confrontational manner. The best known of these extreme environmental organizations was Earth First!, which was led by Dave Foreman. These activists argued that protests and writing letters were not sufficient. "Earth First!ers" sought out more active methods of action, which became known as "eco-radicalism" or "ecoterrorism" (Foreman 1994, 5–9).

Earth First! is an international movement comprising small bioregionally organized groups of supporters. The goal of the movement is for each group to learn about the ecosystem in their region and identify the most immediate and serious threats to that ecosystem. The movement follows many of the ideas of "deep ecology," a branch of ecological philosophy that views humans as an integral part of the environment rather than a superior entity (Naess 1973). Intellectually, Earth First!ers and other extreme environmentalists also introduced ideas more extreme than those broadly entertained by environmentalists of previous eras. In the past, an individual such as Thoreau or Muir presented an extreme philosophical stance and interested Americans steered their minds in that general direction. Now, however, fairly large groups of Americans were willing to entertain concepts such as the need to focus less on human needs and more on those of other portions of nature.

One of the best known of these philosophical stances is referred to as "deep ecology." Subscribers to deep ecology argue that nature contains its own purposes, energy, and matter, and its own self-validating ethics and aesthetics. Calling themselves defenders of wilderness, such thinkers included Arne Ness, David Rothenberg, William Duvall, and George Sessions. For these deep ecologists, even preservation was based on science that grew from values that remained committed to the control of nature by industrial society. Deep ecologists urged environmentalists to turn their backs on society and adopt a radical new position for humans within nature. All decisions begin by asking the question of what benefits the entire natural system. This holistic perspective prioritizes wildness over any utilitarian view of the environment. Earth First! recognizes an intrinsic value in all aspects of the environment and in all living things.

Unlike many mainstream environmental groups, today Earth First! considers itself a movement rather than an organization. Because of this, they do not have members but rather participants that are drawn together by the belief that the life of the earth comes before the comforts, wants, and needs of humans. Additionally, within Earth First!, the only leaders that exist are those participants who are achieving their goals.

Earth First!ers believe in using all types of tactics to achieve their goals. Thus, they can be found working to achieve their goals through educating the public, working through the legal system, and participating in "creative civil disobedience" (Earth First! 2007). This "creative civil disobedience" has historically consisted of not only protests and demonstrations but also a variety of monkey-wrenching activities. Earth First!ers state that they are the most committed and uncompromising of environmental activists and as such may appear radical compared with other environmentalists. Despite this, Earth First! instructs their participants to weigh all options before taking actions that will get them arrested. Earth First! understands that, although being arrested may draw more media attention to the issue at hand, once arrested, the Earth First!er is no longer free and thus will not be able to continue taking actions to help inform others about the issue.

Earth First! was not so much created as named in 1980 when a group of activists were traveling across the southwestern desert from northern Sonora, Mexico, to New Mexico. Exacerbated by mainstream environmental groups and their willingness to make compromises concerning wilderness, the activists developed the idea of a movement to link together individuals working to help preserve ecologies across the United States. Specifically, the movement was meant to address the "selling out" of the mainstream environmental groups. The event that drew particular ire from this group of activists was the U.S. Forest Service's Roadless Areas and Review Evaluation (RARE II) in 1979. The activists wanted to be a part of something that "put the Earth first" (Sierra Nevada Earth First! 2007). Drawing from the ideas of Rachel Caron, Aldo Leopold, and Edward Abbey, this group of activists moved forward with their desire for ecological preserves across the United States.

Carrying the banner of antiestablishment direct action, Earth First! claimed the slogan "no compromise in defense of Mother Earth." Initially, the organization adopted many of the tactics of the American Civil Rights movement. They were inspired by the writing of Abbey, who, in his 1975 book *The Monkey Wrench Gang*, wrote of a band of activists wreaking havoc on development efforts in the American West. Earth First!ers swiftly intensified their actions to include ecotage (environmental terrorism), which became standard fare, particularly in the American West. During the first years, Earth First! took the initiative to propose biocentric-focused wilderness proposals, beyond the ideas being advocated by mainstream environmentalists, along with more creative actions such as putting a depiction of a "crack" onto Glen Canyon Dam. At this time, the movement also published *Earth First! The Radical Environmental Journal*, a periodical created to convey Earth First!ers' perspectives on environmental issues. Over time, Earth First! participants became involved in more protests, particularly against logging in the northwest. It was during one of these protests, in Oregon, that activist Mike Jakubal devised and carried out the first tree sit (Davis 1991). The event lasted for less than a day, but it started a new form of struggle against the destruction of wildlife habitats and wilderness.

After the creation of tree sitting, Earth First! participants became mainly associated with actions to prevent logging and the construction of dams. This change attracted a new type of activist to Earth First!, bringing the movement closer to the ideas championed by anarchists and the counterculture. This change in direction made many of the original Earth First!ers uncomfortable. Thus, by 1990, many of the original members and most of the group of activists that created the movement had severed ties from Earth First!. It was at this time when Earth First! changed from an organization with a leadership to a movement greatly influenced by anarchist philosophy. Another change to Earth First! occurred in 1992 when some participants pushed to have recognition as a mainstream movement. The Earth First! activists who did not want to give up their criminal acts started an offshoot focused more on monkey-wrenching than Earth First! This group became known as the Earth Liberation Front (ELF) (Fattig 2003).

By 1992, the Earth First! movement appeared to be made up of small groups and individuals who attended or set up protests and educational campaigns, along with supporting civil disobedience in the form of tree sitting, road blockades, and occasionally "ecotage" (monkey-wrenching). However, more recently the movement, led by Judi Bari, has begun to renounce ecotage and work with small logging businesses to defeat large-scale corporate logging, particularly in Northern California.

Of course, perspectives such as deep ecology made it impossible for subscribers to also support middle-class, American ideas. Therefore, quite inadvertently, the radical end of environmentalism functioned to make the mainstream movement seem more reasonable. In 1992, for instance, some members of Earth First! grew frustrated that their organization had become too mainstream and then formed the international group known as ELF. ELF's uncompromising goal was to "inflict economic damage on those profiting from the destruction and exploitation of the natural environment."

Opponents of Earth First! have always focused on the radical element of the movement, pointing out the terrorism-like activities in which some of the members partake. These actions are responsible for economic losses to industries such as logging and mining, mainly as a result of protests that stop the extraction of resources. Proponents of Earth First! counter their critics with claims that the value of the environment they protected is greater than the economic losses taken by the industries. In addition, supporters of Earth First! also point to their more recent change to work with small, local business, as a means to show a move away from their more radical periods.

Over time, the Earth First! movement has changed. The members' desire to put Earth above humans comforts has remained, but the actions taken to support this effort have been adjusted to fit the people and issues involved.

Sources and Further Reading: Cronon, *Uncommon Ground*; Davis, *The Earth First! Reader: Ten Years of Radical Environmentalism*; Foreman, *Confessions of an Eco-Warrior*; Lee, *Earth First!: Environmental Apocalypse*; Naess, "The Shallow and the Deep, Long-Range Ecology Movement"; Nash, *Wilderness and the American Mind*; Opie, *Nature's Nation*; Price, *Flight Maps*; Scarce, *Eco-Warriors: Understanding the Radical Environmental Movement*; Steinberg, *Down to Earth*; Wall, *Earth First! And the Anti-Roads Movement: Radical Environmentalism and Comparative Social Movements*.

DEFINING PRESERVATION FOR THE MODERN ERA AT DINOSAUR

Time Period: 1920 to the present
In This Corner: Preservationists, environmental organizations
In the Other Corner: Private property advocates
Other Interested Parties: Middle-class environmentalists
General Environmental Issue(s): National parks, preservation

Grassroots environmentalism relied on the expansion of the middle class after World War II. Of course, it is ironic that this society based on conspicuous consumption and the production to supply it fueled new interest in the environment. The conservative impulse proclaimed by Pinchot and others at the turn of the twentieth century had little bearing on a post-1950 America that will likely be recalled as one of the most wasteful civilizations in world history. Even so, middle-class environmentalists drove new ideas about preservation in the twentieth century through environmental organizations.

Including Scientific Information

At the root of this type of modern environmentalism was more advanced scientific knowledge on the public's part and a much higher degree of expectation for personal safety. Samuel P. Hays wrote that this era "displayed demands from the grass-roots, demands that are well charted by the innumerable citizen organizations" that grew out of such public interest. Within growing suburbanization, middle-class Americans expected health and home safety, as well as economic expansion. These versions of the American dream were not seen to be mutually exclusive in the late twentieth century (Hays 1993, 30).

Although there was as yet little regulative authority available, grassroots environmentalists would demand that their government intercede and ensure community safety. Such a groundswell of interest in the environment mobilized with the counterculture movements of the 1960s.

As Americans rethought the ethics of everyday human existence, some explored a more ecologically sustainable way of life. Alternative fuels, whole grain, natural foods, and communal living are only a few examples of life-changing options that Americans chose to pursue. Most importantly, a national stage linked scientific data with environmental concern. National parks, which had begun without much controversy, became a major forum in which this debate would play out throughout the late twentieth century. As a national symbol, these parks also represented the nation's environmental ethic during a time of reappraisal and open contest.

Of course, at the outset, we also must consider the problem of definitions. Historian Hal Rothman observed that, in 1968 when the Brookings Institute surveyed the nation about which issues most concerned them, their list did not even include any environmental issues. Only a year later, however, "environmental crisis" was the top issue in the same survey (Rothman 1998, 125–30). This is a dramatic shift, but what exactly did the term "environmental" and "crisis" mean to each American? When considering nature's role in his or her everyday life, would each American really be willing to alter their preferences and choices?

Defining National Parks

The mainstay within environmental politics was the ideal: the national park. The earliest national parks possessed little if any unifying philosophy or ethic. A remedy to this began with the Hetch Hetchy episode and then with passage of the National Park Service Act in 1916. This act created the NPS as a unit of the Department of the Interior, staffed no longer by military personnel but now specially trained rangers (although this change would not be truly noticeable until later in the century). In his popular 1917 book on the parks, Enos Mills stated their mission as follows: "A national park is an island of safety in this riotous world. Within national parks is room—glorious room—room in which to find ourselves, in which to think and hope, to dream and plan, to rest, and resolve" (Nash 1982, 189).

Stephan T. Mather, a businessman, was made the first NPS director. In addition to creating a unifying mission based in preservation, Mather also sought to develop parks as certifiable tourist attractions. By midcentury, some critics had even come to criticize over-crowding in the parks. Preservationist organizations such as Muir's Sierra Club and the Land Conservancy would argue for as little use as possible; others argued that national parks were a trust open for the use of any citizen. This, of course, meant Americans had every right to use the sites as they saw fit.

Environmental policy also continued to move forward from 1950 through the 1960s. The initial interest of the public in the 1940s and 1950s was garnered through an event similar to Hetch Hetchy. The Bureau of Reclamation, an agency developed by applying Pinchot's idea of conservation to waterways of the American West, set out to construct the Echo Park Dam along the Utah-Colorado border, and within a little-used national monument, named Dinosaur, although most of its fossils and bones had been stolen. As Congress neared a vote on the issue in 1950, seventy-eight national and 236 state conservation organizations expressed their belief that national parks and monuments were sacred areas.

David Brower, executive director of the Sierra Club, and Howard Zahniser of the Wilderness Society used the opportunity to create a model for environmental lobbyists to follow. Direct-mail pamphlets asked, "What Is Your Stake in Dinosaur?" and "Will You DAM the Scenic Wildlands of Our National Park System?" Additionally, a color motion picture and a book of lush photos, each depicting the Echo Park Valley's natural splendor, were widely viewed by the public. Such images and sentiments forced Americans to react. With mail to Congress late in 1954 running at 80 to 1 against the dam, the bill's vote was suspended and the project was eventually abandoned. The issues had been packaged by environmentalists to connect concerns with romantic images of the American past. Americans loved this idea and reacted more positively than ever before (Gottlieb 1993, 36–41).

Conclusion: National Parks as Centerpiece

Built on this example, national parks became the nation's greatest tourist attractions during the twentieth century. The eighty million acres of national park in the year 2000, however, was just one segment of the federally owned lands in the United States. In 2000, acreage for the BLM was 270 million acres, the U.S. Forest Service held 191 million acres, and the Fish and Wildlife Service administered ninety-one million acres. In each case, the administration of these federally owned lands could be guided and controlled through federal legislation. This controversial fight offered the environmental movement a unique opportunity to express a new ethic for the entire nation.

Sources and Further Reading: Gottlieb, *Forcing the Spring: The Transformation of the American Environmental Movement*; Nash, *Wilderness and the American Mind*; Opie, *Nature's Nation*; Pinchot, *Breaking New Ground*; Price, *Flight Maps*; Reisner, *Cadillac Desert*; Rothman, *Greening of the Nation?*; Steinberg, *Down to Earth*; White, *It's Your Misfortune and None of My Own*.

WILDERNESS ACT SHAPES ENVIRONMENTAL IDEALS FOR A GENERATION

Time Period: 1935 to the present
In This Corner: Preservationists, some scientists, NGOs, environmental organizations
In the Other Corner: Conservationists, some scientists

Other Interested Parties: Policy makers, American citizens
General Environmental Issue(s): Wilderness, preservation

The concept of wilderness was not always part of environmentalism. In fact, some conservationists were openly critical of the extreme ideas that wilderness preservation embodied. However, as modern environmentalism matured in the mid-twentieth century, wilderness was at its core.

Conceiving of Wilderness

As Aldo Leopold and other visionaries applied a new understanding on human's relationships with the natural world, many grew increasingly frustrated with the limits of what was known in the 1930s as "conservation." Although the New Deal initiated many new ways of applying ideas of conservation to the landscape, the alternative ethic of preservation also found new energy. Leopold and others combined the new ideas of ecology with Muir's conception of preservation to organize their efforts around the idea of wilderness.

Robert Marshall, a young wealthy outdoorsman, led the formation of the Wilderness Society. He joined efforts with two other visionaries who were frustrated with their work for New Deal conservation: Robert Sterling Yard and Benton MacKaye. In early 1935, these three environmentalists joined with Leopold to form an organization committed to the preservation of an wondrous idea: wilderness. Utopian in its conception, the Wilderness Society embodied the ideas expressed by Marshall when he said, "There is just one hope of repulsing the tyrannical ambition of civilization to conquer every niche on the whole earth. That hope is the organization of spirited people who will fight for the freedom of the wilderness." The concept of wilderness prioritized a lack of human impact; however, there was otherwise little definition to the concept or the organization's efforts. This would change during the next three decades.

Basing Policy in Ecology

After experiencing success at Dinosaur National Monument, Howard Zahniser of the Wilderness Society identified this moment as the best to press for the environmental movement's greatest goal: a national system of wilderness lands. Based on the idealistic notion of pristine wilderness espoused by Theodore Roosevelt and others, such a system had been called for beginning with Aldo Leopold in the 1910s. With increased recreation in parks and public lands, argued Zahniser, it had become even more crucial that some of the land be set aside completely (Gottlieb 1993, 41–44).

The ideal of wilderness had received scientific definition through the growing science of ecology and the related development of the concept of ecosystems. Building on the ideas of Clements and others, Henry Cowles and scientists including Aldo Leopold, who wrote *A Sand County Almanac, and Sketches Here and There* during the 1940s, developed a much deeper understanding of human impact on the natural environment. The term ecosystem is credited to Arthur George Tansley, who in the 1940s argued that nature occurred in self-sufficient (except for solar energy) ecological systems. He would go on to add that such systems could overlap and interrelate. The existence of such systems, of course, began to suggest that the human agent existed as the interloper in any system (Worster 1994, 307–11).

This concept of the human surrounded by biological systems independent from it became a basic realization to the modern environmental movement. The concept used by environmentalists to link together high-level ecological theories for the general public was wilderness. By the early 1960s, when many Americans had begun to be aware of the fact that many human activities were injuring the environment, wilderness became the term used to describe a nature free from human influence.

Zahniser's bill, which was introduced to Congress in the 1950s, precluded land development and offered recreational opportunities only for a few rather than for the great mass of travelers. Such an ideal goal required great salesmanship, and Zahniser was perfect for the job. As the political climate shifted in the early 1960s, lawmakers became more interested in wilderness. Finally, in 1964, President Lyndon Johnson signed the Wilderness Act into law. The United States had taken one of the most idealistic plunges in the history of environmentalism: nearly ten million acres were immediately set aside as "an area where the earth and its community of life are untrammeled by man, where man himself is a visitor who does not remain." Additional lands would be preserved in a similar manner by the end of the decade (Harvey 2005, 23).

Although the concept of wilderness forced the general American public to begin to understand ecosystems and the webs of reliance operating within natural systems, the application of scientific understanding to environmentalism occurred most often in other realms. Defeating the dam at Echo Park and the passage of the Wilderness Act set the stage for a 1960s shift in environmental thought that combined with NIMBY culture of the 1970s to create a federal mandate for policy action. From a vague ideal, wilderness became a structuring agent for the administration of federal lands (Gottlieb 1993, 41–48).

John Saylor Crosses the Political Divide for the Environment

As this revolutionary legislation came forward, it required the support of a broad political base. Inspired by Republican Theodore Roosevelt, Representative John Saylor of Pennsylvania balanced a commitment to the conservation of natural resources that was unafraid to favor whole-scale preservation of specific areas. Wrote biographer Thomas Smith, "Saylor believed that once national parks and monuments had been established, they became sacrosanct." His efforts on behalf of the environment also helped him to emphasize earth stewardship with a strong religious base. "Protecting natural splendors," Smith wrote, Saylor believed "would bring present and future generations closer to the Creator" (Smith 2006, 2). Beginning from a fairly equal-part conservation of natural resources and resistance to preferential treatment of a region other than his own, Saylor was identified as a friend by environmental organizations. In two decades of lawmaking and litigation, Saylor toed the line for national park preservation. He fought his battles publicly and, more important, privately in the decision-making bodies on Capitol Hill.

The Echo Park proposal prompted Saylor to visit Dinosaur National Monument in 1953. The trip was arranged by an opponent of the scheme, Joe Penfold, a Coloradan who served as western representative of the Izaak Walton League. The excursion took a float trip on the Yampa River through remote, rugged canyon lands. That experience, Penfold later related, made a lasting impression and may have heightened Saylor's commitment to wilderness, undisciplined rivers, and the inviolability of national monuments and parklands. Shortly after the trip, Bus Hatch wrote David Brower of the Sierra Club that he believed "Congressman Saylor is going to help us out" (Smith 2006).

Working with Brower, the Sierra Club, and others, Saylor led the defeat of this project. The mobilization of the American public over this single controversy helped to define the organization and mechanisms of modern environmentalism. Most remarkable, when the Wilderness Society followed this effort with a push for a federal law to preserve wilderness, Saylor led that cause as well. The Wilderness Act of 1964 is one of the most idealistic, forward-looking environmental documents in human history. Smith reported that Brower believed that, because of his hard work on its behalf, the wilderness system created by the act should be named after Saylor (317). The following is from the Wilderness Act, September 3, 1964:

> To assure that an increasing population, accompanied by expanding settlement and growing mechanization, does not occupy and modify all areas within the United States and its possessions, leaving no lands designated for preservation and protection in their natural condition, it is hereby declared to be the policy of the Congress to secure for the American people of present and future generations the benefits of an enduring resource of wilderness. For this purpose there is hereby established a National Wilderness Preservation System to be composed of federally owned areas designated by Congress as "wilderness areas", and these shall be administered for the use and enjoyment of the American people in such manner as will leave them unimpaired for future use and enjoyment as wilderness; and no Federal lands shall be designated as "wilderness areas" except as provided for in this Act or by a subsequent Act....

> Definition of Wilderness

> (c) A wilderness, in contrast with those areas where man and his own works dominate the landscape, is hereby recognized as an area where the earth and its community of life are untrammeled by man, where man himself is a visitor who does not remain. An area of wilderness is further defined to mean in this Act an area of undeveloped Federal land retaining its primeval character and influence, without permanent improvements or human habitation, which is protected and managed so as to preserve its natural conditions and which (1) generally appears to have been affected primarily by the forces of nature, with the imprint of man's work substantially unnoticeable; (2) has outstanding opportunities for solitude or a primitive and unconfined type of recreation; (3) has at least five thousand acres of land or is of sufficient size as to make practicable its preservation and use in an unimpaired condition; and (4) may also contain ecological, geological, or other features of scientific, educational, scenic, or historical value.

Conclusion: An Environmental Ideal or a Problem

By constructing the criteria for wilderness, environmentalists helped to define an era of modern environmentalism that strove to have areas included in the national wilderness system. By the end of the century, however, some environmentalists claimed that the wilderness ideal was a false goal.

For instance, historian William Cronon wrote the following:

But the trouble with wilderness is that it quietly expresses and reproduces the very values its devotees seek to reject. The flight from history that is very nearly the core of wilderness represents the false hope of an escape from responsibility, the illusion that we can somehow wipe clean the slate of our past and return to the tabula rasa that supposedly existed before we began to leave our marks on the world. The dream of an unworked natural landscape is very much the fantasy of people who have never themselves had to work the land to make a living—urban folk for whom food comes from a supermarket or a restaurant instead of a field, and for whom the wooden houses in which they live and work apparently have no meaningful connection to the forests in which trees grow and die. Only people whose relation to the land was already alienated could hold up wilderness as a model for human life in nature, for the romantic ideology of wilderness leaves precisely nowhere for human beings actually to make their living from the land. (Cronon 1996)

Although contemporary environmentalists will continue to debate the importance of the wilderness designation, it is clear that the concept helped to direct environmental thought during its formative era.

Sources and Further Reading: Nash, *Wilderness and the American Mind*; Rothman, *The Greening of a Nation*; Runte, *National Parks: The American Experience*; Sellars, *Preserving Nature in the National Parks: A History*; Smith, *Green Republican: John Saylor and Preservation of America's Wilderness*; Sutter, *Driven Wild: How the Fight Against Automobiles Launched the Modern Wilderness Movement*.

NATURAL BACKLASH FUELS THE NATIONAL ENVIRONMENTAL POLICY ACT AND ENVIRONMENTAL POLICY OF THE 1970S

Time Period: 1969
In This Corner: Nature
In the Other Corner: American consumers, politicians
Other Interested Parties: Environmentalists, industrial leaders
General Environmental Issue(s): Pollution, policy

In the 1970s, many Americans favored the establishment of environmental regulation because they felt that the American ethic for development had pushed nature too far. If any additional evidence were needed to prove the need for such legislation, nature seemed to be in open rebellion against humans in 1969. These events, which became large-scale media events, functioned to radically change the role of nature in everyday American life.

Together, these environmental catastrophes changed the American conscience. Each of the following events cycled directly into the new concern for the environment and became prominent headlines in American newspapers and news media (Rothman 1998, 96–99).

Lake Erie Is Dead

In the 1960s, with little visible life in its waters, Lake Erie was declared "dead." Scientists quickly learned that the opposite was true; the lake was full of life but not a correct balance of life forms. Primarily, an excessive amount of algae, created by pollution and excessive nutrients, had created eutrophication in Lake Erie. As the excessive algae expanded, it soaked up the lake's supply of oxygen, making it impossible for other species to survive.

Lake Erie's situation was particularly acute because it is the shallowest and warmest of the five Great Lakes. It had been used extensively for decades, enduring runoff from agriculture, urban areas, industries, and sewage treatment plants along its shores. The worst of the pollutants coming into the lake was phosphorous, which entered from agricultural fields. Unfortunately, the phosphorous did just what the farmers intended when they spread the fertilizer on their fields: only it also stimulated vegetative growth once it ran off into Lake Erie. This phosphorous was one of the primary reasons for the growth of the algae.

In response to Lake Erie's "death," the Great Lakes Water Quality Agreement (GLWQA) was signed by the United States and Canada in 1972. The agreement emphasized the reduction of phosphorous entering Lakes Erie and Ontario. Coupled with the U.S. and Canadian clean water acts, the GLQWA made significant progress toward reducing the phosphorus levels in Lake Erie. Through GLQWA, two nations for the first time committed themselves to creating common water quality goals (Ashworth 1986, 130–37).

The Cuyahoga River Is on Fire

Throughout American industrialization, many rivers were exploited and left heavily laden with pollutants. Rivers near industrial centers such as Pittsburgh and Cleveland became symbols of this degradation. This legacy attracted national attention in 1969 when the Cuyahoga River did something that it had done many times before: caught on fire. This time, however, the American public viewed the occasion as a profound statement about the impact of pollution on our natural environment (Steinberg 2002, 239–40).

The fire was actually very brief in duration. It began at 11:56 A.M. and lasted for approximately twenty minutes. The area of the river that caught fire was fairly out of the way: just southeast of downtown Cleveland. The flames damaged two railway bridges and were estimated to reach heights of roughly five stories. The cause was nothing out of the ordinary for the Cuyahoga: a slick of highly volatile petroleum derivatives had leaked from one of the refineries located along the river. That turned out to be the real issue: national disdain that such an event was fairly routine for such industrial locales.

The river fire served as a reminder of the importance of continued support for cleanup of Lake Erie and the Cuyahoga River. On the day after the fire, Cleveland Mayor Carl Stokes stood on the damaged Norfolk and Western bridge and called for the public to rally support for an effort to cleanup the Cuyahoga and promised to sue the state of Ohio and the individual polluters. Stokes referred to the polluted state of the river as "a long-standing condition that must be brought to an end."

The Cuyahoga fire was most important as a symbol for a new era. The river fire received major national media attention with a *Time* magazine article on August 1, 1969 (approximately one month after the fire). In October of 1969, federal officials passed a bill that would grant states more assurances that projects aimed at improving water quality would receive federal support.

Oil Spills off Santa Barbara, California

On the afternoon of January 29, 1969, a Union Oil Company platform stationed six miles off the coast of Summerland suffered what is referred to as a blowout. Oil workers had drilled a well down 3,500 feet below the ocean floor. Riggers began to retrieve the pipe to replace a drill bit when the "mud" used to maintain pressure became dangerously low. Natural gas blew out the pipe, sending oil into the surrounding water. When the drillers successfully capped the well, pressure built up and five breaks appeared in an east-west fault on the ocean floor, releasing oil and gas from deep beneath the earth (Rothman 1998, 101–5).

For eleven days, oil workers struggled to cap the rupture. During that time, 200,000 gallons of crude oil bubbled to the surface to form an 800-square-mile slick. Incoming tides brought the thick tar to beaches from Rincon Point to Goleta, covering thirty-five miles of coastline and coastal life. The slick also moved south, tarring Anacapa Island's Frenchy's Cove and beaches on Santa Cruz, Santa Rosa, and San Miguel Islands.

Many lessons were learned from this first oil spill of the environmental era. Rapid cleanup was crucial. For this spill, it took oil workers eleven and a half days to control the leaking oil well. The cleanup had to be multifaceted: skimmers scooped oil from the surface of the ocean while air planes dumped detergents on the spill to try and break up the slick. Meanwhile, on the beaches and harbors, volunteers spread straw to soak up the tar and oil, and rocks were steam cleaned.

Just days after the spill occurred, activists founded Get Oil Out in Santa Barbara. Founder Bud Bottoms urged the public to cut down on driving, burn oil company credit cards, and boycott gas stations associated with offshore drilling companies. Volunteers also helped the organization to gather 100,000 signatures on a petition calling for the ban of offshore oil drilling. Although drilling was only halted temporarily, new laws eventually tightened regulations a bit more on offshore drilling.

Environmental Protection In the 1970s

Although many Americans remained skeptical about the federal government's ability to regulate the natural environment and about the prescience of any controls over industrial development, these events helped to convince politicians and voters that action was needed. Many Americans simply had enough. Others very clearly felt shameful at the present state of America's natural environment. During the next decade, the social and cultural change initiated by the 1960s and fed by a barrage of demonstrations of environmental degradation and the need for conservation created a deluge of environmental legislation.

The public outcry would be so severe that even a conservative such as Richard Nixon might be deemed "the environmental president" as he signed the NEPA in 1969, creating the EPA:

National Environmental Protection Act (NEPA), 1969
(a) The Congress, recognizing the profound impact of man's activity on the interrelations of all components of the natural environment, particularly the profound influences of population growth, high-density urbanization, industrial expansion, resource exploitation, and new and expanding technological advances and recognizing further the critical importance of restoring and maintaining environmental quality to the overall welfare and development of man, declares that it is the continuing policy of the Federal Government, in cooperation with State and local governments, and other concerned public and private organizations, to use all practicable means

and measures, including financial and technical assistance, in a manner calculated to foster and promote the general welfare, to create and maintain conditions under which man and nature can exist in productive harmony, and fulfill the social, economic, and other requirements of present and future generations of Americans.

(b) In order to carry out the policy set forth in this Act, it is the continuing responsibility of the Federal Government to use all practicable means, consistent with other essential considerations of national policy, to improve and coordinate Federal plans, functions, programs, and resources to the end that the Nation may—

1. fulfill the responsibilities of each generation as trustee of the environment for succeeding generations;
2. assure for all Americans safe, healthful, productive, and aesthetically and culturally pleasing surroundings;
3. attain the widest range of beneficial uses of the environment without degradation, risk to health or safety, or other undesirable and unintended consequences;
4. preserve important historic, cultural, and natural aspects of our national heritage, and maintain, wherever possible, an environment which supports diversity, and variety of individual choice;
5. achieve a balance between population and resource use which will permit high standards of living and a wide sharing of life's amenities; and
6. enhance the quality of renewable resources and approach the maximum attainable recycling of depletable resources.

(c) The Congress recognizes that each person should enjoy a healthful environment and that each person has a responsibility to contribute to the preservation and enhancement of the environment....

The Congress authorizes and directs that, to the fullest extent possible: (1) the policies, regulations, and public laws of the United States shall be interpreted and administered in accordance with the policies set forth in this Act, and (2) all agencies of the Federal Government shall—

(A) utilize a systematic, interdisciplinary approach which will insure the integrated use of the natural and social sciences and the environmental design arts in planning and in decisionmaking which may have an impact on man's environment;
(B) identify and develop methods and procedures, in consultation with the Council on Environmental Quality established by title II of this Act, which will insure that presently unquantified environmental amenities and values may be given appropriate consideration in decisionmaking along with economic and technical considerations;
(C) include in every recommendation or report on proposals for legislation and other major Federal actions significantly affecting the quality of the human environment, a detailed statement by the responsible official on—

(i) the environmental impact of the proposed action,
(ii) any adverse environmental effects which cannot be avoided should the proposal be implemented,

(iii)   alternatives to the proposed action,

(iv)   the relationship between local short-term uses of man's environment and the maintenance and enhancement of long-term productivity, and

(v)   any irreversible and irretrievable commitments of resources which would be involved in the proposed action should it be implemented.

The public entrusted the EPA as its environmental regulator to enforce ensuing legislation monitoring air and water purity, limiting noise and other kinds of pollution, and monitoring species to discern which required federal protection. The public soon realized just how great the stakes were (Rothman 1998, 115–21).

During the 1970s, nearly every industrial process was seen to have environmental costs associated with it. From chemicals to atomic power, long-believed technological "fixes" came to have long-term impacts in the form of wastes and residue. Synthetic chemicals, for instance, were long thought to be advantageous because they resist biological deterioration. In the 1970s, this inability to deteriorate made chemical and toxic waste the bane of many communities near industrial or dump sites.

Among the assorted catastrophes, Love Canal stood out as a new model for federal action. The connection between health and environmental hazards became obvious throughout the nation. Scientists were able to connect radiation, pollution, and toxic waste to a variety of human ailments. The "smoking gun," of course, contributed to a new era of litigation in environmentalism. Legal battles armed with scientific data provided individuals armed with only NIMBY convictions with the ability to take on the largest corporations in the nation.

Rapidly, this decade instructed Americans, already possessing a growing environmental sensibility, that humans, just as Carson had instructed, needed to live within limits. A watershed shift in human consciousness could be witnessed in the popular culture as green philosophies infiltrated companies wanting to create products that appealed to the public's environmental priority. Recycling, requiring the use of daylight savings time, carpooling, and EISs became part of everyday life after the 1970s.

Conclusion: Earth Day 1990

How far could the influence of environmental thought come in twenty years? One of the culminating symbols of the changes brought by events and expectations of 1969–1970, the first Earth Day on April 22, 1970, marked an important moment for modern environmentalism. Therefore, assessing changes in Earth Days over the next decades might allow us some understanding of the persistence of changes in environmental thought and policy from the 1970s.

April 22, 1990 may have seemed like an ordinary day; however, instead of the regular television schedule listings, the *New York Times* listed "TV and the Environment." Most impressive, there was a host of programs from which to choose! However, many Americans were not taking time to watch TV; instead, they performed cleanup enterprises on behalf of Mother Earth. Internationally, more than forty million humans marked some kind of celebration on Earth Day 1990.

For American culture, however, Earth Day 1990 marked a day of broader recognition. American society celebrated a new relationship with the natural world surrounding it. Today, polls reveal that nearly seventy percent of Americans refer to themselves as "environmentalists." Such developments are simply the latest in a watershed shift in Americans' awareness of the human impact on the natural environment. The 1990s marked a maturing period for the environmental movement, which had been evolving in the United States since the nineteenth century.

Sources and Further Reading: Andrews, *Managing the Environment, Managing Ourselves*; Ashworth, *The Late, Great Lakes: An Environmental History*; Nash, *Wilderness and the American Mind*; Opie, *Nature's Nation*; Rothman, *The Greening of a Nation*; Steinberg, *Down to Earth*.

INVOLVING THE FEDERAL GOVERNMENT IN PUBLIC HEALTH

Time Period: 1900 to the present
In This Corner: Health officials
In the Other Corner: Believers in small government, pharmaceutical companies
Other Interested Parties: American public, political officials
General Environmental Issue(s): Health

Throughout American life in the 1970s, new expectations were placed on the federal government. In some cases, however, new legislation simply picked up on trends begun in an earlier generation. With each major change during this period, the political landscape of healthcare changed. At the close of the twentieth century, high healthcare prices and malpractice suits contributed to significant changes. Because of the massive money involved, healthcare became one of the most active areas of special interest lobbying.

Bring in the Federal Government

When progressive reformers demanded that the federal government become more involved in Americans' health, for instance, the PHS was established in the early 1900s as part of the Department of Health and Human Services. By the end of the twentieth century, the PHS had grown to be the major health agency in the federal government, with more than 51,000 civil service employees and a budget of more than $20 billion.

NIMBY expectations significantly altered the expectations placed on the PHS. With a broad mission to protect and advance the nation's health, today the PHS creates programs to help control and prevent diseases, fund biomedical research, protect Americans against unsafe food and drugs, and try to make new technologies and medicines available to as many Americans as possible. The major changes for the PHS began in the 1950s, when President Dwight D. Eisenhower reorganized the federal health agencies to ensure that the important areas of health, education, and social security be represented in the president's cabinet. The newly created Department of Health, Education, and Welfare eventually gave way in 1979 to the Department of Health and Human Services.

The eight major agencies that make up the PHS and that do this work are the Centers for Disease Control and Prevention, the Agency for Toxic Substances and Disease Registry, the National Institutes of Health, the FDA, the Substance Abuse and Mental Health Services Administration, the Health Resources and Services Administration, the Agency for Health Care Policy and Research, and the Indian Health Service.

Recent events and debates have involved the agency in some of the most controversial issues of our age, including national security, abortion rights, and the morality of bioengineering. A similar effort to call into question the basic values of post–World War II American consumption came from a genre of writing known as consumer advocacy. Ralph Nader helped define a field known as consumer rights with the publication of *Unsafe at Any Speed* in 1965. In 2002, similar titles, including *High and Mighty* and *Fast Food Nation*, criticized the sport utility vehicle (SUV) and American nutrition. *Fast Food Nation* suggested that restaurant chains such as McDonald's contributed to Americans becoming the most obese people in world history.

The Contested Realm of Federal Healthcare

The federal role in citizens' healthcare was the source of great debate at the end of the twentieth century. Although most Americans wished good health on other citizens, paying for such services became enmeshed in political philosophy. For instance, in the 1990s, the administration of President William Clinton placed the federal government at the forefront of research to explore the use of DNA in medical procedures and, especially, to map the human genome. Similar to space travel, such scientific frontiers were often considered to merit the organization and funding of the federal government. In addition, many scientists believed the project was best kept in the public sector. The administration of George W. Bush, in contrast, took a stand against medical research related to stem cells and human cloning and argued that the federal government should not be involved in such scientific efforts.

In efforts such as food labeling, the federal government has been consistently involved in helping to give Americans nutritional information about what they eat and drink. Responding to reports of nutritional deficiencies, the FDA in 1973 adopted voluntary labeling that emphasized vitamins, minerals, and proteins. Although nutritional deficiencies are now uncommon, problems with labeling became apparent. As consumers became concerned about the link between diet and disease, the food industry began adding phrases such as "light" and "healthy" or "low fat" to labels. What these phrases meant was unclear. The Nutrition Labeling and Education Act of 1990 required food labels, and the FDA set standard serving sizes. In 1994, requirements for a standardized food label took effect, making it easier for consumers to check the nutritional content of packaged foods. The nutrition pyramid released by the FDA is part of a national effort to reduce obesity, particularly among younger Americans.

One of the major controversies of the 1980s and 1990s was over the government's response to the AIDS (acquired immunodeficiency syndrome) crisis. When AIDS emerged in the 1980s, gays and lesbians were already a well-organized political force. Their experiences had often led them to be skeptical of the medical and scientific communities. AIDS activist groups, such as the New York-based ACT UP (AIDS Coalition to Unleash Power) and the San Francisco–based Project Inform, pressured the FDA to provide promising but still experimental drugs to people with AIDS on a parallel track with standard clinical trials required before FDA approval of drugs.

Despite resistance, the FDA moved toward such a policy and established accelerated approval procedures in 1992. Accelerated approval is intended to get promising but still unproven drugs for life-threatening diseases to patients as quickly as possible. The drugs must still be shown to be safe, but the usual standards of efficacy are relaxed. Although drug companies have long resisted increasing regulation, this particular case did not meet with resistance. Unlike other consumer groups, AIDS activists had an interest in less stringent regulations, and, by the late 1980s, prominent researchers became convinced that there was a moral obligation to provide promising therapies as early as possible.

The public expected protection and action from the federal government. These NIMBY expectations stimulated federal action on a number of issues related to public health by the end of the 1970s. Although these initiatives varied with each presidential administration, the expectations of middle-class Americans have made certain that they never are lost entirely.

Conclusion: But Not THAT Much Federal Involvement

Although Americans clearly believed that the federal government possessed an important role to play in monitoring and regulating healthcare, there was a clear line to be drawn: they did not want the government to be actively involved in providing healthcare. The concept of socialized medicine became a hot-button topic during the 1990s. The Clinton Administration immediately set out to reform U.S. healthcare, even including the first lady, Hillary, to oversee the project, which became known as the Task Force on National Health Care Reform.

The outcome was remarkable. The involvement of the first lady clearly politicized the volatile issue even further. During hearings on Capitol Hill, the first lady received significant criticism for this plan. Meanwhile, Democrats, instead of uniting behind the president's original proposal, offered a number of competing plans of their own. Some criticized the plan from the left, preferring a Canadian-style single payer system. After an attempt to compromise was defeated, the 1994 election swung Congress heavily to Republicans. The Clinton effort at healthcare reform would not be revived.

As candidates prepare for the 2008 presidential elections, one of them, of course, is Senator Hillary Clinton. Most candidates for the Democratic nomination argue that healthcare reform is an urgent need. Having learned from the experiences of the 1990s, however, none of the candidates are willing to provide specifics before they are elected.

Sources and Further Reading: Opie, *Nature's Nation*; Rothman, *The Greening of a Nation*; Steinberg, *Down to Earth*.

THE ALL-AMERICAN CANAL

Time Period: 1934 to today
In This Corner: The United States, California, United States agriculture producers
In the Other Corner: Mexico, Baja California, Sonora, Mexican farmers
Other Interested Parties: Environmentalists
General Environmental Issue(s): Ground water, drought, habitat destruction

The United States' arid southwest is a region of great developmental and agricultural potential, but only if the region has water. Long ago, this region recognized its need and the resulting competition that ensues when the issue of water is addressed, competition among individual stakeholders and between the United States and Mexico. The past, current, and future competition for the region's most important natural resource, water, has shaped and continues to shape the region's structural and political development. The construction and management of the All-American Canal has been and continues to be one of the most important elements that impact the region.

The eighty-two-mile (132 kilometers) long All-American Canal conveys water from the Colorado River in southeastern California through the Colorado Desert, to southwestern California's Imperial and Coachella Valleys. Used by nine municipalities and more than 500,000 acres of agricultural land, the All-American Canal carries approximately 3.1 million acre-feet to Colorado River water, 26,155 cubic feet per second of water, making it the largest irrigation canal in the world (Imperial Irrigation District 2007). The canal is the only source of water for the Imperial Valley and is an essential source in the Coachella Valley. With Colorado River water carried to the Imperial and Coachella Valleys, the increased yield of farm crops has been exceptional, but for the Imperial Valley to remain the United States' fourth most productive agricultural region and for the Coachella Valley to remain productive, the area needs to continue receiving water from the All-American Canal.

The United States Bureau of Reclamation built the All-American Canal between 1934 and 1942. Located totally within the United States (thus the name All-American), it was designed to replace the Alamo Canal or Inter-California canal, which served the same function but passed through Mexico. Politicians and agricultural producers in the United States wanted sole control of the flow of Colorado River water for the state of California rather than sharing management duties with their Mexican counterparts, thus the creation of the All-American Canal.

The All-American Canal is a part of the federal irrigation system that also includes the Hoover Dam, the Imperial Dam (the structure that delivers the water from the Colorado River into the Canal), the East Highline Canal, the Central Canal, the Coachella Canal, and the Westside Main Canal, in addition to a number of smaller canals that move the water into and throughout the Imperial Valley and Coachella Valley. In addition to providing water, the All-American Canal also provides a limited amount of electricity to the Imperial and Coachella Valleys. Eight relatively small hydroelectric power plants exist along drops in the canal system; the hydroelectric energy production is determined by water delivery needs. Although the All-American Canal is owned by the United States Bureau of Reclamation, it is the Imperial Irrigation District that manages both the water and energy production of the canal.

Supporters of the All-American Canal praise the water and energy provision. These resources have aided in the development of both the urban and rural agricultural areas of the region. Thus, there are many economic benefits that have stemmed directly from the construction and functioning of the All-American Canal. In addition, supporters also point to the security that stems from the United States decision to keep Colorado River water within the United States rather than continuing to use the Alamo or Inter-California Canal that transported the water into Mexico before returning it to California. This decision not only protects the water from contamination, it also eliminates the potential for Mexican stakeholders to use the water before it reaches California.

Critics of the All-American Canal focus on water quality and quantity issues. Functionally, the main criticisms of the All-American Canal focus on salinity and water seepage. The

Colorado River carries a substantial amount of sediment. To prevent the clogging of sections of the canal and to avoid subsequent expensive and difficult repairs, sediment is removed at desilting basins. Surveillance of these facilities allows for a fairly easy resolution to these criticisms. The second main criticism with the function of the All-American Canal deals with the level of salinity. Two rivers, the Alamo and New River, gain most of their flow from runoff from farmland irrigated by the All-American Canal. Although these rivers would have substantially less flow without the water from the All-American Canal, the water that would exist would not contain the high levels of salts, fertilizers, and pesticides. Critics of the All-American Canal suggest that the canal has facilitated an environmental disaster by providing the water that allows the agricultural industry to grow products that demand the fertilizers and pesticides that pollute the region's soil and water.

A third critique of the All-American Canal addresses the siphoning of water from the Colorado River to provide for the Imperial and Coachella Valleys. Thus, water is being taken from stakeholders farther downstream and from the Colorado River Delta to supply agriculture and municipal needs farther west. Individuals living downstream, in addition to environmentalists who have recorded damage done because of lack of water, take opposition to supply needs in the Imperial and Coachella Valleys while ignoring the needs of those downstream.

A final criticism of the All-American Canal deals with the issue of water loss from seepage. The United States Bureau of Reclamation concludes that, along one twenty-three-mile section, the canal loses an estimated 70,000 acre-feet of water each year (California Department of Water Resources 2007). In an attempt to limit seepage, the United States government authorized the U.S. secretary of the interior to construct a lined canal or to line the previously unlined portions of the canal. Although some critics of the All-American Canal see lining the canal as the solution to their concerns, others view the lining as appropriating water that should go to replenish the aquifer. Water from the aquifer both maintains springs and supplies farmers in northern Mexico with water. By lining the canal and thus minimizing the water that will seep back into the aquifer, both the natural world and the citizens of northern Mexico are likely to suffer the consequences.

Overall, the All-American Canal successfully achieves its goal of providing water from the Colorado River to the Imperial and Coachella Valleys in California. Supporters note the agricultural and municipal benefits that come from supplying more water to these arid regions. Critics, conversely, look at the harm done to the regions that are not receiving the water.

Sources and Further Reading: California Department of Water Resources, "All-American Canal Lining Agreement Signed"; California Department of Water Resources, "Coachella Canal and All-American Canal Lining Projects"; http://wwwdpla.water.ca.gov/sd/environment/canal_linings.html; Culp, *Restoring the Colorado Delta with the Limits of the Law of the River*; Frisvold and Caswell, "Trans-boundary Water Management. Game-Theoretic Lessons for Projects on the U.S.-Mexico Border"; Hayes, "The All-American Canal Lining Project: A Catalyst for Rational and Comprehensive Groundwater Management on the United States–Mexico Border"; Imperial Irrigation District, "Water Department"; Jones et al., "Assessing Transboundary Environmental Impacts on the U.S.-Mexican and U.S.-Canadian Borders"; Pitt et al., "Two Nations, One River: Managing Ecosystem Conservation in the Colorado River Delta"; Sanchea Munguia, *El Revestimiento del Canal Todo Americano: Competencia o cooperacion por el agua en la frontera Mexico-Estados Unidos?*

BALD EAGLE AND THE IMPORTANCE OF THE ENDANGERED SPECIES ACT

Time Period: 1973 to the present
In This Corner: Biologists and wildlife managers
In the Other Corner: Property-rights advocates
Other Interested Parties: American public, American politicians
General Environmental Issue(s): Wildlife management, policy

If the Wilderness Act marks the beginning of the political expressions of America's new environmental sensibility, the high point of this new legislative push came in the early 1970s, particularly in 1973 with the passage of the ESA. This act illustrated how far the ecosystem realization might carry Americans in reconfiguring their view of humans' role in nature: this act prioritized keystone or endangered species over the wants and needs of human society.

Such a concept would have been unthinkable a century before, or even during the early years of progressive environmentalism. The concept of biological integrity, enforced by scientific understanding, presented a revolution to human life on Earth. Ecology had come to demonstrate that humans were not the most important species on Earth (Rothman 1998, 126–28). The historic text read as follows:

(a) FINDINGS—The Congress finds and declares that—

   (1) various species of fish, wildlife, and plants in the United States have been rendered extinct as a consequence of economic growth and development untempered by adequate concern and conservation;

   (2) other species of fish, wildlife, and plants have been so depleted in numbers that they are in danger of or threatened with extinction;

   (3) these species of fish, wildlife, and plants are of aesthetic, ecological, educational, historical, recreational, and scientific value to the Nation and its people;

   (4) the United States has pledged itself as a sovereign state in the international community to conserve to the extent practicable the various species of fish or wildlife and plants facing extinction, pursuant to-

      (A) migratory bird treaties with Canada and Mexico;

      (B) the Migratory and Endangered Bird Treaty with Japan;

      (C) the Convention on Nature Protection and Wildlife Preservation in the Western Hemisphere;

      (D) the International Convention for the Northwest Atlantic Fisheries;

      (E) the International Convention for the High Seas Fisheries of the North Pacific Ocean;

      (F) the Convention on International Trade in Endangered Species of Wild Fauna and Flora; and

      (G) other international agreements; and

   (5) encouraging the States and other interested parties, through Federal financial assistance and a system of incentives, to develop and maintain conservation

programs which meet national and international standards is a key to meeting the Nation's international commitments and to better safeguarding, for the benefit of all citizens, the Nation's heritage in fish, wildlife, and plants.

(b) PURPOSES—The purposes of this Act are to provide a means whereby the ecosystems upon which endangered species and threatened species depend may be conserved, to provide a program for the conservation of such endangered species and threatened species, and to take such steps as may be appropriate to achieve the purposes of the treaties and conventions set forth in subsection (a) of this section.

(c) POLICY—

(1) It Is further declared to be the policy of Congress that all Federal departments and agencies shall seek to conserve endangered species and threatened species and shall utilize their authorities in furtherance of the purposes of this Act.

(2) It is further declared to be the policy of Congress that Federal agencies shall cooperate with State and local agencies to resolve water resource issues in concert with conservation of endangered species.

(http://www.fws.gov/Endangered/esa.html#Lnk02)

Such a policy inspired almost constant challenge and debate over the next decades. To environmentalists, the ESA was one of the basic building blocks of further environmental regulation for the federal government. To critics, it was a symbol of wrong-headed policy that favored other portions of nature over humans. Three case studies represent these perspectives.

Tennessee Valley Authority v. Hill, 437 U.S. 153 (1978) (The Snail Darter Case)

With laws such as the ESA now enforced, how would the courts react to the needs of nature conflicted with those of human development? One of the first public controversies along these lines occurred when the TVA began construction on the Tellico Dam on the Little Tennessee River in 1967, and initially it appeared that the project would proceed smoothly. Typical of most TVA projects, the goal was multiuse, including the creation of hydroelectric power, shoreline, and recreational opportunities, as well as flood control. During much of the twentieth century, such projects were considered progressive improvements, even called conservation. In the new era of environmental consciousness, however, such projects endured new scrutiny (Stine 1993, 20–32).

In 1973, an ichthyologist conducting sampling in the area that would be flooded by the Tellico Dam discovered a previously unknown species of fish: a three-inch, tannish-colored perch called the snail darter. Studies of this small fish showed that its entire population lived in one small part of the Little Tennessee River, an area, of course, that was destined to be turned into the Tellico Dam Reservoir. To protect the snail darter and its habitat, the secretary of the interior used the new environmental legislation and listed it as an endangered species.

This lawsuit then was brought against the TVA in an effort to force a halt to construction on the Tellico Dam. While the courts debated the issue, the project approached eighty percent completion. Despite this stage of near completion, the Supreme Court ruled that the construction must be halted. However, just as environmental activists were learning exactly how to use these new legal tools, other interests were perfecting ways to thwart them.

In mid-1979, Senator Howard Baker, who was also the Republican Senate minority leader, and Congressman John Duncan, both from Tennessee, buried a small provision into a larger piece of legislation pending before the Congress. This provision undid the Supreme Court's decision in *TVA v. Hill* and provided that the Tellico Dam project could be completed without further legal delay. The project was completed in late 1979 (Rothman 1998, 127–28).

The case at once represented a success and a failure for the nascent environmental movement. No snail darters survived in the specified portion of the Little Tennessee River impacted by the Tellico Dam project. Small populations of the snail darter, however, were subsequently discovered. As a result, the Department of Interior now lists the species as "threatened" rather than "endangered."

Spotted Owl v. Logging Interests

When President Richard Nixon signed the ESA into law in 1973, it was intended to conserve plant and animal species in danger of extinction. Specifically, the ESA sought to ensure the species in question access to a significant portion of their range within their typical ecosystem. In the late 1980s and early 1990s, the ESA was the focus of a series of public, political, and legal controversies. One of the most memorable was the northern spotted owl on national forest lands in Washington and Oregon.

The debate over the spotted owl played across newspapers across the country and led to hostilities in many of the Pacific Northwest's small towns. Although the issues were in fact far more complex, many reports pitched the controversy as a struggle between loggers' jobs and protection of the owls' ancient forest habitat.

The debate began in June of 1989 when the U.S. Fish and Wildlife Service advised the spotted owl be placed on the threatened list and protected under the ESA. With the prospect of a halt in cutting and lost jobs, the timber industry reacted quickly. "Save Timber" groups started in many places in the Pacific Northwest. Towns united to fight the listing of the spotted owl, urging that a single species was not worth the disruption to countless families. Some logging companies responded with their own "threatened" listing for the jobs, towns, and lives that saving the spotted owl would alter. By 1989, the government had begun to loosen its stand and to allow cutting. This resulted in a suit filed by the Seattle Audubon Society against the Forest Service.

On March 7, 1991, U.S. Federal District Court Judge William L. Dwyer ruled on this lawsuit and directed the Forest Service to revise its standards to protect owls and their habitat, "to ensure the northern spotted owl's viability." Such victories for environmentalists, however, seem to have often helped anti-environmental groups learn how to better present their future arguments to the public.

Bald Eagle: An American Symbol in More Than One Way

By the 1990s, there were examples that the initiatives of the 1960s and 1970s had made changes in the nature of everyday American life. Possibly the best example of the possibilities of modern environmentalism was the bald eagle. By the 1990s, our nation's symbol had also become a symbol of the possibilities of environmental regulation.

In the early 1900s, eagles inhabited every large river and concentration of lakes in the interior United States. It was estimated that eagles nested in forty-five of the lower forty-eight states. Within a few decades, however, the population of eagles had plummeted. Scientific explanations for this decline are complex and include a number of different factors.

Essentially, eagles and humans were in competition for the same food, and humans, with guns and traps at their disposal, had the upper hand. By the 1930s, public awareness of bald eagles and their plight began to increase, and in 1940 the Bald Eagle Act was passed. This act reduced the direct pressures caused by humans (such as hunting), and eagle populations began to rebound. However, this was also the era when American agriculture began to rely more than ever on DDT and other pesticides. Part of Rachel Carson's research for *Silent Spring* directly discussed the effect of these chemicals on eagle populations. Primarily, the chemicals ran off the land and into waterways where they infected fish that would eventually be eaten by eagles and other birds of prey.

Scientific research demonstrated definitively that the chemicals caused a softening of eagles' eggs, which caused them to break prematurely and not reach birth. With their population plummeting, bald eagles were officially declared an endangered species in 1967 in all areas of the United States south of the 40th Parallel. This law was then followed by the ESA of 1973. Federal and state government agencies, along with private organizations, successfully sought to alert the public about the eagle's plight and to protect its habitat.

Because of its symbolic importance to the United States, the eagle's struggle represented the general over-exploitation of resources by American society. If the eagle could be saved, then it could be a symbol of a new era in environmental management and resource conservation.

Conclusion: Success for the Eagle But Continued Debate over ESA

The ESA will continue to arouse debate on either side of the environmental argument, but at least some of the results are quite indisputable. Although only a handful of species have fought their way back from the U.S. endangered species list, the eagle regained its population stability in the 1990s.

Once endangered in all of the lower forty-eight states, the bald eagle's status was upgraded to "threatened" in 1994, two decades after the banning of DDT and the passing of laws to protect both eagles and their nesting trees. In the early twenty-first century, experts considered upgrading the eagle's status further. Clearly, Americans had learned valuable lessons about their role in managing the natural environment.

Sources and Further Reading: Andrews, *Managing the Environment, Managing Ourselves*; Gottleib, *Forcing the Spring: The Transformation of the American Environmental Movement*; Nash, *Wilderness and the American Mind*; Opie, *Nature's Nation*; Rothman, *The Greening of a Nation*; Steinberg, *Down to Earth.*

GREEN CULTURE SETS NEW STANDARDS FOR AMERICAN ENVIRONMENTALISM

Time Period: 1970s to the present
In This Corner: Mainstream environmental organizations
In the Other Corner: Scientists, extreme environmentalists, anti-environmental organizations

Other Interested Parties: Politicians, members of the media
General Environmental Issue(s): Green culture, popular culture

As policy makers reacted to grassroots demand for environmental reform in the 1970s and 1980s, many Americans sought ways to integrate their new-found environmental ethics into their everyday lives. More than at any other time in American history, the living patterns of everyday American life in the 1980s included a thought or awareness of humans' impact on the world around them.

Once this environmental awareness made it into basic patterns of American mass culture, it often held little identifiable connection to its roots in the ecological principles of Cowles, Clements, Leopold, and Carson. However, many Americans clearly had added impact on the environment to their list of considerations when they made choices about which product to buy, where to eat, and what to do in their free time.

When these choices reflected a bit of environmental conscience or reflection, it can be grouped with a cultural pattern termed "green culture." Often, this change was marked by alterations to tradition and practices already ingrained in American life, including residential patterns, leisure culture, and film preferences. Many scientists and active environmentalists decried such efforts as depthless attempts to exploit environmental greenness without understanding the real issues. They described green culture as a consumer America's example of "green washing" seen in corporate America.

Regardless of the debate over its true meaning, green culture represents the manner in which the late 1970s and the 1980s mark a period of applied cultural change in American ideas about nature. Of course, once one begins to study green culture, one finds that there are examples of this throughout the twentieth century. In certain cases, green culture can derive from forms that seem incongruous with the environment. However, there are also plenty of clear patterns within American mass culture that suggest a growing interest in the environment, possibly even a more widespread environmental ethic.

Imaging Conservation: Smokey and Woodsy

Smokey Bear, the guardian of our forests, has been a part of the American popular culture for more than sixty years. Although Smokey's story is interesting for a variety of reasons, it especially symbolizes an era when resource conservation issues could be taken directly to an interested and educated American public. Today, Smokey Bear is one of the most famous advertising symbols in the world and is protected by federal law. He has his own private zip code, his own legal council, and his own private committee to ensure that his name is used properly. Smokey Bear is much more than a make-believe paper image; he exists as an actual symbol of forest fire prevention (Pyne 1982, 170–75).

Smokey Bear and the interest in bringing fire prevention to the public actually relate directly to events of World War II. As one historian explains, in the spring of 1942, a Japanese submarine surfaced near the coast of Southern California and fired a salvo of shells that exploded on an oil field near Santa Barbara, very close to the Los Padres National Forest. Americans throughout the country were shocked by the news that the war had now been brought directly to the American mainland. There was concern that further attacks could bring a disastrous loss of life and destruction of property. One of the areas that seemed ripe for destruction were valuable forests on the Pacific coast. With experienced firefighters and other able-bodied men engaged in the armed forces, the home communities had to deal with the forest fires as best they could. Protection of these forests became a matter of national importance, and a new idea was born. If people could be urged to be more careful, perhaps some of the fires could be prevented (Smokey 1982). This very genuine security concern eventually combined with the emerging American interest in the environment.

During World War II, the Forest Service worked with the Wartime Advertising Council to create a marketing campaign in which posters and slogans proclaimed "Forest Fires Aid the Enemy" and "Our Carelessness, Their Secret Weapon." Throughout the campaign, the suggestion was clear: more careful practices by humans could prevent many forest fires. Indirectly, fewer fires could help the war effort by maintaining timber supplies and minimizing necessary man power.

Disney also got into the act when it released the motion picture *Bambi* in 1944. The company allowed the character to be used in the forest fire prevention campaign for one year. The "Bambi" poster was a success. Although the Forest Service was contractually bound not to use a fawn after Bambi's one-year run, they did feel certain that an animal mascot had helped to raise public interest in fire prevention. After internal debates, the Forest Service settled on using a bear as its new mascot.

The first Smokey Bear poster was released later in August 9, 1944. The poster, which depicted a bear pouring a bucket of water on a campfire, was a hit. Within a decade, Smokey Bear was a moneymaker for the federal government. His character was officially moved from the public domain and placed under the secretary of agriculture's control. Whether Smokey was a stuffed toy or a poster, the federal government could now collect fees and royalties while also publicizing (Pyne 1982, 178–80).

Smokey was joined a few decades later by Woodsy Owl. Created by the USDA's National Forest Service in celebration of Earth Day 1970, Woodsy Owl told Americans, "Give a Hoot! ... Don't Pollute!" These ad campaigns were especially effective with schoolchildren, and Woodsy was used on many informational programs developed to educate American children about these environmental issues.

Interpreting Green Culture

Green culture did not only emanate from conservation agencies. The dissemination of greener ethics has also greatly impacted the popular culture created by mass media. This development seems to have occurred at a pace with the growing interest of Americans in the environment; therefore, one can argue that the popular images fed the evolving desire of many Americans to be environmentally aware.

Most prevalent might be the genre of culture that seeks to give viewers access to the natural world, which lay quite distant from the professional worlds of most viewers. Mutual of Omaha's Wild Kingdom began this tradition in the 1970s. In the tradition of National Geographic, Marlon Perkins created adventure from far-off locations that was based in the unknown secrets of the natural world. Breeding an entire genre of television, and even an entire network, Wild Kingdom continues production and has spawned a great many programs, particularly for young viewers. Finally, Perkins's search for showing animals in their natural surroundings contributed to the interest in "ecotourism," in which the very wealthy now travel to various portions of the world not to shoot big game but only to view it.

Zoos and wildlife parks have also seized on this interest and attempted to manufacture similar experiences for visitors. Possibly the most well-known cultural manifestation of environmental themes is Sea World, the marine theme park that first opened in 1964 and now includes parks in Florida, California, and Ohio. Unlike Disneyland and other amusement parks, Sea World carries a full-blown theme: the effort to bring visitors into closer contact with the marine world. As this agenda has become more routine since 1980, performing mammals have taken center stage. The most famous of these performers is Shamu the killer whale. In the highly competitive amusement industry, Sea World has exploited its niche by focusing since 1990 on environmental themes deriving from threats to marine life (Davis 1997, 68–69).

Such cultural interest in natural history and science is most clearly evident for children. Although entire school curricula have been altered to include environmental perspectives, juvenile popular culture has guided such interest. From Disney's *Bambi* to Dr. Seuss's *The Lorax*, artists have clearly identified a sensitivity in juveniles. Each of these tales stresses overuse, mismanagement, or cruelty toward the natural world. The typical use of easily recognizable examples of good and evil that support children's media have been radically expanded.

Mixing science with action, environmentalism proved to be excellent fodder for American educators. More importantly, however, the philosophy of fairness and living within limits merged with cultural forms to become mainstays in entertainment for young people, including feature films such as *Lion King*, *Free Willie*, and *Fern Gully*, environmental music, and even clothing styles. Contemporary films such as *Lion King* bring complex ecological principles of balance between species to the child's level.

Many parents find children acting as environmental regulators within a household. Shaped by green culture, a child's mindset is often entirely utopian, whereas parents possess more real-world stress and knowledge. Even so, many adults long for such simplicity and idealism, and scholars say that children awaken these convictions in many adults. In fact, a growing number of adults hold jobs that are involved with the environment. Environmental regulation and green culture created a mandate for a new segment of the workforce: technically trained individuals to carry out new ways of managing waste and consumption.

Imaging Environmental Cataclysm: *The Late, Great Planet Earth*

Throughout the history of mass and popular culture, creators of popular reading material and film, especially science fiction, have sought to strike American anxieties by choosing topics about which the public has great trepidation. During the 1970s, one of the most popular paradigms revolved around human exploitation of natural resources that then resulted in widespread human suffering. This form of tragic environmental drama continues to be popular in the twenty-first century.

One of the first examples of this line of thought was Hal Lindsey's *The Late, Great Planet Earth*, which was originally published in 1970. In this text, Lindsey offered readers a guide to finding the future in the text of the Bible. With fifteen million copies in print, this bestseller obviously struck a nerve in the modern world. Specifically, Lindsey offered order to the chaotic close of the twentieth century by arguing that many of the predictions of the Old and New Testament have come true. Such a connection offered hope to many Judeo-Christians that the Bible and the morality that it imposes had resonance in contemporary life. It also made many readers turn to the Bible to prophesy future events. In this manner, Lindsey spurred contemporary readers to read the Bible with care and helped to reenergize Christianity. Many critics, however, suggest that few of his predictions for the 1980s came true and that he preyed on readers' hopes and fears. Regardless, the prophetic rhetoric of *The Late, Great Planet Earth* made it one of the most popular books of the 1970s.

Although there have been many science fiction films designed to depict cataclysm and the end of the world, one of the most recent blockbusters was more specific than all of the others. *The Day After Tomorrow* (2003) marked a new approach and a new awareness of the potential impact of human living patterns.

The Whole Earth Catalogue and Green Consumption

Aspects of the early environmental movement's roots in the 1960s counterculture persisted in alternate forms of consumption during the late twentieth century. The roots of this movement reach back to 1968 when Stewart Brand's *The Whole Earth Catalogue* (*TWEC*) introduced Americans to green consumerism. Winner of the National Book Award and a national bestseller, *TWEC* quickly became the unofficial handbook of the 1960s counterculture. The book contained philosophical ideas based in science, holistic living, and metaphysics as well as listings of products that functioned within these confines. As many Americans sought to turn their backs on America's culture of consumption, *TWEC* offered an alternative paradigm based in values extending across the counterculture. Even if Americans did not choose these values, they garnered a valuable lesson in discerning consumption.

As many of the participants in the 1960s counterculture entered communes or "returned to the land" in other manners, Brand's book became an instruction manual. The first page declared that "the establishment" had failed and that the catalogue aimed to supply tools to help an individual "conduct his own education, find his own inspiration, shape his own environment, and share his adventure with whoever is interested." The text offered advice about organic gardening, massaging, meditation, and do-it-yourself burial: "Human bodies are an organic part of the whole earth and at death must return to the ongoing stream of life." Many Americans found the resources and rationale within *TWEC* to live as rebels against the American "establishment." Interestingly, however, Brand did not urge readers to reject consumption altogether. *TWEC*'s enlightened philosophy had a significant impact on patterns in American mass culture.

In particular, *TWEC* helped to create the consumptive niche known today as green consumerism, which seeks to resist products contributing to or deriving from waste or abuse of resources, applications of intrusive technologies, or use of non-natural raw materials. *TWEC* sought to appeal to this niche by offering products such as recycled paper and the rationale for its use. As the trend-setting publication of green consumption, *TWEC* is viewed by many Americans as having started the movement toward whole grains, healthy living, and environmentally friendly products. Today, these products make up a significant portion of all consumer goods, and entire national chains have based themselves around the sale of such goods.

The original catalogue combines the best qualities of the *Farmer's Almanac* and a Sears catalogue, while merging wisdom and consumption with environmental activism and expression. Today, although green culture has even infiltrated mass society, *TWEC* continues as a network of experts who gather information and tools to live a better life and, for some, to construct "practical utopias." *The Millennium Whole Earth Catalog*, for instance, claims to integrate the best ideas of the past 25 years with the best for the next, based around the *TWEC* standards, such as environmental restoration, community building, whole systems thinking, and medical self-care. Despite the increased environmental awareness of the American public, *TWEC* continues to find a niche for its unique ideas about soft living and careful consuming. It has also led to the growth of organic products and markets.

Nature Company Empowers Green Consumption

As with many aspects of green culture, some of the forms appear to be false attempts to exploit consumer interest in the environment. A revealing example of this green culture occurred during the 1980s when the American shopping mall, the quintessential example of artifice, played host to green consumerism. The Nature Company, which sold scientific and naturalistic gadgets as well as holistic and third-world crafts, had originally been founded in Berkeley, California, in 1973 by Tom and Priscilla Wrubel, who had been members of the Peace Corps in the 1960s.

By 1994, the Wrubels had sold their interest to a corporation that specialized in newer models of consumption targeting "yuppie" consumers. Now, there were 124 stores in the United States and approximately twenty more worldwide. Since then, Nature Company's fortunes have declined. However, the interest in this type of green consumption has not diminished. The niche that the Nature Company identified has now become part of mainstream consumption and can be found in many different types of stores (Price 2000, 195–200). In this niche, Americans unapologetically mix consumptive practices with symbols and forms of the natural environment.

Conclusion: What Is the Environmental Ethic of the Plastic Pink Flamingo?

How willing are we to accept elements of green culture into the environmental movement? Another way of posing this question is to ask, can a pink, artificial, plastic decoration actually connect the everyday life of Americans to the natural forms around them? Some scholars argue that the pink flamingo placed in many American lawns does just that.

The pink flamingo was first sold to the public in 1957. By the 1970s, the flamingos became a prevalent part of the landscape of middle-class suburbs. For suburbanites, the plastic bird signaled a hint of disconformity in the homogeneous world of suburbia. Jenny Price, however, wrote that the flamingo swiftly came to also signal definitions of nature.

The baby boomers did not invent the bird, but as with television, we were born with it and grew up with it and we appropriated it for ourselves. Through the 1970s, we used the pink flamingo as a ubiquitous signpost for crossing the various, overlapping boundaries of class, taste, propriety, art, sexuality, and nature (Price 2000, 146).

During the 1980s and 1990s, the flamingo became more popular than ever. Price adds that, during this ever more fluid era of boundaries and definition, the flamingo became a symbol of the connection between nature and art or artifice (Price 2000, 161).

Whether it be through lawn ornaments or through the environmental themes contained in many feature films or even the remarkable popularity of *An Inconvenient Truth*, green culture became one of the most noticeable portions of environmentalism to mass culture Americans. In fact, green culture might be the only environmentalism many of them ever get to know.

Sources and Further Reading: Cronon, *Uncommon Ground*; Nash, *Wilderness and the American Mind*; Opie, *Nature's Nation*; Price, *Flight Maps*; Rothman, *The Greening of a Nation*; Smokey, http:// www.smokeybear.com/vault/wartime_prevention.asp; Steinberg, *Down to Earth.*

LOVE CANAL AND THE CONTROVERSIAL MANDATE FOR SUPERFUND

Time Period: 1970s

In This Corner: Homeowners in Niagara Falls area, environmentalists

In the Other Corner: Industrial interests, litigators

Other Interested Parties: Suburban homeowners elsewhere, public officials

General Environmental Issue(s): Toxic waste, modern environmentalism, Superfund, NIMBY

The meeting was similar to any number of gatherings of concerned citizens in the late twentieth century. The subject might be the routing of a new highway, a problematic new bookstore around the corner, or a resident who failed to sufficiently manicure her gardens and lawn. However, this 1978 meeting of the Love Canal Homeowners Association (LCHA) focused the group on the concerns of one member, Lois Gibbs.

Gibbs lived in the neighborhood and her children attended the 99th Street School located on toxic ground. She explained to the group the story that she had already told them individually. She and her family had never been told about the chemical dump that was buried before developers constructed the residential neighborhood and school. When she learned of the existence of the chemical waste from newspaper articles describing the contents of the landfill and its proximity to the 99th Street School, she grew particularly concerned, for her own sickly child attended the school. She became convinced that the waste may have poisoned her child. So, Gibbs approached the school board with notes from two physicians who recommended that her son be transferred to another school for health reasons.

The board refused on the grounds that it could not admit that the health of one child had been impacted without first admitting the possibility that the entire school population had been affected as well. Enraged, Gibbs began talking with other parents in the neighborhood to see if they were having problems with their children's health. After speaking with hundreds of people, she realized that the entire community was concerned about the potential problems from having their children interact with toxic waste. The parents wanted definitive answers.

From this start, the LCHA took shape and committed itself to alerting the entire nation, if necessary, to the dangers of chemical waste. Members of this grassroots organization came from approximately 500 families residing within a ten-block area surrounding Love Canal. Most of the members of the LCHA were workers at one of the local chemical plants, with an average annual income of $10,000–$25,000. They risked jobs and security to better understand the issue of waste disposal in their community.

Mark Zanatian, one of the children endangered by the Love Canal chemicals under 99th Street, waves a banner in protest during a neighborhood protest meeting on August 5, 1978. AP Photo/DS.

Ultimately, the LCHA became a model for countless similar organizations that gave voice to individual claims of environmental carelessness and injustice.

A Story of Chemical Progress

The LCHA was one of the first examples of grassroots action on environmental matters. Motivated activists from the middle class were products of a shift in attitudes toward the use of technology in American life. In the early twentieth century, Americans identified chemicals, like most technological innovations, almost blindly with future progress. There was very little knowledge about possible health issues related to the storage and dumping of wastes made in the production of these progressive chemicals. When, in the case of Love Canal, the Hooker Chemical Company dumped 21,800 tons of waste into an abandoned canal in New York between 1942 and 1953, it was practicing disposal methods seen throughout industrial America. It was business as usual in the United States.

The empty canal had been part of an elaborate plan to create a futuristic metropolis in the area of Niagara Falls in the late 1800s. When these plans were abandoned, the partially dug canal was left unfilled, a large hole in the ground. For the Hooker Chemical Company fifty years later, the thick clay walls of the canal provided a strong dump site for chemical waste. After disposal of their chemical wastes, Hooker Chemical Company covered the land with more clay. Eventually, the land was sold and developed into a small town. This became the town of Love Canal. It is likely that this residential development weakened the clay walls of the dump and allowed chemical waste to leach into the soils of the area.

In 1953, this new, growing community needed a school. The Niagara Falls Board of Education bought the land, including the dumping site, for $1. Although there is some debate about the builders' knowledge of the dump, the school was built without giving consideration to the possibility of ill effects from toxic waste. As early as the 1950s, residents complained of strong chemical odors; however, they were attributed to the chemical factories that were active in town. There were also some reports of children and dogs who developed skin irritations after spending time in the fields around the school. Some residents even talked about stray rocks that exploded when dropped to the ground. With these unproven stories swirling about, the *Niagara Gazette* finally took some sludge from a neighbor's sump pump and sent it to a laboratory for analysis. The results showed that the sludge was definitely from what the Hooker Chemical Company had disposed of in the ground decades before.

These results proved to be the "smoking gun"—or smoking sludge, in this case—that was necessary to attract attention and get results. This scientific evidence was publicized by the LCHA, and ultimately the Department of Environmental Conservation of New York completed additional investigations. The statistics of health problems stemming from the sludge dump were overwhelming. A few days after the report was released, the state agreed to purchase the 239 homes closest to the canal. These unprecedented actions were evidence of a new era in the American attitude toward technology and in our expectations of personal health and safety. One of the best tools for bringing about action proved to be the formation of strong united citizens' organizations (University at Buffalo Libraries, *Love Canal Collection*, http://ublib.buffalo.edu/libraries/projects/lovecanal/).

Development of the Superfund and the Legacy of Industrialization

NEPA was a watershed change in the role of nature in American life. Although the degree of its involvement would continue to vacillate, the federal government was now entrusted with the responsibility of enforcing and maintaining the health of American citizens. Overnight, for instance, government projects in nearly all branches required an EIS. These reports required that, before any construction or natural disruption, federal agencies needed to first have consultants assess the project's impact on the natural ecosystem (Opie 1998, 452–56).

As a reaction to environmental catastrophes and to NEPA, environmental legislation became of interest to nearly all legislators during the 1970s. One of the most interesting pieces of legislation considered the unwieldy problem that began this chapter. After it had been made clear that Love Canal was a serious health threat, actions were taken to create a program that would fund the cleanup of this and similar sites. Chemical companies resisted new legislation that forced them to clean up their own wastes. In December 1980, on the heels of the tenth anniversary of the U.S. Earth Day, and amid toxic waste fires in New Jersey and contamination at Love Canal, President Jimmy Carter signed the Comprehensive Environmental Response, Compensation and Liability Act of 1980 (CERCLA). President Carter stated that CERCLA was "landmark in its scope and impact on preserving the environmental quality of our country" and that it "fills a major gap in the existing laws of our country."

CERCLA, or Superfund as it is more commonly known, was formed to primarily deal with cleaning up hazardous waste sites where owners had shirked responsibility but also allowed injured parties to sue in federal court for damages. The U.S. EPA administers the Superfund program, in cooperation with states and tribal governments. Nothing like Superfund had ever before existed. EPA was expected to immediately extinguish the fires, stop tank leaks, and clean up sites. Over time, the program has evolved to creating a regulatory framework to protect human health and the environment from the dangers of hazardous waste. The original bill created a $1.6 billion fund and allowed participating companies to escape liability from private citizens who wanted to sue for property and personal damage caused by the dumps or spills. Owners and operators of disposal sites and producers or transporters of hazardous wastes were made liable for up to $50 million in cleanup costs. EPA's Superfund program identifies contaminated sites and assesses their risks to human health and the surrounding environment.

The sites are placed on a National Priority List (NPL) to determine when they will receive further investigation and long-term cleanup actions. The first NPL was announced in 1983, with 406 priority sites identified. The NPL is updated regularly based on the evaluation of both new sites and the progress of clean up at sites already on the NPL. Over the years, in addition to completing remedial construction at more than 750 sites, EPA has deleted 219 sites from the NPL.

Funding for Superfund, however, is tied to politics. Given the costs that such regulation can add to industrial development, many private interests have been against the program since the beginning. Politicians influenced by this point of view have reduced funding for the program and also tried to do away entirely with the program. This point of view argues that Superfund was intended to provide temporary emergency federal funding for the cleanup of chemical waste if responsible parties could not be found or were unable to pay.

Other serious arguments against the program follow issues such as the following: retroactive liability, the difficult issue that claims that companies are responsible for wastes legally deposited years or decades ago that cause damage to present owners; joint and several liability, which means that costs might be partly divided according to the percentage of waste at a given site; or strict liability, which argues that companies must pay regardless of fault and whether or not they used legally mandated disposal technologies in a good faith effort to properly dispose of wastes.

Critics argue that Superfund's liability rules generate endless litigation. From 36 to 60¢ of every dollar spent on Superfund has gone for legal and other transaction costs. Over the past decades, it appears that the financial costs proved too great for the Superfund to succeed. In addition, different presidential administrations fluctuated the political and financial support for this project (Opie 1998, 458–60). Still, there appears to be no better method for remediating infected sites that are already burdening many regions of the United States.

Sources and Further Reading: Andrews, *Managing the Environment, Managing Ourselves*; Environmental Literacy Council, *Superfund*, http://www.enviroliteracy.org/article.php/329.html; Gottleib, *Forcing the Spring: The Transformation of the American Environmental Movement*; Nash, *Wilderness and the American Mind*; Opie, *Nature's Nation*; Price, *Flight Maps*; Rothman, *The Greening of a Nation*; Steinberg, *Down to Earth*; University at Buffalo Libraries, Love Canal Collection, http://ublib.buffalo.edu/libraries/projects/lovecanal/; U.S. Environmental Protection Agency, *The Birth of Superfund*, http://www.epa.gov/superfund/20years/ch2pg2.htm.

NATIVE RIGHTS AND NUCLEAR WASTE DISPOSAL

Time Period: Post-1950
In This Corner: Native peoples with health concerns
In the Other Corner: Native peoples with developmental concerns, companies and agencies seeking dump sites
Other Interested Parties: Policy makers, American citizens
General Environmental Issue(s): Native rights, nuclear, the West

Although the issues related to mining on Native American reservations remain hotly debated, they are a fairly straightforward matter of law. By contrast, some tribes in recent years have become embroiled in a most complicated manner of construing their unique political status: inviting the paid disposal of hazardous waste, particularly nuclear material. Debate surrounding the ethics and health hazards of such arrangements have divided some native communities. Environmentalists and others have accused industry and corporate dumpers of exploiting the difficult economic times in many reservation communities. However, the offer of income from hosting solid, hazardous, and nuclear wastes is too promising an opportunity for many reservations to resist. The use of tribal land for this purpose provides a bitter irony when one considers the traditional environmental ethics of all native groups.

In 1982, Congress passed the Nuclear Waste Policy Act (NWPA), which provided a comprehensive policy for dealing with the nuclear waste problem, including "science-based" approaches to siting both a repository and a monitored retrievable storage (MRS) facility. However, strong public and political opposition limited the practical viability of the NWPA. In response, Congress amended the NWPA in 1987 to create a bifurcated approach around the siting impasse. The permanent repository was to be sited by congressional fiat, while a voluntary process was stipulated for the MRS facility. Furthermore, no MRS facility could be built until a permanent repository was issued a license. The voluntary siting process for the MRS facility was to be implemented by the Office of the Nuclear Waste Negotiator created by the 1987 amendments to the NWPA.

There is a long list of tribes considering or implementing such offers. In 1990, the Pine Ridge Sioux rejected proposals by subsidiaries of O & G Industries to build a landfill, but the neighboring Rosebud Sioux council approved a 5,700-acre facility that would pay them $1 per ton of trash accepted. The Campo of California agreed to host a 600-acre landfill, and the Kaibab-Paiutes of Arizona and the Kaw of Oklahoma each accepted incinerators that burned hazardous waste. Recent debates over accepting nuclear waste have involved the Mescalero Apaches, Skull Valley Goshutes, and others. In each case, the sovereignty of the reservations circumvents restrictions and regulations of federal, state, and local laws. The lack of proper inspection and oversight, however, promises that these decisions may pose long-term environmental problems that could outweigh the short-term benefits.

The bitter irony, of course, is the stereotype that native people were the original stewards of Earth and that they practice ethics that place them in closer communion with the natural world than do Americans who are most concerned with the profits and potential of capitalism. In the 1960s and 1970s, Indians became symbols for the counterculture and conservation movements, for example, Iron Eyes Cody shedding a tear in television ads as he looked over a polluted landscape; an apocryphal speech attributed to Chief Seattle became the litany of true believers. Many observers argued that native people possessed an elaborate land ethic based on use, reciprocity, and balance.

In Utah, the Skull Valley Band of Goshute Indians has been targeted for a very big nuclear waste dump. Private Fuel Storage (PFS), a limited liability corporation representing eight powerful nuclear utilities, wants to "temporarily" store 40,000 tons of commercial high-level radioactive waste (nearly the total amount that presently exists in the United States). Only approximately twenty-four tribal members live on the reservation that would be adjacent to the storage site. This group has established Ohngo Gaudadeh Devia (or OGD, Goshute for "Mountain Community") to publicly oppose the dump.

Native groups claim that the federal government has targeted reservations for waste storage since the late 1980s. In 1987, the U.S. Congress created the Office of the Nuclear Waste Negotiator in an effort to open a federal MRS site for high-level nuclear waste. Negotiator reportedly sent letters to every federally recognized tribe in the country, offering hundreds of thousands and even millions of dollars to tribal council governments for first considering and then ultimately hosting the dump. After this initial contact, Negotiator eventually focused its effort on a few tribes.

In response, many native groups fought Negotiator's efforts from the start. Some formed organizations such as OGD; others took advantage of large organizations, including Indigenous Environmental Network and Honor the Earth. Eventually, the government set its sights on Mescalero Apache Reservation in New Mexico. When this plan was resisted, the Negotiator's office was dissolved in 1994.

The effort to negotiate then became the work of Northern States Power (now Xcel Energy), eight nuclear utility companies that formed a coalition that attempted to overcome the resistance at Mescalero. After failing, the coalition turned to the Skull Valley Goshutes in Utah. At approximately this same time, federal authorities developed a permanent underground dump for high-level atomic waste at Yucca Mountain, Nevada, land owned by the Western Shoshone.

An agreement was reportedly struck with members of the tribal council late in the 1990s. Critics continue to protest, however, arguing that the agreement is a violation of environmental justice. OGD and its legal representatives must now navigate the complex legal and bureaucratic labyrinth of the NRC's Atomic Safety and Licensing Board.

The huge financial costs of this litigation is a significant burden for any native group. However, the outcome of the PFS fight may set important precedents for tribal sovereignty and environmental protection on reservations. The tribal critics argue that the nuclear power industry is attempting to evade environmental regulations and to sidestep the opposition of the State of Utah with the shield of tribal sovereignty. If successful, this could threaten to undermine tribal sovereignty itself.

Sources and Further Reading: White, *It's Your Misfortune and None of My Own;* http://www.nir-s.org/factsheets/pfsejfactsheet.htm; Gowda and Easterling, "Nuclear Waste and Native America: The MRS Siting Exercise."

RECYCLING BECOMES AMERICA'S MOST ACTIVE ENVIRONMENTAL PRACTICE

Time Period: 1970s to the present
In This Corner: Mainstream environmental activists, American consumers
In the Other Corner: Extreme environmental activists, initially reluctant industry
Other Interested Parties: Politicians
General Environmental Issue(s): Waste, recycling

During World War II, Americans experimented with conservation and recycling as a matter of national security. Afterward, 1950s middle-class life unapologetically adopted the ethics of expansion and newness. However, as more and more middle-class Americans began to express environmental attitudes, the wastefulness of modern consumption became obvious to more and more consumers.

More Americans than ever became willing to integrate such practices into their lives as part of a commitment to the environment. For instance, most children born after the 1980s assume the "recycle, reduce, and reuse" mantra has been part of the United States since its founding. In actuality, it serves as a continuing ripple of the cultural and social impact of Earth Day 1970 and the effort of Americans to begin to live within limits, albeit at their most conservative roots.

Earth Day 1970 suggested to millions of Americans that environmental concern could be expressed locally. Through organized activities, many Americans found that they could actively improve the environment with their own hands. Many communities responded by organizing ongoing efforts to alter wasteful patterns. Recycling has proven to be the most persistent of these grassroots efforts. Although the effort is trivialized by extremist environmentalists, trash and waste recycling now stands as the ultimate symbol of the American environmental consciousness.

Recycling, of course, did not begin as an expression of environmentalism; instead, it grows out of a conservative impulse to reduce waste. The effort to make worthwhile materials from waste can be traced throughout human society as an application of common sense rationality. The term became part of the American lexicon during wartime rationing, particularly in World War II. Scrap metals and other materials became a resource to be collected and recycled into weaponry and other materials to support fighting overseas. Ironically, it is also World War II that aids Americans in disavowing such an effort during times of peace.

Historians point to the climactic conclusion of World War II and its commensurate growth in the American middle class as defining points in the American "culture of consumption," which became prevalent in the 1950s but certainly extended in some form through the 1980s. The prodigious scale of American consumption quickly made the nation at once the most advanced and the most wasteful civilization in the world. It was only a matter of time before a backlash brought American patterns of consumption into question. As the 1960s counterculture imposed doubt on much of the American "establishment," many Americans began to consider more carefully the patterns with which they lived everyday life.

Belittled by many environmentalists, recycling often seems like busywork for kids with little actual environmental benefit. However, such a minor shift in human behavior suggests the significant alteration made to many humans' view of their place in nature. This change in worldview, caused by many political, social, and intellectual shifts in the 1960s, forced Americans to question their lack of restraint. The culture of consumption of post–World War II America reinforced carelessness, waste, and a drive for newness. Environmental concerns contributed to a new ethic within American culture that began to value restraint, reuse, and living within limits. This ethic of restraint, fed by overused landfills and excessive litter, gave communities a new mandate to restrain the wasteful practices of their population. Reusing products or creating useful byproducts from waste offered application of this new ethic while also offering new opportunity for economic profit and development.

Green, or environmental, industries have taken form to facilitate and profit from this impulse, creating a significant growth portion of the American economy. Even more impressively, the grassroots desire to express an environmental commitment has compelled middle-class Americans to make recycling part of an everyday effort to reduce and better manage waste.

Sources and Further Reading: Cronon, *Uncommon Ground*; Gottlieb, *Forcing the Spring: The Transformation of the American Environmental Movement*; Nash, *Wilderness and the American Mind*; Opie, *Nature's Nation*; Price, *Flight Maps*; Rothman, *The Greening of a Nation*; Steinberg, *Down to Earth*.

HOW GREEN IS THE AMERICAN SUBURB?

Time Period: 1950s to the present
In This Corner: American landscape taste, turf companies, sports proponents
In the Other Corner: Planners, designers, natural scientists
Other Interested Parties: Developers, public officials
General Environmental Issue(s): Planning, human environment

Americans over the past 150 years have created a uniquely built living environment known as the suburb. Census 2000 showed that the slow transition had engulfed the entire nation when it reported that 56 percent of Americans now resided in what would be defined as suburban areas.

In the last decade, however, many Americans have sought to revise the 1950s vision of the suburb as America's ideal living environment. Instead, some Americans now argue for planners to reach back to forms of design and town layout that predate the burb.

Giddy over Grass

Golf courses helped to shape the aesthetic with which Americans measured and defined their preferred environments. By association, creating better species of turf grass became an interest of many Americans. The agriculture departments of state land-grant colleges and universities studied turf-grass growing with subsidies from the Golf Association of America. The most significant developments came from the combined forces of the U.S. Golf Association and the USDA to create a heat-, drought-, and disease-resistant grass for use on the modern American landscape. Specifically, the American aesthetic required a grass that could be green year round. Such priorities also extended to the American home, where the golf course aesthetic contributed to the American idealization of the lawn surrounding most homes (Teysott 1999, 121–22).

The popularity of golf and the use of turf grass in urban parks linked it into the American aesthetic. This portion of American taste combined with home-building technology of the twentieth century to make green grass the context for nearly every American home. The lawn can often seem more closely related to an artificial area than to an organic creation such as a garden; however, placed within its historical context, this landscape form can be viewed for its "natural" significance in the lives of many Americans.

A greensward generally surrounds nearly 60 percent of American homes, offering a border between public and private space. Although there are certainly utilitarian purposes for the lawn, particularly for children's play, it remains largely an aesthetic creation. Imported from France and England, the lawn was normally a transitional zone into manicured gardens. Although lawns were not uncommon in the United States at the turn of the eighteenth century, the more purposeful design was not devised until the mid-1800s. Starting with the rural cemetery movement of the 1830s, Andrew Jackson Downing and other landscape architects created a general aesthetic that relied on the green space as a multipurpose setting that helped to civilize the wild vista beyond one's property. This was particularly important in the country estates that Downing normally designed (Teysott 1999, 44–49).

More than a century later, when Arthur Levitt and other builders streamlined the suburban model of construction, the lawn remained an integral part of the American landscape. As nondescript housing developments swept the nation after World War II, the lawn became one of the few aesthetic staples of the design. Today, many homeowners parcel their lawn into different zones depending on patterns of use, including a front lawn that is most heavily manicured and managed, intended most to frame the home's presentation, and the backyard, which is less manicured and chemically managed and is more of a personal space.

Making a Better Blade

The evolution of the lawn has not been without a scientific presence. The lawn industry got its start in 1901 when the U.S. Congress allotted $17,000 to ascertain the best turf-grass species for lawns and pleasure grounds. In 1920, the U.S. Golf Association began working with the USDA to devise a species of grass that would remain green all year round. Today, grass research centers are available at most agricultural universities. The industry now mixes science, aesthetics, and marketing. Interestingly, advertising is most often carried out by successful golfers who represent lawn-care companies. Also composed of elements such as a garden, the lawn has clearly become a social landscape for Americans, one that provides important statements about one's standing in society. In many communities, the pursuit of the perfect lawn is reinforced by peer pressure. The physical aesthetic is relatively simple: the space contains only healthy, green grass and few weeds and is consistently cut so that it maintains a low, even height (Jenkins 1994, 160–65).

Of course, such a lawn is not attuned to particularities of place. Significant technology is required to insert such a garden site in areas of varied climate and rainfall. Using marketing skills and advertising, lawn and garden companies were able to sell their products and gradually shape the concept of the lawn to meet their desire for increased profit. Michael Polan wrote that the artificial green space that he refers to as the "industrial lawn" also has become a significant symbol in American life. He wrote, "Since we have traditionally eschewed fences and hedges in America, the suburban vista can be marred by the negligence—or dissent—of a single property owner. This is why lawn care is regarded as such an important civic responsibility in the suburbs" (Pollan 1992, 10–11).

The industrial lawn has also spurred the exportation of the lawn mystique to other parts of the world. Although many nations included gardens and pasture near homes, the standardized lawn is considered an American creation. Today, the image of American prosperity has aided the dissemination of the lawn aesthetic globally, particularly into wealthier homes. This is also demonstrated by the popularity of other related forms, such as golf courses in Asia and elsewhere. Regardless of the nation's garden preferences, such lawn design is directly modeled after the American form.

The lawn is considered so uniquely American that Canadian scholars created a museum exhibit in 1998 titled "The American Lawn," which contained historical developments as well as design replicas. Although Americans may prefer that other nations acknowledge our natural wonders or technological developments, for many international observers, the lawn has become a symbol of America's successful era of consumption. To many Americans, caring for and maintaining the aesthetic of the lawn may be their closest relationship to the natural world. Clearly, however, grass forms an aesthetic preference for Americans whether in their own home space or in the sporting events they choose to patronize.

A Nearly Green Era in Suburban Development

By 1970, most Americans lived in suburbs. These developments were often designed around the aesthetic of the lawn, which meant that, for most Americans, their section of turf grass represented their most consistent nature/human relationship. The homes themselves also reflected new preferences for many Americans.

When the United States went through one of the largest sweeps of new home construction in history after World War II, it fed the movement toward prefabrication and homogeneous suburbs. However, many builders actually argued for green technology well before it was prevalent in other parts of American life. In the 1950s, wrote historian Adam Rome, solar and atomic technologies vied for the home heating market in a way that would never be seen again in the twentieth century (Rome 2001, 53). The September 1943 *Newsweek* offered readers a postwar dream house that would "hedge against future fuel shortages" by describing a tour through a solar house in Chicago, Illinois.

In fact, the 1950s marked the high point for research into creating an American home that did not rely on oil or gas for heating. This research sector interest, however, could not compete in the marketplace. Rome wrote, "The predictions of a bright new day in housing proved false, however. From 1945 to 1970, the energy consumption of the average American household increased precipitously." In the 1960s alone, household energy consumption jumped by 30 percent (Rome 2001, 46).

The housing market, of course, was governed by the interest in this era to make homes as inexpensive as possible. In the end, the price of the mortgage was more important than that of utilities. Another influence on this outcome was a specific technical innovation: air conditioning.

The efficiency of the home became unnecessary with the sale of inexpensive air conditioning by the end of the 1940s. Sales of room air conditioners rose from a few thousand during World War II to 43,000 in 1947 and over one million in 1953. Thoughts of energy conservation literally went out the window with the window air conditioner (Cooper 2002, 24–34).

Ironically, during this era of peaked interest for solar and other green designs, the home-building industry produced a vast new demand for energy consumption. Rome suggests that the spread of air conditioning, for instance, became a symbol of American abundance during the Cold War, but the implications to home construction held long-term significance. To keep prices low and to still install central air conditioning, wrote Rome, "builders eliminated traditional ways of providing shade and ventilation" (Rome 2001, 85).

When President Richard Nixon's Council for Environmental Quality took a look at criticism of the lack of energy efficiency in American tract housing during 1970s, they emphasized zoning and land use. The primary outcome would be state and federal regulations designed to regulate home building to minimize erosion, runoff, and the like; however, no regulations considered the energy conservation of the design.

Everyday Setting: The Xeriscaped Home

Something looks incongruous at 555 Bottonfield Lane. The single-level home is nearly identical to the two dozen homes on either side of the block. However, each of the other homes is cast on the typical American setting: a backdrop of green. Number 555, however, has no green. It is surrounded by pebbles and packed dirt. Where other residents have placed trees that must be watered twice a day, 555 has cacti, sagebrush, and other plants indigenous to the Colorado area. None of these require water or extra care.

Here, in the arid lands of western North America, however, that backdrop of green relies on the region's precious supply of water. Nationally, communities have been faced with increased demands on existing water supplies. Consequently, there is a greater focus on water conservation, not just in times of drought, but in anticipation of future population growth. Water can no longer be considered a limitless resource. A philosophy of water conservation through creative landscaping has engendered the new term, xeriscape.

Early on, a landscape such as that at 555 Bottonfield Lane often resulted in complaints from neighborhood associations, but with Americans' increasing understanding of the limits of arid environments, xeriscape was seen as a form of environmental planning. The term xeriscape derives from the Greek word xeros, meaning dry, combined with landscaping. Therefore, xeriscaping is an aesthetically pleasing landscape planned for arid conditions. The term is said to have been used first by the Front Range Xeriscape Task Force of the Denver Water Department in 1981. The stated goal of a xeriscape is visual appeal with plants selected for their water efficiency. Properly maintained, a xeriscape can easily use less than one-half the water of a traditional landscape. Once established, a xeriscape should require less maintenance than turf landscape.

In many of these areas, homeowners have implemented other ideas while still maintaining a public outdoor buffer. One of the most interesting gardening methods is xeriscaping in which one uses species native to the natural surroundings. Instead of turf grass, the use of cacti and scrub or prairie grasses reduces maintenance significantly.

Clearly, the lawn of green grass is a social and cultural product of the age of conspicuous consumption. Its roots, however, fall within the realm of gardening. Ideas such as xeriscaping reintroduce the ethic of gardening, while requiring that such a human-managed environment still function within the natural ecosystem.

New Urbanism Tries to Control Sprawl

How did planners and architects apply the ethics of modern environmentalism and move Americans from Levittown toward xeriscaping? One of the most important guides in this intellectual progression was Ian McHarg, who was one of the true pioneers of the environmental movement but in a very applied manner. Whereas many activists argued for "green" methods of constructing or laying out towns, McHarg took the idea of planning and reconfigured the entire process to reflect ecological considerations.

In terms of spreading his ideas, McHarg was responsible for the creation of the department of landscape architecture at the University of Pennsylvania. In addition, however, McHarg sought to tell the public about his ideas of green planning. In 1960, McHarg hosted his own CBS television program, *The House We Live In,* in which he discussed ways of creating inviting landscapes for humans that did not overlook the surrounding environment. McHarg argued that well-designed architecture and landscape helped to define humans' relationship with the natural world. Therefore, the planning process of designing such spaces required knowledge about ecology and the nature of the site. In 1969, McHarg published his ideas in a landmark book, *Design With Nature.* He specifically called for urban planners to consider an environmentally conscious approach to land use and provided a new method for evaluating and implementing it.

According to Ian McHarg, "the task [of design] was given to those who, by instinct and training, were especially suited to gouge and scar landscape and city without remorse—the engineers." He urged those trained in design to demand control over such projects from the earliest points. "[The engineer's] competence is not the design of highways," McHarg explained, "merely of the structures that compose them—but only after they have been designed by persons more knowing of man and the land" (McHarg 1992, 1–5).

McHarg's ideas moved through the architecture community for a few decades before coalescing into a new philosophy of planning and land use. Clearly, the automobile had radically reorganized American society and its landscape after 1950. Certain aspects of this structure are undeniable: license plates, drivers' licenses, and parking meters and lots, just to name a few. The auto landscape culminated in the shopping mall, which quickly became a necessary portion of developers' plans for the roadside strips that would connect housing with shopping areas. By the 1970s, developers' initiative clearly included regional economic development for a newly evolving service and retail world. Incorporating suburbs into such development plans, designs for these pseudo-communities were held together by the automobile. The marketplace for this culture quickly became the shopping mall. Strip malls, which open on to roadways and parking lots, were installed near residential areas as suburbs extended farther from the city center. Developers then perfected the self-sustained, enclosed shopping mall.

Try as they might, such artificial environments could never recreate the culture of local communities. Beginning in the 1960s, Ralph Nader and other critics would begin to ask hard questions of the auto's restructuring of American society. Shopping malls became the symbol of a culture of conspicuous consumption that many Americans began to criticize during the 1990s.

Critics such as Jane Jacobs and Jim Kunstler identified an intrinsic bias on the American landscape in the 1970s. Kunstler wrote, "Americans have been living car-centered lives for so long that the collective memory of what used to make a landscape or a townscape or even a suburb humanly rewarding has nearly been erased" (Kunstler 1993, 21). The 1990s closed with the unfolding of the new politics of urban sprawl. "I've come to the conclusion," explained Vice President Al Gore on the campaign trail in 1999, "that what we really are faced with here is a systematic change from a pattern of uncontrolled sprawl toward a brand new path that makes quality of life the goal of all our urban, suburban, and farmland policies." One form of these new plans became known as "new urbanism" as designers attempted to design for humans not just for automobiles.

Sources: Jackson, *Crabgrass Frontier: The Suburbanization of the United States*; Jacobs, *The Death and Life of Great American Cities*; Kunstler, *The Geography of Nowhere: The Rise and Decline of America's Man-Made Landscape*; Rome, *Bulldozer in the Countryside: Suburban Sprawl and the Rise of American Environmentalism*.

GREENWASHING DEMONSTRATES MAINSTREAM ENVIRONMENTAL EXPECTATIONS AND CONCERNS

Time Period: 2001 to the present
In This Corner: Anti-environmental groups
In the Other Corner: Environmentalists
Other Interested Parties: Consumers, politicians
General Environmental Issue(s): Culture, greenwashing, policy

Although many corporations have attempted to implement environmentally friendly measures, others sought to appeal to green culture while doing business as usual. The term "greenwashing" is the term used for corporations who try to present an image of being environmentally friendly without necessarily making any changes in their actual business practices. In such instances, a company, an industry, a government, a politician, or even an NGO creates a pro-environmental image to sell a product or a policy or to try and rehabilitate their standing with the public and decision makers after being embroiled in controversy.

The use of greenwashing demonstrates how prevalent environmental perspectives have become. However, it also leads to suspicion and doubt about the true interests or capabilities of companies to alter their ethics.

Businesses Go Green, Or So It Seems

Joshua Karliner of CorpWatch traces this trend back to the late-1960s and the rise of environmentalism. Initially, a few ad executives, including Jerry Mander, who observed the trend, referred to it as "ecopornography." Karliner wrote the following:

> It seemed that everyone was jumping on the bandwagon. It was a time when the antinuclear movement was coming into its own. In response, notes Mander, the nuclear power division of Westinghouse ran four-color advertisements "everywhere, extolling the anti-polluting virtues of atomic power" as "reliable, low-cost ... neat, clean, safe." ... Meanwhile, in the year 1969 alone, public utilities spent more than $300 million on advertising—more than eight times what they spent on the anti-pollution research they were touting in their ads. Overall, Mander estimated that oil, chemical, and automobile corporations, along with industrial associations and utilities, were spending nearly $1 billion a year on "ecopornography" and in the process were destroying the word "ecology" and perhaps all understanding of the concept. (Karliner, *A Brief History of Greenwash*)

The prevalence of greenwashing in companies peaked each year on Earth Day as many companies felt responsible to note the holiday in some manner. By the twenty-first century, however, efforts to placate environmentally minded consumers occurred year round. Karliner is careful to point out that greenwashing occurs hand-in-glove with globalization. Often, green-speak seemed to be the primary trans-border language. For instance, one of the most notable efforts in this arena has been that of the giant oil company BP. Formally known as British Petroleum, in 2000, BP Amoco officially made its initials stand for "Beyond Petroleum."

Greenwash History

In the year 1969 alone, public utilities spent more than $300 million on advertising—more than eight times what they spent on the anti-pollution research they were touting in their ads. Greenwash advertisements became even more numerous and more sophisticated in the 1970s and 1980s, reaching new heights in 1990 on the twentieth anniversary of Earth Day. This annual event gave businesses the opportunity to appear green, at least temporarily.

For instance, one-fourth of all new household products that came on to the market in the United States around the time of "Earth Day 2000" advertised themselves as "recyclable," "biodegradable," "ozone friendly," or "compostable."

Evidently, these temporary initiatives influenced consumers. In the early 1990s, one poll found that 77 percent of Americans said that a corporation's environmental reputation affected what they bought. In 1985 Chevron launched its "People Do" advertisements aimed at a "hostile audience" of "societally conscious" people. When the UN held its Rio Conference in 1992, Secretary General Maurice Strong created an Eco-Fund to finance the event. The Eco-Fund franchised rights to the Earth Summit logo to the likes of ARCO, ICI, and

Mitsubishi group member Asahi Glass. Obviously, it was an association that many corporations desired.

By the twenty-first century, many corporations saw "greening" as an important part of public relations. For instance, BP, the world's second largest oil company and one of the world's largest corporations, advertised its new identity as a leader in moving the world "Beyond Petroleum." It touted its $45 million purchase of the largest Solarex solar energy corporation. The company touts this image instead of the $5 billion over five years that it will spend on oil exploration in Alaska alone.

Shell, the world's third largest oil company, developed a clever but misleading ad series "Profits or Principles," which touted Shell's commitment to renewable energy sources while displaying photos of lush green forests. However, Shell spends a miniscule 0.6 percent of its annual investments on renewables. For Earth Day 2000, Ford Motor Company announced that all corporate brand advertising will have an environmental theme. It expects to spend as much on this greenwashing as it does to roll out a new line of cars (Karliner, *A Brief History of Greenwash*). Green business has become good business in the twenty-first century.

Wise Use and Greenwashing Politics

In the realm of politics and planning a national movement has been organized around the name and idea of wise use, which was borrowed from conservationists such as Gifford Pinchot. This is a well-financed right-wing movement that blossomed in the 1980s, and most observers agree that the use of the conservation terminology derived more from greenwashing than from any genuine ethic. Wise users sought to use public frustration with government interference to attract the support of middle-class voters.

Wise use became one of the first organized responses to environmentalism. Prioritizing use, particularly of federally owned resources, wise users fought against federal regulation and environmental limitations. Some support came from property rights advocates. Other supporters of wise use argued that environmentalists were antihumans. Wise use continues to have an active influence on efforts to mitigate or abolish environmental regulations

Increasingly a matter of individual definitions, many Americans move into the twenty-first century unclear about what it means to be an environmentalist. In the presidential election of 2004, for instance, when a questioner asked President Bush to assess his view on the environment, the president responded simply, "I'd say that I am a good steward of the land." Many environmental organizations would never say this about the Bush administration. However, history will most certainly show that Bush's policies introduced an entirely new phase in the American use of policy to administer and enforce adherence of environmental legislation. Part of this new phase in legislation is a lack of clarity in presenting this agenda to the American people. In short, politicians factor in the green expectations of the public and adjust their rhetoric accordingly.

The Bush administration stresses an effort to achieve quantifiable results when discussing its no-nonsense approach to overseeing America's natural environment. A brief list of accomplishments claimed by the administration includes "Clear Skies Initiative," cuts in mercury emissions, a "growth-oriented" approach to climate change, tax incentives for renewable energy and hybrid and fuel-cell vehicles, and the development of domestic energy sources, including hydrogen. In each of the cases, however, initiatives and even the names of the policies demonstrate little about the ethical intent of the policy and are more an example of greenwashing. Are Americans being misled? Many critics say so.

One of the Bush administration's most vocal critics is attorney and business consultant Robert F. Kennedy, Jr. In his book *Crimes Against Nature*, Kennedy wrote, "You simply can't talk honestly about the environment today without criticizing this president. George W. Bush will go down as the worst environmental president in our nation's history" (Kennedy 2004). Clearly, however, Americans continue to be able to agree that nature has played a tremendously important role in the development of the nation during the twentieth century.

Senator Patrick Leahy, a Democrat from Vermont, made the following statement on the floor of the Senate on April 26, 2004:

> Just recently the *New York Times* reported on the creative White House fact-spinning of the Administration's proposed retreat from strong mercury controls on power plants.
>
> Of course, we all recognize that [the Bush administration's] favorite tactic is just giving one of their environmental assaults a green name and hoping the American public believes it. "Clear Skies" and "Healthy Forests" are just about as accurate as "No Child Left Behind."
>
> The Administration has used all of these tactics when it comes to misleading the public about wetlands protections. Last January, on a Friday, the Administration announced one of its most sweeping rollbacks to take away protections under the Clean Water Act for 20 million acres of wetlands.
>
> The policy created such a groundswell of opposition from hunters, anglers, environmental groups and others that the President finally withdrew the proposed rulemaking last December. Unfortunately, what the Administration did not tell the public is that they were not revoking the underlying instructions to federal agencies to follow the same policy that leaves 20 million acres of wetlands at risk.
>
> That is why I found it so interesting that the President would start his election year attempts to greenwash his Administration's anti-environmental record by talking about wetlands....
>
> While the President was touting his plan to restore 1 million acres of wetlands, he made no mention of his policy to revoke protection for 20 million acres....
>
> The Administration's retreat from aggressive mercury controls on power plants is just the most recent missile in his all out environmental assault.
>
> Again, the President did get some nice photo ops in Maine and Florida, but his record on the environment is too mired in reversals and rollbacks for any greenwash to last for long.
>
> Greenwash—like whitewash—doesn't stick.
>
> Despite all their public relations maneuvering, the public recognizes the enormous and long-term affect of these Bush policies on our environment and our health.
>
> They will mean more pollution in our rivers and streams, more toxics in our air and less natural resources to pass on to the next generation. (Leahy 2004)

Conclusion: Separating True Environmental Ethics

In 1999, "greenwash" was added to the Oxford English Dictionary, in which it is defined as "Disinformation disseminated by an organization so as to present an environmentally responsible public image." Although many observers have become skilled at discerning authentic environmental statements from greenwashing, the mass of American consumers are assumed to accept greenwash as reality.

At times, environmental organizations seem to complicate matters. For instance, to combat what it viewed as misleading corporate practices, Greenpeace USA began publishing the *Book of Greenwash* during the 1990s. Some critics even argue that community recycling programs also should be grouped under the heading of greenwash because they prevent calls for reducing consumption and economic growth.

Therefore, the admission by nonenvironmental political and corporate actors that greenness appeals to the general public has contributed to a confusing era in environmental politics and marketing. By obscuring the connection between corporations or policies and environmental degradation, the public often is misled about the primary agenda of such practices. The web has become an important resource for savvy consumers who have sought outlets to clarify green initiatives by Wal-Mart and others, particularly in relation to the use of the term "organic."

Overall, however, these variations in thought demonstrate how the nature of environmentalism had changed dramatically by the end of the twentieth century. Regardless of one's perspective, however, environmentalism had clearly gained a new connection to everyday patterns of living instead of being restricted to the annual dues sent to conservation organizations. Although false or misleading advertising of any type is unfair, greenwashing represents an awareness by companies, politicians, and others that environmental concerns must be addressed or at least recognized.

Sources and Further Reading: Andrews, *Managing the Environment, Managing Ourselves*; Karliner, *A Brief History of Greenwash*, http://www.greenwashing.net/; Leahy, http://leahy.senate.gov/press/200404/042604a.html; Nash, *Wilderness and the American Mind*; Opie, *Nature's Nation*; Rothman, *The Greening of a Nation*; Steinberg, *Down to Earth*.

ELECTRIC VEHICLES RISE AGAIN?

Time Period: 1890s to the present
In This Corner: Electric proponents, environmentalists
In the Other Corner: Petroleum interests, industry related to internal-combustion engine
Other Interested Parties: Policy makers, American public
General Environmental Issue(s): Automobiles, transportation, energy, electricity

Energy management has emerged as one of the hot-button topics of the twenty-first century. Although the interested parties have changed a great deal over the past century, the energy conundrum has been a part of American life since before 1900. Transportation makes up one of the most important sectors of energy consumption and, therefore, is one of the most combustible areas of debate.

For instance, city planners and New Urbanists saw the automobile as the greatest symbol of American waste by the end of the 1970s. Even so, for the American public, the lesson of the oil shortage of the 1970s did not radically alter energy use. Americans increased their use of electricity even more than oil, and utilities struggled to meet the new demand. Rejecting initiatives under the Carter administration, utility companies dismissed energy conservation. Eventually, utility companies even managed to defeat efforts at cleaner generation of power that were initiated during the Clinton administration.

As the twenty-first century began, however, high oil prices at last appeared to force Americans to give serious consideration to devising a greener automobile. At the close of the twentieth century, when Americans began to more fully consider more efficient autos, one of the proposals, ironically, was the path not taken in the 1890s: electric cars.

Historical Precedents for Electrics

Electric or hybrid cars were introduced to the United States in 1905 when H. Piper applied for a patent on a vehicular power train that used electricity to augment a gasoline engine. Piper's technology actually followed the work of French inventors. From 1897 to 1907, the Compagnie Parisienne des Voitures Electriques (Paris Electric Car Company) built a series of electric and hybrid vehicles, including the fairly well-known 1903 Krieger. Although these vehicles used electric power, some also ran on alcohol (Motavalli 2001, 9–14).

General Electric (GE) even dabbled in the vehicle business after 1900. The GE laboratory built a hybrid with a four-cylinder gasoline engine. In Chicago, the Woods Motor Vehicle Company produced the 1917 Woods Dual Power, which was a parallel hybrid engine that could use its electrical and the gasoline-burning sources of power simultaneously. Another Chicago firm, the Walker Vehicle Company, built both electric and gasoline-electric trucks, from around 1918 to the early 1940s. These examples show us that it was not at all a foregone conclusion that automobiles must run on gasoline when early manufacturing began in the 1900s.

However, during the early 1900s, great new supplies of petroleum transformed gasoline into an abundant and inexpensive source of energy. This spurred inventors to focus on perfecting gasoline-powered engines. In addition, Rockefeller and other titans of American business lobbied hard for the use of the internal-combustion engine in American automobiles. Under this influence, the path to take was clear, and we have burned petroleum in our cars ever since. However, that could be changing.

The new surge in hybrid development occurred in 1993 when the Clinton administration announced the formation of the Partnership for a New Generation of Vehicles consortium. This group consisted of the "Big Three" automobile manufacturers (GM, Ford, and Chrysler) and about 350 smaller technical firms. Their goal was to create operating prototypes by 2004, and then high gasoline prices further stimulated new hybrid designs (Motavalli 2001, 109–17).

Hybrids of the Twenty-First Century

In the early twenty-first century, the electric car has reemerged as a possible supplement to the internal-combustion engine. Similar to the first designs in the 1890s, these hybrid autos combine gas power with the use of an electric battery that allows the vehicle to use much less petroleum. New designs look for opportunities to minimize power and shift off the gasoline engine. For instance, most vehicles require their full wheel power for only very short bursts. Whereas the maximum power for most new cars sold in the United States today is more than 100 kilowatts, the average power actually used during city and highway driving is only about 7.5 kilowatts. During these lulls, the hybrid runs on battery power and burns no gasoline (Motavalli 2001, 154–57).

The hybrid rebirth began when the Toyota Prius went on sale in Japan in 1997, making it the world's first volume-production hybrid car. Today, the five-passenger Prius is the world's most popular hybrid. The Prius uses a separate generator to keep the electric battery pack charged. For this reason, owners never need to plug in their cars for recharging. During braking, the electric motor also acts as a generator to recapture energy and further charge the batteries. That means that, even in stop-and-go city driving, the Prius can travel approximately fifty-two miles per gallon of gas.

Other hybrids include the Honda Insight, first sold in the United States as a 2000 model, and the Honda Civic Hybrid, which came to market in the United States as a 2003 model. A variation of the Civic sedan, the Civic Hybrid is a mild hybrid, meaning that the electric motor assists the gasoline engine during acceleration or times of heavy load but does not move the car on its own.

Honda and Toyota are leading the movement toward hybrids, but both the Mitsubishi and the Nissan Corporations released hybrids in 2006. Of the "Big Three" American manufacturers, only Ford has moved into the hybrid sector with vigor (Motavalli 2001, 159–70).

Timeline: Life and Death of the Electric Car

1832–1839	Scottish inventor Robert Anderson invents the first crude electric carriage powered by non-rechargeable primary cells.
1835	American Thomas Davenport is credited with building the first practical electric vehicle: a small locomotive.
1859	French physicist Gaston Planté invents the rechargeable lead-acid storage battery.
1891	William Morrison of Des Moines, Iowa, builds the first successful electric automobile in the United States.
1893	A variety of electric cars are exhibited in Chicago at the World's Columbian Exposition.
1897	The first electric taxis hit the streets of New York City early in the year. The Pope Manufacturing Company of Connecticut becomes the first large-scale American electric automobile manufacturer.
1899	Setting his sites on powering America's future electric fleet of autos, Thomas Alva Edison begins work on a long-lasting, powerful battery for commercial automobiles. Although his research yields some improvements to the alkaline battery, he ultimately abandons his quest a decade later.
1900	The high point of American use of the electric automobile: of the 4,192 cars produced in the United States, 28 percent are powered by electricity. In cities such as New York City, Boston, and Chicago, one-third of all autos are electric powered.
1908	Henry Ford introduces the mass-produced and gasoline-powered Model T, which swings the market from the electrics for good.

1912	Charles Kettering's electric automobile starter makes gasoline-powered autos more practical by eliminating the unwieldy hand-crank starter.
1966	Congress introduces the earliest bills recommending use of electric vehicles as a means of reducing air pollution. A Gallup poll indicates that thirty-three million Americans are interested in electric vehicles.
1972	Victor Wouk, the "Godfather of the Hybrid," builds the first full-powered, full-size hybrid vehicle out of a 1972 Buick Skylark provided by GM for the 1970 Federal Clean Car Incentive Program.
1974	Vanguard-Sebring's CitiCar makes its debut at the Electric Vehicle Symposium in Washington, DC. The CitiCar has a top speed of more than thirty miles per hour and a reliable warm-weather range of forty miles.
1975	The U.S. Postal Service purchases 350 electric delivery jeeps from AM General, a division of American Motors Corporation (AMC), to be used in a trial program.
1976	Congress passes the Electric and Hybrid Vehicle Research, Development, and Demonstration Act. The law is intended to spur the development of new technologies, including improved batteries, motors, and other hybrid-electric components.
1988	Roger Smith, chief executive officer of GM, agrees to fund research efforts to build a practical consumer electric car. GM teams up with California's AeroVironment to design what would become the EV1.
1990	California passes its Zero Emissions Vehicle (ZEV) Mandate, which requires 2 percent of the state's vehicles to have no emissions by 1998 and 10 percent by 2003. The law is repeatedly weakened over the next decade to reduce the number of pure ZEVs it requires.
1997	Toyota unveils the Prius, the world's first commercially mass-produced and marketed hybrid car, in Japan. Nearly 18,000 units are sold during the first production year.
1997–2000	A few thousand all-electric cars (such as Honda's EV Plus, GM's EV1, Ford's Ranger pickup EV, Nissan's Altra EV, Chevy's S-10 EV, and Toyota's RAV4 EV) are produced by large manufacturers. Each of the major automakers' advanced all-electric production programs will be discontinued by the early 2000s.
2002	GM and DaimlerChrysler sue the California Air Resources Board to repeal the ZEV mandate first passed in 1990. The Bush Administration joins that suit.
2003	GM announces that it will not renew leases on its EV1 cars, saying it can no longer supply parts to repair the vehicles and that it plans to reclaim the cars by the end of 2004.
2005	On February 16, electric vehicle enthusiasts begin a "Don't Crush" vigil to stop GM from demolishing seventy-eight impounded EV1s in Burbank, California. The vigil ends twenty-eight days later when GM removes the cars from the facility. In the film *Who Killed the Electric Car*, GM spokesman Dave Barthmuss states that the EV1s are to be recycled, not just crushed.

2006   A few pure electric cars and plug-in hybrids are in limited production and
              new ones are on the horizon. The success of the gasoline hybrid Toyota
              Prius is a promising sign.

(from http://www.pbs.org/now/shows/223/electric-car-timeline.html)

Conclusion: Mandating Change

In the twenty-first century, obvious changes have indicated a new dawn for electric-powered automobiles. The major stimulus for this change has, of course, been high petroleum prices. These costs coupled with new awareness of the detrimental effect of the internal-combustion engine on the environment have made many consumers interested in purchasing one of the hybrid or "flex-fuel" models in production. In addition, federal and state tax incentives have been used to help offset the higher costs of hybrid vehicles. The result has been that hybrids have received more popularity than at any other time in history.

For Americans, purchasing and selecting a vehicle is tightly linked to individual rights of expression and personal choice. For instance, this is part of the reason that corporate average fuel efficiency (CAFE) standards have been structured across a corporation's vehicle fleet, instead of stipulating types and sizes of vehicles to be produced. Although these standards have stimulated auto manufacturers to produce a hybrid model, they have not yet forced them to sell the more efficient vehicles. Experts expect this may begin to change as standards begin to factor in vehicle size and type by applying "footprint" models, which multiplies a vehicle's wheelbase by its track width.

In addition, CAFE standards may receive their first overhaul in more than twenty years. In June 2007, the Senate passed a new standard requiring thirty-five miles per gallon for cars, SUVs, and light trucks by 2020. Although a final agreement needs to be worked out, there is more active discussion about increasing CAFE standards than ever before.

Sources and Further Reading: Gorman, *Redefining Efficiency: Pollution Concerns, Regulatory Mechanisms, and Technological Change in the U.S. Petroleum Industry*; Jackson, *Crabgrass Frontier: The Suburbanization of the United States*; Jacobs, *The Death and Life of Great American Cities*; Kay, *Asphalt Nation*; McShane, *Down the Asphalt Path*; Motavalli, *Forward Drive: The Race to Build "Clean" Cars for the Future*; *Who Killed the Electric Car?* (film).

WHAT IS THE PRIORITY OF AMERICAN AUTO DESIGN?

Time Period: Late twentieth century
In This Corner: American automakers and designers
In the Other Corner: Green designers
Other Interested Parties: Consumers
General Environmental Issue(s): Automobiles, transportation

Automobile consumers are led by taste and symbolism, cultural details molded by everything from film images to gas prices and from attitudes toward safety to attitudes toward wealth ("bling") and, ultimately, the types of vehicles put on the market by car manufacturers.

In moments of anxiety over fuel supplies, such as the 1970s, American attitudes toward vehicles altered significantly. In fact, there have been many adjustments to American vehicle preferences in the late twentieth century. Ultimately, however, many of these shifts have proven to be temporary. With minor adjustments, most Americans seem to always return to certain basic qualities in their vehicles.

Ralph Nader and Designing Cars for Everything but Safety

Beginning in the 1920s, auto manufacturers achieved a nearly untouchable status in the corporate world. For many years, this allowed these massive corporations to disregard innovation that did not help their profits. This was most glaring in the area of vehicle safety; however, it also relates to gas mileage and the environmental impact of emissions, which will be discussed below.

Car safety became an issue almost immediately after the invention of the automobile. However, the first recorded crash tests did not occur until the 1950s. Conducted by Mercedes-Benz, these first tests spurred vehicle safety to become an international trend in 1958 when the U.N. established the World Forum for Harmonization of Vehicle Regulations, an international agency intended to formulate standards for auto safety that would be used worldwide. This organization helped to stimulate life-saving safety innovations, including seat belts and roll cage construction.

In the United States, the effort to allow consumers to balance this playing field begins with a single individual: Ralph Nader. As a student at Harvard Law School in the late 1950s, Nader researched unconventional issues. In 1959, Nader took research that he conducted about the rationale for specific automotive design and turned it into an article in *The Nation* magazine. Titled "The Safe Car You Can't Buy," Nader's article asked hard questions about the choices driving Detroit's design of automobiles. Primarily, he wrote, "It is clear Detroit today is designing automobiles for style, cost, performance, and calculated obsolescence, but not—despite the 5,000,000 reported accidents, nearly 40,000 fatalities, 110,000 permanent disabilities, and 1,500,000 injuries yearly—for safety."

With his knowledgeable concern, Nader carved himself a niche in Washington, DC, as a "consumer advocate." His consulting and research asked the simple question, "Why shouldn't American consumers expect to be safe and healthy?" Obviously, the primary mechanism that could help Nader implement new ethics in American corporations was the federal government. He went to work for Daniel Patrick Moynihan in the Department of Labor and volunteered as an adviser to the U.S. Senate. Having continued his research about automotive design, Nader in 1965 published *Unsafe at Any Speed*. This best-selling book was an indictment of the auto industry in general for its poor safety standards. However, Nader specifically targeted GM's Corvair for its faulty design. Largely because of his influence, Congress passed the 1966 National Traffic and Motor Vehicle Safety Act.

This act established the U.S. Department of Transportation with automobile safety as one of its purposes. Also in 1966, Congress held a series of highly publicized hearings regarding highway safety, which resulted in the passage of legislation that made installation of seat belts mandatory. In 1970, the National Highway Traffic Safety Administration (NHTSA) was officially established, and, in 1972, the Motor Vehicle Information and Cost Savings Act expanded the scope of the NHTSA to include consumer information programs.

Overall, these safety measures are credited with dropping vehicular fatalities by 10,000, to around 40,000 annually, a lower death rate per mile traveled than in the 1960s. In 1996, the United States had about two deaths per 10,000 motor vehicles, comparable with 1.9 in Germany, 2.6 in France, and 1.5 in the United Kingdom. In 1998, there were 3,421 fatal accidents in the United Kingdom, the fewest since 1926.

Nader's influence transcended the automobile market. Activists flocked to Nader's causes. Referred to as "Nader's Raiders," activists involved in the modern consumer movement demanded that the federal government enforce corporate responsibility. They pressed for protections for workers, taxpayers, and the environment. In 1969, Nader established the Center for the Study of Responsive Law, which exposed corporate irresponsibility and the federal government's failure to enforce regulation of business. He founded Public Citizen and U.S. Public Interest Research Group in 1971, an umbrella for many groups of concerned citizens.

From the use of seat belts to product labeling, the influence of Nader's efforts can be seen throughout American everyday life. Crash tests and other consumer information became more and more available as public awareness grew. In addition, environmentally motivated standards for clean air were forcing auto manufacturers to change the way that they did business (Kay 1997, 250).

Making Lightweight Automobiles

Across the entire fleet of American vehicles being manufactured, the design of automobiles reflects lessons of the environmental era. Thanks to the Oil Embargo of the 1970s, Americans, at least temporarily, became interested in cars that burned less gasoline to move. Size mattered through the 1980s, as American drivers bought smaller, lighter automobiles, many of which were manufactured by Japanese companies. Although this interest in smaller cars gave way eventually to the era of semi-powered large pickup trucks, SUVs, and Hummers, the heavy steel used by automobile manufacturers during the twentieth century has given way to increased use of lighter-weight materials such as aluminum.

The use of automotive aluminum has doubled in the past decade. Increasingly, many automakers who are concerned about the environment are choosing to make their vehicles from aluminum. Aluminum allows manufacturers to maintain the size of their vehicles while reducing weight, which, of course, improves fuel economy. Improved fuel economy, then, reduces greenhouse gases emitted from motor vehicles (see below). The weight reduction of using aluminum is quite noticeable: approximately one pound of aluminum typically replaces two pounds of conventional metals. The resultant fuel savings can significantly lower the operating costs of the vehicle over its lifetime.

The use of aluminum is also helpful on the production end of the process. Automotive aluminum can be made from recycled metals. In addition, after use the aluminum, scrap retains its value for other purposes. Today, nearly two-thirds of the aluminum used in automotive applications is recycled metal. Presently, the automotive aluminum recycling rate is 85–90 percent, and two-thirds of the aluminum used in today's vehicles is derived from recycled metal.

Automobiles' Contribution to Air Pollution

As scientists began to understand the complexities of air pollution in the late 1960s, it became increasingly apparent that, in addition to specific toxic emissions such as lead, the internal-combustion engine was a primary contributor to air pollution, which in cities is usually referred to as smog. Emissions from the nation's nearly 200 million cars and trucks account for about half of all air pollution in the United States and more than 80 percent of air pollution in cities. The American Lung Association estimates that America spends more than $60 billion each year on healthcare as a direct result of air pollution (Doyle 2000, 134).

When the engines of automobiles and other vehicles burn gasoline, they create pollution. These emissions have a significant impact on the air, particularly in congested urban areas. This is hard to track or trace, however, because the sources are moving. The pollutants included in these emissions are carbon monoxide, hydrocarbons, nitrogen oxides, and particulate matter. Nationwide, mobile sources represent the largest contributor to air toxins, which are pollutants known or suspected to cause cancer or other serious health effects. These are not the only problems, however. Internal-combustion engines also emit greenhouse gases, which scientists believe are responsible for trapping heat in the earth's atmosphere. These gases are credited by many scientists for intensifying global climate change

In each state, clean air acts have spurred regulations on vehicle emissions, but these levels and policies vary with each state. For instance, California, which has one of the nation's most severe smog problems in Los Angeles, is a national leader in smog regulation. In 1982, California began a vehicle inspection and maintenance (Smog Check) program. The state's Smog Check program has achieved an overall tailpipe emissions reduction of 17 percent in hydrocarbons and carbon monoxide from vehicles repaired after failing a Smog Check test (Doyle 2000, 200). Beginning in 1994, California's Smog Check II took smog regulation to the next step. By using remote sensing devices that are placed on the side of the freeways, state regulators could trace gross polluters (vehicles that pollute at least two times the emissions allowed for that particular model). These vehicles account for 10–15 percent of California vehicles but create more than 50 percent of the vehicular smog.

Reducing Greenhouse Gas Emissions

The difficulties caused by this air pollution are not limited to local environmental impacts. Cars and light trucks, which include SUVs, pickups, and most minivans, emit more than 300 million tons of carbon into the atmosphere each year in the United States. The transportation sector alone is responsible for about one-third of our nation's total production of carbon dioxide, the greenhouse gas that contributes in a big way to global warming. In response to burgeoning consumer demand over the past decade, automakers have shifted their fleets to SUVs and other light trucks, popular vehicles whose fuel economy standards are unfortunately lower than those of cars (Gelbspan 1995, 9–13).

An ever-growing concern among government, industry, and environmental organizations is global climate change. A buildup of carbon dioxide in the atmosphere over the past century has been identified as a possible contributor to climate change. Auto companies are working on a number of initiatives to improve vehicle fuel economy to reduce carbon dioxide emissions.

Although automobile engines have been made more efficient than ever before, Americans continue to use more gasoline than ever. Since 1980, the miles-per-gallon rating of passenger cars has improved 39 percent, yet fuel consumption is up 19 percent. On average, Americans are driving about 50 percent more miles than they did in 1980. In addition, some vehicles have been made heavier and traffic has gotten more congested. That means that less gasoline than ever is actually being used to propel vehicles forward (Gelbspan 1995, 40–45).

Fuel Conservation and Ensuring American Automobility

Given the significant problems associated with the emissions generated by vehicles, common sense, then, follows that every effort be made to increase automobiles' efficiency. The CAFE program was started in 1978 in an effort to stimulate the manufacture of more efficient autos in hopes of reducing American dependence on foreign oil. Each auto manufacturer was required to attain government-set mileage targets (CAFE standards) for all its car and light trucks sold in the United States. In a compromise with manufacturers, the complex standards were calculated as a total for the entire fleet of autos and trucks made by each company. Thus, the manufacture of a few fuel-efficient models could offset an entire line of light trucks that fell below the standards (Doyle 2000, 240).

Originally passed in 1975, the Energy Policy and Conservation Act was a reaction to the Arab oil embargo of the early 1970s. The public demanded that the federal government force auto manufacturers to offer them some assistance in managing the rising price of petroleum. The act quickly influenced many aspects of the industry, including petroleum mixes, automobile design, and vehicle safety.

As a supplement to this original act, the 1978 CAFE standards required eighteen miles per gallon for cars. To spur innovation, the standard increased each year until 1985. With the automobile standard at 27.5 miles per gallon in 1985, lawmakers expected that manufacturers would willingly surpass this goal. Instead, the 1990s saw manufacturers increase the size of most vehicles, particularly in the area of light trucks, including pickup trucks and SUVs, arguably the most popular type of vehicle at the close of the twentieth century. With the fleet standards based on vehicular weight, manufacturers have found ways to sell vehicles such as Hummers if they offset it with enough other fuel-efficient products to allow the overall fleet to meet the CAFE standards.

One other initiative begun in the 1970s was a federally mandated speed limit. In the 1970s, federal safety and fuel conservation measures included a national speed limit of fifty-five miles per hour. Today, consumers have led states to loosen such restrictions; however, concern over fuel conservation continues (Doyle 2000, 251–62).

The End of the Road: The American Junkyard and "Life-Cycle Analysis"

Through a combination of federal regulation, industrial shifts, and consumer awareness, the automobile has addressed many of its most dire environmental impacts. However, there remains one intrinsic problem with the automobile: its end. An eyesore around the nation, junkyards have been the typical conclusion to the life of each auto. Whether it has been totaled in a collision or worn out from use, the end of the road for each automobile in the United States is an ill-prepared and often ill-managed site that we refer to as a junkyard.

Manufacturers have begun to implement a scientific approach for estimating the environmental impacts of their products that is referred to as life-cycle assessment. Used by many European nations, this form of product accounting considers its impact from cradle to grave. Life-cycle analysis quantifies the environmental impact of a vehicle, beginning with its production and continuing through the end of its useful life and through the recycling of vehicle materials back into use. Considerations include the following: price of purchase; insurance and maintenance; fuel economy; environmental impact of manufacture, operation, and disposal; hazardous substance content; transportation of components and cars; road construction and maintenance; storage and packaging; recycling; and health and safety of workers.

When accounted in this manner, portions of the production process such as using environmentally hazardous substances appear very different from in the typical production accounting used by auto manufacturers. Use of hazardous substances tends to have a lower initial purchase price; however, the associated costs of using the hazardous substances (such as management, disposal, and liability) increase the price of using such materials over time. When viewed through this type of accounting, alternatives to hazardous substances wind up saving money in the long run, in addition to being the right thing for the environment. Thus far, the responsibility for implementing these ideas has been placed on manufacturers.

For instance, in 1995, the United States Council for Automotive Research, a research arm of DaimlerChrysler, Ford, and GM, requested that the aluminum industry conduct a life-cycle inventory for the North American aluminum industry. Their objective in doing so was to generate a resource consumption and environmental profile of the industry as a whole that would allow manufacturers to better comprehend the benefits of making changes. This study demonstrated that only 5 percent of the energy required to produce primary aluminum (from ore to finished product) is required to produce product from recycled aluminum. Accordingly, recycling drastically reduces the emissions that would otherwise be emitted from the production of primary aluminum.

Toyota has started its own system, which is called Eco-VAS (Eco-Vehicle Assessment System). This comprehensive environmental impact assessment system allows, throughout the entire vehicle development process, the systematic assessment of the burden a vehicle will have on the environment as the result of its production, use, and disposal.

Under Eco-VAS, the process begins at the very start of planning. The person responsible for designing a particular vehicle sets environmental impact reduction targets for that vehicle. These figures also consider fuel efficiency, emissions and noise during vehicle use, the disposal recovery rate, the reduction of substances of environmental concern, and carbon dioxide emissions throughout the entire life cycle of the vehicle from production to disposal.

Through these processes, and under the direct management of the person in charge, necessary measures to reduce the environmental impact of each vehicle can be devised in the initial stages of development, enabling steady development progress with an eye toward achieving targets and raising a vehicle's overall environmental performance.

Conclusion: What Do American Drivers Really Want?

As additional nations around the world seize on their own automobile age, many Americans look back on our own gluttonous use of cars with chagrin. Simultaneously, however, other Americans are purchasing heavier, less efficient cars than at any time in since the 1940s, before we knew better. Other Americans have helped to make NASCAR one of the nation's most popular "sporting" events, as hundreds of cars furiously burn leaded gasoline in a race against one another. Although many Americans are willing to consider change for the sake of the environment, the cultural roots of our passion for cars run deep. It could be that the best hope for the future is an "adjusted" technology such as hybrid automobiles. However, petroleum will remain a national necessity. That is the basic environmental impact of the automobile industry.

Regardless of the location of future supplies, the fact remains that petroleum is a finite resource. The petroleum age, agree most scientists, will near its end by 2050. Modern technology, unfortunately, allows us to account rather exactly for this certain demise: we have guzzled 800 billion barrels during the petroleum era; we know the location of 850 billion barrels more, which are termed "reserves"; and we expect that 150 billion barrels more remain undiscovered. Simply, there is an end in sight. New energy sources will need to be found if we want to continue our love affair with the automobile.

Sources and Further Reading: Belasco, *Americans on the Road*; Bradsher, *High and Mighty: SUVs: The World's Most Dangerous Vehicles and How They Got That Way*; Doyle, *Taken for a Ride: Detroit's Big Three and the Politics of Pollution*; Flink, *The Automobile Age*; Kay, *Asphalt Nation*; Marcello, *Ralph Nader: A Biography*; Schiffer, Butts, and Grimm, *Taking Charge: The Electric Automobile in America*.

THE EMERGENCE OF AMERICA'S "SUV HABIT"

Time Period: 1986 to the present
In This Corner: Car manufacturers, American consumers
In the Other Corner: Planners, ecologists, architects
Other Interested Parties: Politicians
General Environmental Issue(s): Transportation, planning, energy

In the 1960s, before the lessons of embargoed oil, Watergate, and catalytic converters, a fateful meeting between two very wealthy Americans helped to determine an important portion of the next phase in Americans' relationship with the automobile. Roy Chapin, who descended from some of the nation's earliest automobile tycoons and had come to head the AMC, spent a weekend at his southern Ontario hunting lodge with Stephen Girard, one of the leaders of Kaiser Jeep, maker of one of the world's first four-wheel drive vehicles (Bradsher 2002, 13).

Convinced that Americans wanted to buy cars and not Jeeps, Girard was actively looking to dump the Jeep on another company. Although the deal took a number of years, in 1969 AMC purchased the specialized Jeep for $10 million and millions more in stock options. This came just in time for the 1970s oil crisis and almost pulled AMC, the smallest of the Detroit manufacturers, into bankruptcy. Who would have thought that ultimately, however, the Jeep would be just what Americans yearned for.

In the words of the comedian George Carlin, Americans of the late twentieth century may go down as most memorable for our passion for collecting "stuff." When we have a lot of stuff, when we are a nation prioritizing consumption, it follows that we would prefer to have the option of taking our stuff with us. In an odd, winding road of logic, regulation, and petroleum scarcity, Americans' relationship with petroleum during the late 1900s became tightly wound into economic status and the rise of the trophy-like technological achievement of the American SUV. Ironically, this passion developed simultaneously to strong new calls for conservation and efforts to find ways of using less petroleum.

The SUV as Unintended Consequence

For the United States, a century of energy decadence came to a screeching halt during the 1970s. Politicians could no longer appear to be take their responsibilities seriously without at least proposing ideas on the subject of energy use and conservation. However, few individuals agreed with each other on how to proceed. In the end, this active disagreement might be part of the reason that many of the 1970s attention to energy use disappeared as quickly as it came.

For American manufacturers, the future became the stipulation within the CAFE legislation that the fleet be broken down into cars and light trucks. Which came first, the American consumer's taste for large vehicles or the manufacturers' emphasis of these models? It appears to be a hand-in-glove, synchronistic relationship. The irony, however, is that the policies created to conserve petroleum supplies spurred the increase in vehicle size and weight traveling American roadways. When AMC bought Jeep and began its work as the first American manufacturer to aggressively market such a vehicle, the concept of the SUV did not exist.

Similar to the fictional Dr. Frankenstein, auto manufacturers carefully studied the new guidelines of the 1970s and concocted the best hope for their industry's future. Instead of pursuing the efficiency mandated by the new guidelines (a course they would leave to Japanese manufacturers), American car makers found a loophole and exploited it. Their Frankenstein was the large SUV sought by many Americans in the twenty-first century.

Initially, the primary issue for manufacturers was vehicle weight. This is measured as "gross vehicle weight," which is the truck's weight when fully loaded with the maximum weight recommended by manufacturers. Instead of the 10,000 pounds used for trucks, light trucks were initially set at 6,000 pounds. Automakers realized that they could escape the light truck category all together by increasing the weight of their vehicles, so, as journalist Keith Bradsher wrote, "[they] shifted to beefier, less energy-efficient pickups even in a time of rising gasoline prices rather than try to meet regulations that they deemed too stringent" (Bradsher 2002). In 1977, the maximum for light trucks was raised to 8,500 pounds. In 1981, Ronald Reagan took office and made one of his first priorities to freeze most auto regulations where they now stood. By the late 1990s, one expert defined the ubiquitous SUV in this manner: (1) four-wheel drive available, (2) has an enclosed rear cargo area, (3) has a high ground clearance, (4) uses a pickup-truck underbody, and (5) is designed for urban consumers and is marketed primarily to them.

What began as gimmicky, the small-selling vehicle for a specific purpose morphed into ubiquity through the odd convergence of consumer taste and auto manufacturers' interest in exploiting a specific niche in new vehicle regulations. As defined by the CAFE standard, a light truck is any four-wheel vehicle weighing less than 8,500 pounds that is not a car. Although arbitrary, this category, therefore, includes vans, minivans, pickup trucks, and SUVs. "In the mid- to late 1980s," wrote sociologist James D. Gartman, "upscale demands for functionality and distinction" brought small-market, specialized vehicles into the mainstream (Gartman 1994, 222).

Each of the light trucks had been used for specific activities for many years. Most famous, the Jeep (mentioned above) pioneered the four-wheel-drive design to allow military users flexibility not possible with the animals used in World War I. By World War II, the Jeep was a symbol of the American military. Built originally by Ford and Willys-Overland, the best explanation for the name Jeep derives from G.P., short for "general purpose." Willys began making a Jeep station wagon with four-wheel drive in 1949; however, it still only had one door on each side. During the 1960s, this vehicle grew to be the Jeep Wagoneer, which inspired the Ford Bronco and other vehicles.

It is the Chevrolet Suburban, however, that is credited with being the world's oldest SUV or light truck nameplate, a product continuously in circulation. In 1935, the Suburban was advertised as a delivery truck, whether used as a hearse or for delivering illicit moonshine. Not until the 1960s did Chevrolet add a door on each side, and it was 1967 when a four-door model was offered. (Bradsher 2002, 6) In terms of automobile taste, Gartman suggests that the large trucks contradicted almost everything that typical American consumers desired: in 1970, 14,000 light trucks were registered, whereas nearly 90,000 cars filled America's roadways. Although AMC made the Jeep available to many more American consumers, it remained a rough, difficult-riding truck.

CAFE standards initiated changes in the vehicles made available to American consumers. One of the clearest changes of new legislation was the weight of cars. Since 1978, the average weight of domestic and imported cars dropped nearly 1,000 pounds, from 3,831 to 2,921. Although there are many variables to factor in, we can at least say that, overall, the weight of the cars on American roadways has decreased since the 1970s.

In creating the category of light truck, however, American manufacturers found their safety valve. Of course, this new category of vehicles contained very few vehicles when the standards were set [decades at approximately 10 percent of entire fleet (Volti 2006, 143)]. The light-truck share of the passenger vehicle fleet rose to 20.9 percent in 1975 and to 30 percent in 1987. In 1995, this had risen to 41.5 percent. Remarkably, by the year 2001, there were almost an equal number of cars and light trucks on the road (approximately 8.5 million of each). In a bitter irony, the CAFE standards and ensuing legislation had created the opportunity to build large, heavy, inefficient vehicles, and, to the shock of the owners of AMC and other manufacturers, Americans wanted such vehicles.

The Anti-Conservation Vehicle

In short, under CAFE, large cars are penalized, small cars are subsidized, and light trucks are largely unregulated, and one could expect that both small cars and light trucks would grow in popularity.

Auto manufacturers realize these details only in hindsight. One of the best indicators of this changing taste in vehicle was the evolution of AMC's Jeep as it blazed the way forward in the new SUV category. From the army Jeep to the Commando, CJ5, and even the early Wagoneers, Jeeps were rough work vehicles with a tendency to rollover. AMC designers set out in 1983–1984 to create a more aerodynamic, four-door Jeep. The revolutionary design became the Cherokee.

When gas supplies grew and prices increased mid-decade, the Cherokee became one of the nation's most popular vehicles. Most surprising, the sales were not in rural, specialized markets; instead, it was family-oriented, urban and suburban consumers who opted for the nation's first bona fide SUV. To take advantage of these consumers, AMC created a limited model that sold even more briskly. The cycle was completed in 1987 when Chrysler bought American Motors for $1.5 billion. The main appeal was not the Pacer; it was the Jeep brand.

With the success of Jeep and its acquisition by Chrysler, Ford got through a generation of hesitation over developing a four-door SUV and released the Explorer in 1990. Combining the consumer interest for having such a vehicle with Ford's massive marketing capabilities, the Explorer proved the tipping point for the auto revolution begun by Jeep. Stating the Ford point of view, Stephen Ross explains, "An SUV buyer is almost anti-minivan—this is a buyer who has a family but doesn't want to broadcast a docile family message" (Bradsher 2002, 51). Ford's surveys left no doubt that consumers wanted the vehicle to have four-wheel drive, although the drivers were likely to never need it or, at least, to use it rarely.

During the 1980s, SUV sales rose from 1.79 to 6.49 percent. Gartman finds a striking contrast between the overall style of design behind these vehicles and models of the past. He wrote, "Most of these erstwhile hauling and off-road vehicles could be seen whining along freeways and suburban surface roads, their engines struggling to maintain speed with absurdly low gearing and to overcome the resistance of high, square bodies and knobby, super-traction tires" (Gartman 1994, 223). Merging the attributes of the SUV with the luxury and comfort of large cars would be one of the challenges for automotive designers of the 1990s. Economists estimate that the growth of the light-truck category offset about 75 percent of the improvement that would have been seen from CAFE standards. The standard did not induce consumers to substitute small cars for large cars but to substitute light trucks for large cars.

Conclusion: Super-Size My SUV, Please!

Although the electric initiative was being fought again in locations such as California, American manufacturers presented American consumers with what they really wanted: the exact antithesis of the EV1. Large SUVs stepped into the SUV loophole exploited in the mid-1980s and blew it wide open. Pickup trucks, SUVs, and minivans had each been introduced into the marketplace to provide manufacturers with a way of meeting new CAFE standards. Light trucks, such as the Ford Explorer, provided automakers with an opportunity to slow the pace of their fleet's efforts to increase efficiency. To the manufacturers' surprise, however, the larger vehicles sold well. In addition, during decades of fast-moving wealth, these vehicles, costing well over $50,000, became a symbol of its driver's economic standard. Thanks to the sales of these large vehicles, Detroit's stock prices soared during the 1990s.

Of the late 1990s, automotive industry journalist Keith Bradsher describes the Big Three bringing prototypes for eighty-mile-per-gallon midsize cars to Washington for display, whereas back in Detroit they announced plans for twenty more versions of full-sized SUVs and pickups by 2003 (Bradsher 2002, 79–80). Only a few years previously, auto manufacturers had been sure that size did not matter to American consumers and that efficiency might have its day. The success of the first SUVs in the late 1980s, however, made Detroit take note. Negotiating additional benefits for increasing the size and weight of their fleet, American automobile manufacturers gave Americans what it appeared they wanted: even bigger, less efficient vehicles.

Auto manufacturers set out in the late 1980s to create economic enticements to make SUVs more attractive to American consumers. Light trucks were already exempt from the gas-guzzler tax created in 1978. In 1984, Congress had closed the depreciation law, which allowed self-employed individuals to deduct the entire purchase price of their vehicle from their taxable income over just three years, which reduced their taxes. With this law in place, such individuals were enticed to buy larger, more expensive vehicles, because they would ultimately pay for only a fraction of it. The 1984 restriction made it less attractive to purchase a large car; however, it provided special consideration for trucks with vehicle weight greater than 6,000 pounds. Buyers of these heavy vehicles could still write off the entire purchase price (up to half in the first year and the rest spread over four more years). Because the previous model for depreciation had applied to trucks weighing 13,000 pounds or more when empty, the stage was set for Detroit's next frontier of design and development.

In the mid-1980s, this new exemption applied primarily to full-size pickups, one of Detroit's most exclusive products; with the positive reaction to the Jeep Comanche, however, manufacturers scurried to put more SUVs on the market. It made perfect business sense, then, to model these new vehicles with a weight greater than 6,000 pounds to attract business purchasers. In 1990, when Congress imposed a 10 percent luxury tax on cars costing more than $30,000, the law exempted light trucks with gross weights greater than 6,000 pounds. Therefore, the laws clearly helped to shrink other markets while creating new enticements for consumers to buy some of the largest, heaviest vehicles ever built.

With very few SUVs or minivans weighing more than 6,000 pounds, the next frontier for Detroit was obvious. Even the Chevrolet Suburban, which was the largest SUV on the market, met the weight class restriction but was not costly enough. At approximately $17,000, the Suburban was almost half of the $30,000 price tag on the luxury-tax benefit. The new vehicles needed to combine size, weight, and luxury in a way that vehicles never before had done to fully take advantage of the tax benefits and loopholes. They needed to be symbols of opulence.

Luxury versions of the Suburban, the Chevy Tahoe, and GMC Yukon each became available in 1994. These were followed before the end of the decade by the Ford Expedition and Lexus LX-450 in 1996, Lincoln Navigator in 1997, the Cadillac Escalade in 1998, Ford Excursion in 1999, and Toyota Sequoia in 2000. Unlike the luxury cars that they were replacing, large SUVs brought manufacturers massive profits, normally as much as $15,000 per vehicle. In most cases, these vehicles almost single-handedly saved the Big Three during the 1990s. For instance, Ford's initial plan in 1996 was to stick the Expedition body on top of its F-150 pickup frame and to make 130,000 units and sell them for approximately $36,000 apiece. Their profit on each Expedition was $12,000. Faced with such profits, Ford shifted entire plants over to manufacturing Expeditions so that, by 1998, it could pump out 245,000 units per year. Its stock price rose more than 200 percent by the end of the decade. The impact of such a frenzy was obvious in the consumer market. In the luxury portion of the car market, cars made up 95 percent of all purchases in 1990. By 1996, cars made up only 44 percent of that segment! (Bradser 2002, 154)

SUVs were originally designed for work crews, hunters, residents of snow country, and others needing to travel off-road. By the end of the twentieth century, however, they were the car of choice for soccer moms, secret service teams, business executives, sports stars, and gangster rappers. Thanks to a convergence of changes in air regulation and tax law, largely orchestrated by auto manufacturers and their hired lobbyists, while environmentalists and others attempted to develop new technologies for transportation, Detroit regressed. Paul Roberts wrote, "The SUVs represent the height of conspicuous energy consumption. The extra size, weight, and power of the vehicles are rarely justified by the way their owners drive them. Even though owners and carmakers counter that the SUV's greater size, weight, and capabilities provide an extra margin of safety, studies indicate that SUVs not only are more likely to kill people in cars they hit but, because they roll over more easily, can actually be more dangerous to their occupants as well" (Roberts 2005, 154).

Viewed objectively—for instance, say as a scientist might—we must reflect on the remarkable data of the late twentieth century use of petroleum: from 1960 to 2005, amount of miles driven by Americans quadrupled; the market share of the light trucks grew from 10 percent to nearly 50 percent by 2001; and the largest-selling vehicles in the United States by the year 2001 had become two full-size pickups, the Ford F-150 and the Chevrolet Silverado.

Source: Bradsher, *High and Mighty: SUVs: The World's Most Dangerous Vehicles and How They Got That Way.*

THE AMERICAN SUBURB, SPRAWL NATION, AND THE EMERGENCE OF NEW URBANISM

Time Period: 1920s to the present
In This Corner: Economic and housing developers, American consumers
In the Other Corner: Planners, ecologists, architects
Other Interested Parties: Politicians
General Environmental Issue(s): Housing, sprawl, planning, New Urbanism

Possibly no landscape form has seen its image transformed as quickly as sprawl has since 2000. In just a few years, sprawl has gone from a model of successful economic development to a model of poor community planning that lacks foresight and good taste.

The United States stands as the world's capital of sprawl; however, nations around the world face similar challenges with the arrival of American-inspired economic development. It is the United States, however, that used the image of the suburb to crystallize a standard of living that became the envy of most of the world. For this reason, suburban development was allowed to proceed almost unchecked. By the 1990s, planners, environmentalists, and sociologists were criticizing aspects of the suburban life.

Communities and Spatial Preference

By definition, sprawl is a development pattern that only became prominent in the United States after World War II. Most important, sprawl is based on the decentralizing of the human population. Stylistically, it also ignores historical or ecological precedent and human experience. Although it may sound unappealing, sprawl has come to dominate nations such as the United States simply because of convenience.

From concentrated towns and cities, suburban development has led middle-class residents to construct satellite areas that are now referred to as sprawl. Until recently, this pattern was largely absent from the rest of the world. Throughout human history, most communities were centered about a common area, possibly a market, central structure, or an open area. Most habitation grew outward from this center and created urban areas.

Agriculture was the most common activity outside of the urban or town center. A great deal of early agriculture grew outward from these central areas on land owned and tended commonly. Societies structured around private property ownership began systematic changes to this community structure. Most commonly, agriculturalists constructed homes in outlying areas where they could also tend their own land. Such shifts rarely meant true decentralization. Markets still kept rural inhabitants intimately involved with the town. On the whole, living near the central town in many societies before 1800 was a mark of status; residences more distant were most often relegated to the poor.

Most of these spatial dynamics continued for much of the industrial age, when workplaces located themselves within a downtown business district. The growth of a more defined middle and upper class combined with a growing desire for a cleaner, simpler residential environment to propel Europeans to country estates and summer homes. By 1850, romanticism and other cultural developments had contributed to an aesthetic appreciation of more rustic and primitive living. This intrigue of some wealthy residents stimulated a larger shift in spatial preferences. Ultimately, this shift led many nations to suburbanize by the late 1800s. The effort to link suburbs with necessary services led to the creation of sprawl on the landscape.

The American Suburb as an Architectural Form

When the idea of home-making and house-planning took shape in the United States around the turn of the twentieth century, designers sought a single style that embodied the evolving American ideals in a form that could be dispersed widely. Inspiration for such home design grew from modern sensibilities that were styled after a regressive tradition known as the Arts and Crafts movement. The enduring marriage of this blend was the well-known bungalow style house. Gustav Stickley's *Craftsman* magazine made plans for such homes widely available. When the style was reacquired by modernist designers, Frank Lloyd Wright used it to create a model design for homes that could be mimicked in any residential setting. His designs were grouped within the tradition known as the Prairie School. Accentuating horizontality and organization of internal spaces, the homes of the Prairie School sought to create models to inspire the homes of middle-class Americans.

The homes of such designs played directly into a growing interest in home management, referred to often as home economics. At the turn of the twentieth century, American women began to perceive the home as a laboratory in which one could promote better health, families, and more satisfied individuals with better management and design. The leaders of the movement of domestic science endorsed simplifying the dwelling in both its structure and its amenities. Criticizing Victorian ornamentation, they sought something clean, new, and sensible. The bungalow fulfilled many of these needs perfectly.

The most familiar use of "bungalow," however, would arrive as city and village centers sprawled into the first suburbs for middle-class Americans. These singular homes were often modeled after the original Stickley homes or similar designs from *Ladies' Home Journal*. The design would make it possible for the vast majority of Americans to own their own homes, thereby updating the Jeffersonian image of Americans as a landowning people. Housing the masses would evolve into the suburban revolution on the landscape; however, the change in the vision of the home can be traced to a specific type: the unassuming bungalow (Roth 1970, 198).

Wright merged this style with a specific region when he began designing homes in the Prairie style from 1900 until 1915. The high point came in 1914 and was based in the American Midwest. During its formative years, the architects focused on suburban Chicago, but it would also reach into rural Illinois, Minnesota, Iowa, and Wisconsin. Emphasizing horizontality of design, the Prairie School was a regional manifestation of the more general, international revolt and reform occurring in the visual arts.

For Wright, the horizontal mimicking of the landscape allowed the form to become organic, concealed in the landscape, and satisfied his modernist desire for simplification. This link between structure and landscape was further stimulated by the management of inner and exterior space. The Prairie homes were designed to bring the inside of the home out and the surroundings inside. Patios, gardens, and windows were designed to facilitate this connectedness. Building materials were selected to include natural elements of the surroundings, such as wood, stone, stucco, brick, or the elemental sand, gravel, cement, and water that make up concrete. This was also true of the plants and landscape design of the elongated gardens and courtyards. The linking device between such spaces would often be stone fences that extended wall lines outward, but most often the most noticeable element was the elongated roof lines.

A product of cultural taste, the Prairie style's popularity petered out as preferences changed. *House Beautiful* illustrated its last prairie house in 1914. Stickley's journal ceased publication in 1916 as the Arts and Crafts movement itself also lost popularity. The Prairie style's great achievement is a mode of design universally applicable to every building type. Its influence can be seen in many other types of architecture of the twentieth century, particularly in gardens and courtyard designs. Many suburban homes, including the ubiquitous ranch house, strive for a similar link between the horizontal exterior space and the domestic living environment. Unlike urban residences, suburban settings, at the very least, prioritize space and contact with some form of nature (Wright 1992, 98).

Transportation and Suburbia

Sprawl grew out of the increasing decentralization that accompanied suburbanization after 1900. Transportation served as the most important tool for this shift. By prioritizing transportation corridors, the United States led the way toward today's landscape of sprawl. Early examples of suburbanization were seen in "green cities" found in Britain and France. Such plans rarely incorporated sprawl; instead, their priorities lay in green spaces and rustic style. Providing shelter remained the primary goal of housing, but setting and aesthetics became crucial components of many homes.

By 1930, some suburban planners had begun to look for inspiration less in nature and more in technological solutions. Planners in Germany helped to create the International style of the 1920s, and their modernist designs influenced architecture throughout the twentieth century. Whereas Modernism was most evident in skyscrapers, the ideas inherent in style pioneered by European architects could also be seen in everything ranging from roadways to home design.

As the United States adopted this pattern by 1920, planners and developers combined the impulse to move outward with transportation infrastructure and prefabricated home design. The result was the unprecedented rapid development of suburban areas outside of major cities. Initially, such developments followed railroad and streetcar corridors. However, once automobiles became widely affordable, many land developers and other businesses began marketing a single vision of the American Dream: owning one's own single-family house on a large lot in the suburbs. This marketing appeal resonated with many consumers, who saw the suburbs as an opportunity to escape the noise, crowds, and social problems of the city and to raise children in a safe, clean environment.

American suburbanization was spurred by a number of social and political factors. Through the FHA and Veterans Administration loan programs, the federal government provided mortgages for eleven million new homes after World War II. With the surge in American population called the "baby boom," a new American ideal was born: a new single-family home in an outlying suburb. Policies of the FHA and Veterans Administration programs discouraged the renovation of existing houses and turned buyers away from urban areas.

Planners created home styles that allowed them to develop one site after another with the automobile linking each one to the outside world. The ticky-tack world of Levittown (the first of which was constructed in 1947) involved a complete dependence on automobile travel. This shift to suburban living became the hallmark of the late twentieth century, with over half of the nation residing in suburbs by the 1990s. The planning system that supported this residential world, however, involved much more than roads. The services necessary to support outlying, suburban communities also needed to be integrated by planners.

Auto suburbs spread quickly before 1940, but post–World War II growth dwarfed anything previous. The world had never before seen a spread in the middle-class standard of living like that in the United States after 1945. The symbol of such change was Levittown, NY, built in 1947. Other exact replicas and hundreds of similarly designed communities spread across the United States.

Standardization was the term that governed the Levittowns. Prefabricated construction allowed for thousands of homes to be constructed in a matter of months. As the economy expanded after the war, suburbs could be planted almost immediately to provide uniform shelter and community to inhabitants. No doubt, however, a great deal was lost in the race to house the baby boomers.

Sprawl Binds Together the Nation

In 1950, 33 percent of the nation lived in urban areas, 23 percent in suburban areas, and 44 percent in rural areas. By 2000, more than 50 percent of Americans lived in suburban areas. Historian Clay McShane wrote, "In their headlong search for modernity through mobility, American urbanites made a decision to destroy the living environments of nineteenth-century neighborhoods by converting their gathering places into traffic jams, their playgrounds into motorways, and their shopping places into elongated parking lots."

Instead of the Main Street prototype, the auto suburbs demanded a new form. Initially, American planners such as Jesse Clyde Nichols devised shopping areas such as Kansas City's Country Club District that appeared a hybrid of previous forms. Soon, however, the "strip" evolved as the commercial corridor of the future. These sites quickly became part of suburban development to provide basic services close to home. A shopper rarely arrived without an automobile; therefore, the car needed to be part of the design program. Signs were the most obvious architectural development of this new landscape. Integrated into the overall site plan would be towering neon advertisements that identified services. Also, parking lots and drive-thru windows suggest the integral role of transportation in this new commerce. In short, sprawl had arrived.

During this period of massive home construction, federal and local subsidies also spurred the construction of roads, including a 41,000-mile interstate highway system. The scale of this construction will likely stand as one of the great building feats of human history. Between 1945 and 1954, nine million Americans moved to suburbs and became entirely reliant on roadways for their everyday life. Between 1950 and 1976, central city population in the United States grew by ten million, whereas suburban growth was eighty-five million. Housing developments and the shopping/strip mall culture that accompanied decentralization of the population made the automobile a virtual necessity.

The automobile proved to be the ultimate tool for decentralizing the landscape. With satellite communities constructed miles from downtown resources and shopping, developers seized the opportunity to develop the arteries connecting suburbs to cities. With little thought to livability or other priorities, suburbs incorporated shopping and service areas that would evolve into forms on the American landscape known as the shopping mall and the strip. The common link between each portion of the landscape became the automobile. The landscape designed around automobile linkages became known as sprawl.

These developments culminated in the shopping mall, which quickly became a necessary portion of sprawl. By the 1970s, developers' initiatives clearly included regional economic development for a newly evolving service and retail world. Incorporating suburbs into such development plans, designs for these pseudo-communities were held together by the automobile.

In a world designed around consumption, the marketplace became the shopping mall. Strip malls, which open on to roadways and parking lots, were installed near residential areas as suburbs extended farther from the city center. Developers then perfected the self-sustained, enclosed shopping mall. Sprawl oozed out from these starting points.

Urban sprawl is a twentieth-century phenomenon. It is the expanding of urban and, more frequently, suburban environments into the rural lands at the fringe of these urban areas. The expansion into these areas tends to be sprawling neighborhoods of single-family homes. The neighborhoods have a low population density and are designed around single-use zoning, areas where residential, industrial, and commercial areas are separated from one another rather than integrated. Because of these characteristics, areas considered sprawl are generally not considered to be pedestrian-friendly neighborhoods; walking is not practical, so the area is populated by people who commute by automobile.

Sprawl is occurring in many locations around the globe. Examples include Los Angeles and Atlanta in the United States, Brussels in Belgium, Copenhagen in Denmark, and Mexico City in Mexico. Some of these areas have expanded their sprawling footprint but have lost population. In large part, these trends began as a part of "white flight," middle-class, Caucasian families moving from the city centers to the outskirts. As other minority groups have gained financial security, they too have moved out of the inner cities. Together, this movement has led to a constant expansion of the suburban neighborhoods.

Try as they might, such artificial environments could never recreate the culture of local communities. Shopping malls became the symbol of a culture of conspicuous consumption that many Americans began to criticize during the 1960s. Many Americans began to ask, have we given up our ties to genuine community?

The Ecological Impact of Sprawl

Suburbs and the transportation systems that support them require the subordination of natural elements for those best suited to rapid site preparation. Throughout the twentieth century, the economics of planning and development fueled the standardization of landscape and environment. Whether woodland, prairie, or farmland, the suburban development cleared the land and reshaped it with heavy equipment. Historian Adam Rome wrote that, after World War II, "To speed the work of site preparation, the typical subdivision builder cleared away every tree in the tract, so millions of postwar homes had no shelter against bitter winter winds and brutal summer sun." In addition, regional ecologies throughout the nation had been significantly altered. In its place, builders installed residential human environments, adorned with non-native turf grass and ornamental trees. Natural habitats and regional ecology can rarely be reinstalled, regardless of the "green" intentions of any developer.

Massive suburbanization damaged the ecological integrity of countless locales and placed residential pressures on many regions, such as Los Angeles, California, that were ill suited for such development, and yet American suburbs became an enviable symbol for much of the world. In the famous "kitchen debate" of 1957, Vice President Richard Nixon used the model American home constructed in Moscow as a symbol that democracy and capitalism represented the world's fastest route to happiness and comfort. This symbol had great resonance to leaders around the world for more than a half century.

Standards of living, of course, did not rise uniformly around the world after World War II. Decolonization helped to create vast areas more concerned with starvation than turf grass maintenance. The suburb became one of the great international symbols of the gap between rich and poor nations. The private "palaces" of the American middle class struck many international observers as decadent and wasteful.

As urban areas in Asia, Mexico, and Africa became the most densely populated regions on Earth, planners and human rights agencies searched for sustainable urban plans. Suburbanization requires that residents have a certain level of capital and that developers be free to make homes affordably. Also, open space and transportation links are essential. For many nations, this model of development is impossible. Other nations desire to follow an urban design pattern that wastes fewer resources. Much of the world, however, will live with the lifestyle of suburbia for decades to come.

Questioning Sprawl

Peter Blake's seminal work *God's Own Junkyard* in the early 1970s asked critical questions of the automobile-inspired landscape. Blake and others referred to the landscape as blight, largely from its lack of a design program or architectural style. Robert Venturi and others set out to consider the auto strip as a viable architectural form, but few architects agreed. Inspired and organized by consumption and not living, they argued, sprawl was a horrific symbol of the most decadent and wasteful aspects of twentieth-century life.

Although it seemed clear that sprawl lacked good taste, few scientists had applied new lines of thought such as ecology to suburban planning. In the late 1960s, Ian McHarg published *Design with Nature*, which urged architects to consider the ecology of a site when devising a plan. For popular readers, critics such as Jane Jacobs and Jim Kunstler identified an intrinsic bias on the American landscape in the 1970s. Kunstler wrote, "Americans have been living car-centered lives for so long that the collective memory of what used to make a landscape or a townscape or even a suburb humanly rewarding has nearly been erased."

Such an idea remained foreign until the late 1970s. The modern environmental movement asked hard questions of American consumption. In the 1970s, this movement resulted in new responsibilities for the federal government, including the EPA. The EPA's regulative authority brought new demands on developers. EISs, particularly as suburban plans effected watersheds, became a standard part of planning during the 1980s. Sprawl still happened, but a great deal of the planning needed to funnel through EPA regulators.

The 1990s closed with the unfolding of the new politics of urban sprawl. "I've come to the conclusion," explained Vice President Al Gore on the campaign trail in 1999, "that what we really are faced with here is a systematic change from a pattern of uncontrolled sprawl toward a brand new path that makes quality of life the goal of all our urban, suburban, and farmland policies."

Gore and others had seen the future not in the United States but in Europe, especially nations such as Finland, Norway, and the Netherlands. The international planners were well ahead of American designers. The green movement in Europe spurred significant changes in land planning in many nations. A great deal of these designs were inspired by Postmodernism. This intellectual approach called for abandoning the rigid rules of architectural modernism and showing greater sensitivity to history and local context and for allowing diverse voices, such as minorities, to be heard. The blend was tailored for European communities with a long history but an interest in modernizing.

Critics of urban sprawl include conservationists who point out the amount of land that is taken up by urban sprawl: urban sprawl's low-density, single-family homes, rather than apartments, are separated by lawns and roads. Furthermore, the lots in areas of urban sprawl are usually larger, and, because residents must drive to work, shop, and recreate, there are often large parking lots. This form of development eliminates the region's open space and habitat. Critics also note that people who live in urban sprawl areas tend to use more resources and emit more pollution per person than do those who do not live in the urban sprawl zones (Hirschhorn 2005; Norman, MacLean, and Kennedy 2006). Because urban sprawl is spread out and located away from shopping areas and business districts, people who live within these regions must drive to shop and work.

Another issue critics point out is that urban sprawl affects the health of those who live in the neighborhoods. First, those who live in these areas are less likely to walk or ride a bike for transportation, and, as a result, they are more likely to be obese. Second, the sprawl and increased levels of pollution negatively impact land and water quality, harming the health of both the humans who live in the region and the environment. Finally, opponents of urban sprawl also argue that fast food chains have developed in tandem with sprawl. Fast food architecture and parking lots mimic the tone of sprawl and those who populate the sprawl frequent these restaurants, often furthering health problems (Schlosser 2001).

Defending Sprawl

Supporters of sprawl, or suburban housing developments, claim that the low-density neighborhoods create quieter, safer environments for families and increase privacy for individuals. They also suggest that it does not necessarily increase traffic in the area. To support this claim, they cite research findings that suggest that, in low-density communities, traffic levels are lower, and the traffic moves at a faster pace, actually lowering the air pollution emissions per square mile (Wendell Cox Consultancy 2002).

Advocates also argue that, in many suburban areas, the schools are better and that suburban development brought about the shopping mall. The shopping mall is usually a large building surrounded by parking lots that contains many shops, including "anchor" department stores (Gruen and Smith 1960). Supporters note that shopping malls are in many ways environmentally friendly because people need only drive to one location to buy most things they need. Finally, advocates of suburban development count free marketers among their supporters. Those that support free markets use the argument that sprawl is a matter of consumer choice and freedom and that these things should not be restricted.

Conclusion: Designing Beyond Sprawl

To minimize the sprawling growth and yet use the benefits of the suburban housing development, a "smart growth" movement began in Oregon. In 1973, Oregon enacted a law that limited the geographical area that urban regions could occupy; at the same time, there was a push to focus on improving neighborhoods and schools within these regions. Using the idea of urban growth boundaries, communities such as Portland, Oregon, have created more compact, quality urban areas, protecting wild areas and farm land around the metro region while providing highly desirable urban neighborhoods.

The American version of these plans has come in the form of New Urbanism, one of a handful of responses to sprawl's shortcomings. Instead of single-use developments, New Urbanist communities provide mixed uses within a walkable neighborhood. Additionally, housing options are varied and the automobile becomes an unnecessary part of everyday living. Some examples of New Urbanist communities include, Seaside, Florida, Celebration, Florida, and Kentlands, Maryland. Many European towns and cities were already structured around such pre-automobile models. New Urbanists single out Capri, Florence, and Barcelona as cities that have maintained connections to ideals of these new models of town planning.

Internationally, sprawl has become an issue wherever population density combines with American-inspired residential development. With shopping centers and the spatial organization that they bring, many nations have found that economic development means sprawl. However, efforts are being made by the U.N. and other international agencies to disseminate the ideas of planning and particularly New Urbanism.

These ideas are generally included in the broad mandate to spur sustainable development. Specific programs, such as the U.N.'s Center for Human Settlements–Habitat, seek to link urban planning with environmental understanding all over the world. Also, the U.N. runs a Sustainable Cities Program. Currently, the Sustainable Cities Program operates twenty main demonstrations and twenty-five replicating cities around the world, including cities in China, Chile, Egypt, Ghana, India, Kenya, Korea, Malawi, Nigeria, the Philippines, Poland, Russia, Senegal, Sri Lanka, Tanzania, Tunisia, and Zambia. Activities are planned in Bahrain, Cameroon, Iran, Kenya, Lesotho, Rwanda, South Africa, and Vietnam. Some critics argue that nations must be free to develop strategies of their choice, particularly those that are less costly than urban planning. However, projects such as this one make certain that the information is available for nations desiring it.

Sprawl continues to be the end product when communities allow development to unfold without a plan. However, more and more communities are using new ideas in planning and design to construct a positive human environment and control sprawl.

Sources and Further Reading: Belasco, *Americans on the Road*; Benfield, Terris, and Vorsanger, *Solving Sprawl: Models of Smart Growth in Communities Across America*; Bruegmann, *Sprawl: A Compact History*; Calthorpe, *The Next American Metropolis*; Clark, *The American Family Home*; Duany and Plater-Zyberk, *Suburban Nation: The Rise of Sprawl and the Decline of the American Dream*; Fishman, *Bourgeois Utopias: The Rise and Fall of Suburbia*; Gruen and Smith, *Shopping Towns USA: The Planning of Shopping Centers*; Hirschhorn, *Sprawl Kills: How Blandburbs Steal Your Time, Health, and Money*; Jackson, *Crabgrass Frontier: The Suburbanization of the United States*; Jacobs, *The Death and Life of Great American Cities*; Kunstler, *The Geography of Nowhere: The Rise and Decline of America's Man-Made Landscape*; Norman, MacLean, and Kennedy, "Comparing High and Low Residential Density: Life Cycle Analysis of Energy Use and Greenhouse Gas Emissions"; Relph, *The Modern Urban Landscape*; Rome, *The Bulldozer in the Countryside: Suburban Sprawl and the Rise of American Environmentalism*; Rybczynski, "Suburban Despair"; Wright, *Building the Dream*.

FIELDS OF DREAMS: SPORT GREENSPACES IN AMERICA

Time Period: Twentieth century
In This Corner: Sports and stadium developers, architects
In the Other Corner: Designers, environmental critics
Other Interested Parties: Fans, public officials
General Environmental Issue(s): Landscape architecture, recreation, sports, land use

The historic interest in planned landscapes broke in different directions during the twentieth century. Parks represented a clear environmental sensibility, but some of the other directions were less clear. When planners and landscape designers linked green spaces around leisure activities, including sports and recreational activity, many environmentalists were critical. However, various natural spaces used for sports and recreation represent an important zone in which land planners establish aesthetic tastes that find their way into the lives of most Americans.

Golf Courses and the American Taste for Grass

Although turf grass is not indigenous to most of North America, its presence marks a complicated portion of the American connection to the natural environment. Part of this tradition includes a leisure activity not normally identified with environmentalists: golfing.

With roots in Europe, particularly Great Britain and Scotland, golf was first a hobby of royalty and the very wealthy. The United States broadened this interest to the middle classes but still clearly linked golfing to social and economic status. The most enduring link grew between golf and the new American suburbs of the upper middle class that began to be seen in the late nineteenth century. Today, with nearly 60 percent of Americans living in suburbs, golf's popularity has also increased significantly. No other aspect of the American landscape is as responsible as golf courses for creating an aesthetic bond between twentieth-century Americans and turf grass.

Golf made its appearance in the United States in 1888 and was initially played in close-cropped cow pastures. Despite its appeal to women, golfing was largely limited to a tiny minority of primarily affluent white men who could afford to support the large expanse of carefully managed and highly maintained acreage needed for a golf course. Public links were established in parks in New York City, Boston, Cincinnati, Philadelphia, and Providence, Rhode Island, in the 1890s, with eighty courses in place by 1894 and 982 by 1900. Many of these courses were less than picturesque, however. As American courses became more competitive, the best lured greens keepers from Europe to recreate the American courses as beautiful parks. Of these newer courses, the most influential were Myopia, north of Boston and the Garden City Golf Course and National Golf Links of America, each on New York's Long Island.

More public courses followed in the 1920s, and a few elite suburbs were laid out around golf courses so that the fairways seemed to flow into the lawns around the houses. Charles Hugh Alison, the first course architect to be recognized internationally, was the most active of approximately twelve architects presiding over this golden era. Construction costs, real estate values, and interest rates were low and the public was prioritizing leisure. Six hundred new courses were opened each year between 1923 and 1929! By 1929, there were nearly 6,000 courses in the United States. Only long-established courses were able to weather the Depression and war years. The post-war years, however, would bring significant changes to golf course design and use (Jenkins 1994, 34–40).

The greatest revolution in course construction after 1945 was technology, particularly the availability of massive earth-moving machines. The appearance of courses now could include cultivating and carefully smoothing fairway seed beds. Robert Trent Jones and other well-known designers combined such methods with the expansion of the middle class to recreate golf as a male standard through the late twentieth century. By 1980, there were more than 400 Trent Jones courses located in forty-two states. Additionally, many suburban communities included courses in their overall designs. Clearly, golf was a part of life for many Americans.

The growth in golf's popularity during the latter half of the twentieth century includes strategic marketing through the mass media. As types of labor gave way to service and management, middle-class men had a bit more leisure time as well as an automobile to carry them where they wanted. Historian Virginia Scott Jenkins finds that a connection was firmly made between the aesthetic of the golf courses and the suburbs in which players lived. One advertisement urged homeowners to "ask the Greens-keeper at your own Club what he thinks of TORO Equipment." Golfers such as Sam Sneed and Jack Nicklaus were used in advertisements to instill the connection between homeownership and lawn care (Jenkins 1994, 61–64).

Sports Fields: Green Cathedrals

One of the most ubiquitous versions of our pastoral ideal is the tradition of American sports fields. Of course, athletic stadiums are utilitarian structures that also offer major cities the opportunity to possess a cutting-edge stadium that will earmark a city on the move. The precedent for such structures differs significantly from the private, corporate development of contemporary stadium planning.

The earliest modern stadiums emphasized utility and monumentality. Olympic stadiums, including the White City stadium in London for the 1908 games and the Berlin Olympic stadium for the 1936 games, carried on these traditions while also stressing flexibility in the facility's use. After World War I, the United States broke new ground with a series of pioneering stadiums, including the Yale Bowl at New Haven in 1914, the Rose Bowl at Pasadena, and Ann Arbor stadium. Using grass playing fields as their setting, these and other stadiums built before 1950 simply organized and systematized an audience's ability to watch sports. The nature of stadiums, however, changed dramatically after 1950. Multipurpose facilities became the rage for modernist planners who created urban designs that allowed for stadiums to be surrounded by massive parking lots so that fans could arrive by automobile.

The multipurpose form was extended by a development that seemed foreboding for the American commitment to turf grass: the enclosed dome. The originator of this form was the Houston Astrodome, which opened in 1964. Judge Roy Hofheinz, with the quirky idea to combine attending a sports event with going to a cocktail party, designed the dome around the idea of skyboxes. These private boxes allowed high-paying clients to attend games without interacting with other fans. A young pitcher for the Houston Astros bounded into the stadium in April 1965, taking in the miracles of the dome: air conditioning, grass growing indoors (artificial turf would be laid in 1966), the translucent roof (greenhouse by day, a planetarium by night), and seating for 66,000. "It was," he says, "like walking into the next century" (Jenkins 1994, 143–44).

Many traditionalists viewed attending baseball in air-conditioned splendor as a travesty. The players, however, were most critical of the roof, the panels of which created a glare that made it impossible to see the ball. The league tried changing the color of the ball but to no avail. The team painted over the roof panels, banishing the sun and killing the grass, Tifway 419 Bermuda, which had been specially developed by scientists in Georgia. For the rest of the season, the Astros simply painted over the dead grass. After the season, Monsanto's new artificial turf, renamed AstroTurf, was installed. By 1973, five more stadiums would have synthetic surfaces and many others would follow.

Turf proved the high point of the artificial stadium. Other domes would follow, including the Louisiana Superdome, the largest dome when it was built in 1975 with a seating capacity of 95,000. Although traditionalists would wage war against domes and turf, there was practical value to the controlled environment. Particularly when sports became more concerned with money making, these technologies reduced the dependence on weather and made the games more appealing for family and business groups. The effort to reconcile these needs led to a few innovations, including the retractable roof, which was first installed in the Toronto Skydome in 1989. On the whole, however, the domes have fueled a return to a more traditional model of stadiums seen in the early 1900s.

Many current stadium projects have followed the model of Camden Yards in Baltimore and the new Comiskey Park in Chicago. Returning to the one-dimensional parks, these forms fuse modern convenience with nostalgic detail. The postmodern fusion has been a universal success, even functioning to attract entire families to baseball games. Many stadiums have built on this to include amusement and shopping facilities within the park for those less enamored with sports. Among the nostalgic details is also a return to turf. Natural grass has been found to be much kinder on athletes' bodies. Today, very little artificial turf is installed at the professional level.

From being viewed as a utilitarian structure, stadiums appear now to be viewed as an attraction of their own. The mixture of traditional nostalgia and business concerns created a great stadium boom in the final years of the twentieth century.

Conclusion: A Green Commons

For Americans in the era of mass consumerism, sports fields helped to shape a whole series of preferences, ranging from the desire for virile, active lifestyles and for certain picturesque landscapes. Caring for these highly manicured areas created an entire realm of education, planning, design, and labor. In arid areas, the use of aesthetic land forms that demanded the use of scarce water incurred the wrath of many environmentalists. However, such spaces continue to represent a familiar natural form for most Americans.

> **Sources and Further Reading:** Jackson, *Crabgrass Frontier: The Suburbanization of the United States*; Jacobs, *The Death and Life of Great American Cities*; Jenkins, *The Lawn*; Kunstler, *The Geography of Nowhere: The Rise and Decline of America's Man-Made Landscape*; Rome, *Bulldozer in the Countryside: Suburban Sprawl and the Rise of American Environmentalism*.

THE GREEN REVOLUTION CHANGES RURAL SOCIETIES IN THE DEVELOPING WORLD

Time Period: 1940s to 1980s

In This Corner: Agricultural companies, humanitarian interests

In the Other Corner: Defenders of traditional methods and culture

Other Interested Parties: United Nations, agricultural scientists, citizens of less-developed nations

General Environmental Issue(s): Hunger, agriculture, development

The Green Revolution, a term coined in 1968, was an agricultural and technological revolution that increased agricultural production between the 1940s and the 1980s. The primary objective of the Green Revolution was to increase crop yields and augment aggregate food supplies to alleviate hunger and malnutrition. During the Green Revolution, agricultural programs of research and extension and infrastructural development brought major social and ecological impacts to, particularly, developing nations. The results of the Green Revolution have drawn both praise and criticism.

During the 1940s and 1950s, United States programs to improve seeds (high-yield grains), develop farm technology, improve irrigation, and finance agrochemicals made great progress. Judged as successful by the major funding agencies, including the Rockefeller Foundation and the Ford Foundation, there was a feeling that these programs could be important to the future of other nations. Thus began the movement to spread the techniques to other countries. Beginning in Mexico in the 1940s, collaboration between the Rockefeller Foundation and Mexican President Manuel Avila Camacho's administration brought the expectation that Mexico's developing agriculture would aid in the nation's industrial development and economic growth.

The Green Revolution then moved to India, bringing programs of plant breeding, irrigation, and agrochemicals, including the use of pesticides and synthetic fertilizers that created substantial increases in grain production. In India, the Green Revolution raised rice yields by

30 percent, giving the government time to curb its population growth without having to endure the periodic famines they had previously experienced. By the 1960s, high yielding (so-called "miracle seed") varieties of wheat, corn, and rice were growing in many non-Soviet bloc countries in Asia, Latin America, and North Africa. As a part of the expanding programs, the U.S. Agency for International Development helped subsidize rural infrastructure development and agrochemical shipments.

Supporters of the Green Revolution believe that the revolution has helped to avoid widespread famine by feeding billions of people. Most of these supporters accept the Malthusian principle of population, believing that, in many countries, population would outgrow food production, causing vast famine and malnutrition. The increased yield crops within developing countries have provided for the growing population. Supporters also note that many of the new varieties of crops are fortified with vitamins and minerals that improve the health of the people who consume them. For example, golden rice, grown in an area where there is a shortage of dietary vitamin A, was developed as a humanitarian tool to help combat the irreversible health conditions related to a vitamin A deficiency. Finally, supporters suggest that, by improving the nutrition, one result would be a healthier workforce who could then work to improve the industrial and economic growth of their country.

Critics dispute a number of the supporters' claims. They suggest that, although the numbers show that grain production steadily increased from the 1960s to the 1990s as a result of the Green Revolution and its programs, this does not translate to an increase in overall food production. This claim is backed with the notion that Green Revolution agriculture follows the idea of monoculture, whereas traditional agriculture often incorporates other edible plant species. Thus, despite the fact that more grains are being produced with the techniques from the Green Revolution, there is a simultaneous reduction in other food varieties. Furthermore, some traditional agriculture practices displaced by the Green Revolution are highly productive and would likely compete with the Green Revolution production. For example, systems such as "chinampas," growing crops on the fertile arable land in shallow lake beds, have been replaced by the agricultural practices taught during the Green Revolution. Critics suggest that, through these changes, the Green Revolution may have decreased food security for some of the poorest people, depleting the traditional foods of many peasants through monoculture and the use of pesticides that eliminate other plants and animals within the ecosystem.

Critics also point out the dependence of modern agriculture on petroleum products. As petroleum supplies shrink, programs from the Green Revolution may well become prohibitively expensive for many of the developing nations that they were intended to help. Furthermore, the pesticides used as a part of the Green Revolution were necessary to deal with the large amount of pest damage that inevitably occurs in monocultures. These chemicals manage pests but they do not easily break down in the environment, and thus many environmentalists believe that they accumulate in the food chain and spread throughout ecosystems. These outcomes can lead to water contamination and the evolution of resistance in pest organism populations.

Another area of criticism is that the Green Revolution plant varieties were developed as hybrid seeds. The developers of these seeds hold the intellectual property rights for the seeds and thus require the farmers to purchase new seeds each season rather than saving seed from the last harvest. Although this process pays the companies for their research and development, it also increases the farmers' production costs. Critics suggest that this process goes against the original goals of the Green Revolution, finding a solution to hunger, poverty, and underdevelopment. Additionally, studies of the impact of the Green Revolution show that incomes for the larger farms in rural society have increased with the introduction of technology, whereas incomes of the smaller farm and poorest strata have tended to fall. Critics suggest that this reveals that the purchasing of the seeds and necessary agrochemicals is financially difficult for smaller farmers, often pushing them into debt, whereas the larger farms are benefiting.

Finally, there are also accusations that the Green Revolution is a political program more than a humanitarian program. For example, the name itself, "Green Revolution," is said to be a contrast to the Soviet "Red Revolution" and the Iranian "White Revolution." Additionally, United States journalist Mark Dowie, a major critic of the Green Revolution, suggests that the primary objective behind the programs was increasing social stability in non–Soviet bloc developing countries, thus creating beneficial relationships between these countries and the United States. These connections would not only create positive relationships for the United States but would also function as an alternative to socialist policies of expanding agrarian reforms initiatives.

The Green Revolution has made changes to countries within the developing world. Because of both the praise and the criticism, institutions involved in Green Revolution programs now attempt to take a more holistic view of agriculture. The scientists are working to better understand the problems faced by farmers and are now involving the farmers in the development process. Likewise, international organizations involved in the Green Revolution, such as the Food and Agriculture Organization of the United Nations, are working to formulate a more equitable and sustainable Green Revolution, aiming in particular at improving the standard of living for the people involved in producing, providing, and managing food supply within the poorest rural households: women farmers. To date, overall results have been both positive and negative, guaranteeing continued controversy surrounding the Green Revolution.

Sources and Further Reading: Brown, *Seeds of Change*; Conway, *The Doubly Green Revolution*; Dowie, *American Foundations: An Investigative History*; Food and Agriculture Organization of the United Nations, "Women and the Green Revolution"; Perkins, *Geopolitics and the Green Revolution: Wheat, Genes, and the Cold War*; Shiva, *The Violence of the Green Revolution: Ecological Degradation and Political Conflict in Punjab*; Wright, *The Death of Ramon Gonzalez: The Modern Agricultural Dilemma*.

THE GREEN REVOLUTION INITIATES AN ERA OF INTERNATIONAL COOPERATION

Time Period: 1950 to the present
In This Corner: Nations of the world
In the Other Corner: Other nations
Other Interested Parties: Political leaders, voters
General Environmental Issue(s): Globalization

The evolution of global cooperation on behalf of the environment has occurred most gradually over the past half century. Although some scientists and environmentalists hypothesized that the human species was doomed because of its inevitable need to act in individual self-interest, a culture of international cooperation slowly took shape. Although this required political agencies such as the U.N., the movement for international cooperation also drew from evolving science that assisted in establishing international issues that confronted all humans.

The Green Revolution Ushers in New Cooperative Era

Most scholars date the interest in globalization to the world wars and the Cold War. In addition, the world's worst recorded food disaster occurred in 1943 in British-ruled India. Known as the Bengal Famine, an estimated four million people died of hunger that year alone in eastern India (that included today's Bangladesh). The international community blamed poor food production for the shortage and began to consider ways of helping India and other less-developed nations. The gap separating developed and less-developed nations had become most pronounced by 1950, when modern conveniences and technologies made the standard of living in the United States and many other developed countries leap forward, while other nations lagged further behind.

When British occupiers left India in 1947, the Indian government set about trying to close this gap in terms of food production. However, two decades later, India realized that its efforts at achieving food self-sufficiency were not entirely successful. This awareness led to the importation of new agricultural technology from around the world. Referred to as the Green Revolution, the effort to bring new agricultural know-how to developing nations radically altered the possibility of famine while also attacking the primary issue dividing developed nations, such as the United States, and less-developed nations, such as India and much of Africa.

Headed by Norman Borlaug, a plant breeder from the University of Minnesota, American agricultural technology was given to many Third World nations. Although there were many successful examples, Borlaug was awarded the Nobel Prize for his work on a high-yielding wheat plant. Work on this wheat had begun in the mid-1940s in Mexico as Borlaug and others developed broadly adapted, short-stemmed, disease-resistant wheats that excelled at converting fertilizer and water into high yields.

The "winter wheat," as it was called, could be grown much more easily in areas such as Mexico and India. The impact of this plant was monumental. In 1944, Mexico imported half its wheat, but by 1956 it was self-sufficient in wheat production, and by 1964 it was growing enough of a surplus that it exported approximately one-half million tons of wheat. In India, wheat production increased four times in twenty years (from twelve million tons in 1966 to forty-seven million tons in 1986). This success inspired rice experiments in the Philippines and elsewhere.

By 1992, the system included eighteen centers, mostly in developing countries, staffed by scientists from around the world, supported by a consortium of foundations, national governments, and international agencies. Some critics argue that the Green Revolution depends on fertilizers, irrigation, and other factors that poor farmers cannot afford and that may be ecologically harmful, and that it promotes monocultures and loss of genetic diversity. Clearly, however, the spread of agricultural technology has greatly assisted less-developed nations in feeding their populations.

Garrett Hardin and the "Tragedy of the Commons"

After 1960, Americans' worldview changed considerably, whether influenced by photos from the moon or scientific concepts. Our lives, it became clear, were tied to many other creatures and systems. Therefore, our choices and actions had broad impacts.

Following Rachel Carson, in 1968, Garrett Hardin wrote an article that developed the ecological idea of the commons. This concept and his argument of its tragic (undeniable) outcome in depletion gave humans new rationale with which to view common resources such as the air and the ocean. He wrote the following:

> The tragedy of the commons develops in this way. Picture a pasture open to all. It is to be expected that each herdsman will try to keep as many cattle as possible on the commons. Such an arrangement may work reasonably satisfactorily for centuries because tribal wars, poaching, and disease keep the numbers of both man and beast well below the carrying capacity of the land. Finally, however, comes the day of reckoning, that is, the day when the long-desired goal of social stability becomes a reality. At this point, the inherent logic of the commons remorselessly generates tragedy.
>
> As a rational being, each herdsman seeks to maximize his gain. Explicitly or implicitly, more or less consciously, he asks, "What is the utility to me of adding one more animal to my herd?" This utility has one negative and one positive component....
>
> Adding together the components ... the rational herdsman concludes that the only sensible course for him to pursue is to add another animal to his herd. And another.... But this is the conclusion reached by each and every rational herdsman sharing a commons. Therein is the tragedy. Each man is locked into a system that compels him to increase his herd without limit—in a world that is limited. Ruin is the destination toward which all men rush, each pursuing his own best interest in a society that believes in the freedom of the commons. Freedom in a commons brings ruin to all. (Hardin 1968, 243–48)

Fostering Mechanisms for Cooperation

Although a global perspective seemed inherent in the web of life put forward by Rachel Carson and others, it would take global issues such as the Chernobyl nuclear accident in 1986 and shared problems such as greenhouse gasses and global warming to bind the world into a common perspective. Organizations, including Greenpeace and the U.N., assisted members from many nations to shape a common stand on issues.

The U.N. presented the leading tool for facilitating global environmental efforts. With its first meeting on the environment in 1972, the global organization created its Environmental Program, which was referred to as UNEP. This organization would sponsor the historic Rio Conference on the Environment in 1992 and the conference on global warming in 1997. UNEP would also be the primary institution moving the U.N.'s environmental agenda into the areas that were most in need of assistance, primarily developing nations. In response to such activities, the U.S. federal government declared the environment a genuine diplomatic risk in global affairs by creating a State Department undersecretary for the environment in 1996.

What began as an intellectual philosophy in the early 1800s had so impacted the human worldview that it would now influence global relations. Of course, a primary portion of this environmental worldview was scientific understandings that were communicated to the public after the 1960s. During the 1990s, a global agenda for action took shape that was organized around environmental improvement. Such an agenda, however, was not without its difficulties.

The most difficult portion of this global debate may be the fairness of each nation to be allowed to develop economically. Many less-developed nations resent the environmental efforts of nations that have already industrialized, including many European nations and the United States. Such nations believe they are being denied their own opportunity to develop economically simply because of problems created by industrial nations. This sentiment has been part of major demonstrations in recent meetings of global organizations, such as the World Trade Organization (WTO). Any global agreements will need to balance the basic differences of these constituencies

Forming a Discourse of International Cooperation

During the 1980s and 1990s, the international discourse on environmental issues took a more organized and systematic form. On the twentieth anniversary of the first UNEP meeting in Stockholm, the U.N. hosted the 1992 Rio Conference on Environment and Development, the "Earth Summit," in which world leaders agreed on Agenda 21 and the Rio Declaration.

The summit brought environment and development issues firmly into the public arena. Along with the Rio Declaration and Agenda 21, it led to agreement on two legally binding conventions: Biological Diversity and the Framework Convention on Climate Change.

Agenda 21, in particular, functioned to place an important new idea into mainstream international environmentalism: sustainable development. Agenda 21 was a 300-page plan for achieving sustainable development in the twenty-first century. The U.N. Commission on Sustainable Development (CSD) was created in December 1992 to ensure effective follow-up of the U.N. Conference on Environment and Development and to monitor and report on implementation of the Earth Summit agreements at the local, national, regional, and international levels. The CSD is a functional commission of the U.N. Economic and Social Council, with fifty-three members. A five-year review of Earth Summit progress took place in 1997 by the U.N. General Assembly meeting in special session, followed in 2002 by a ten-year review by the World Summit on Sustainable Development.

The CSD was established as a functional commission of the Economic and Social Council by council decision 1993/207. Its functions are set out in General Assembly resolution 47/191 of December 22, 1992. The commission is composed of fifty-three members elected for terms of office of three years, meets annually for a period of two to three weeks, and receives substantive and technical services from the Department of Economic and Social Affairs/Division for Sustainable Development. The commission reports to the Economic and Social Council and, through it, to the Second Committee of the General Assembly.

The CSD now focuses on assessment and education. Most important, it must contain representatives from the nations that most need assistance with development strategies. In 2005, the membership of the CSD included thirteen members elected from Africa, eleven from Asia, ten from Latin America and the Caribbean, six from Eastern Europe, and thirteen from Western Europe and other areas. Such a balanced membership is a priority of most U.N. efforts; however, on CSD it may be even more important because representation can ensure that nations receive the help that they need most.

However, global environmental initiatives remain difficult. There simply is no authority that can enforce environmental policies and regulations across political lines everywhere in the world. However, great strides have been made in creating regional, cooperative initiatives.

NAFTA and Regional Agreements

Outside of the U.N., the new international connections that were inherent in globalization altered relations between nations. For the United States, global free trade remained a priority, but regional development required new agreements with neighbors. The 1993 NAFTA defined a vast new region for free trade. Organizations such as the WTO also helped to prioritize free trade over environmental protection.

Opposed to such agreements, public interest activist Ralph Nader argued that such deals would erode environmental and social legislation. The Clinton administration negotiated at least two environmental side agreements. One of these created the North American Commission for Environmental Cooperation, which emphasized cross-border initiatives. The second agreement specifically focused on the Mexican border.

The success of such efforts to limit the environmental impact of such initiatives and agreements has been limited by factors, including a difficulty in regulating users and accessing documents that would allow investigators to bring suit. Most observers believe that the environmental problems along the United States–Mexico border have intensified under NAFTA. Green NGOs have joined with American labor organizations to call for the United States to renegotiate or pull out altogether from the 1993 agreement.

Masked protesters lead the flag-waving march from Hong Kong's Victoria Park toward the Hong Kong Convention and Exhibition Center, December 17, 2005, in a protest against the 6th WTO Ministerial Conference in this former British colony. This is one example of the protests that occur at many meetings of international trade organizations in the twenty-first century. AP Photo/Bullit Marquez.

Protesting the WTO and Globalization

A policy infrastructure for globalization had taken shape by the mid-1990s. The nature of the American worldview expanded to consider the interests of other nations as well as other ecological entities. However, just as many proponents began to think that they had created the agencies and initiatives to close the gap separating developed and less-developed nations, activists, many of them associated with environmental causes, altered their view of initiatives such as the WTO and the World Bank.

Whereas instruments of change such as the World Bank, WTO, and even the U.N. had been created to promote global peace and stability, critics began to argue that such organizations exerted the will of powerful nations on those of the less powerful and less-developed nations. Many American environmentalists argued that, not only were the agencies' actions heavy handed, but they also fueled less-developed nations to follow the less sustainable paths to progress used and favored by Western nations.

By the end of the 1990s, every meeting of the WTO became a gathering for activists denouncing its activities. In only a few short years, the attitude toward international assistance had undergone a radical shift. Although there was a clear history of cooperation on some global environmental issues, consensus seemed to shatter at the start of the twenty-first century.

Conclusion: Global Warming

In the late twentieth century, the environmental issue referred to as global warming seemed to be the next great issue in the global unification of efforts to diminish the impact of humans on the earth. However, by the beginning of the twenty-first century, it was obvious that, instead of bringing the world together, global warming was going to cause even more division. Unlike the issues behind the Green Revolution, the science behind global warming was difficult to verify, and, most important, acting on global warming could potentially hurt the U.S. economic dominance of the globe.

What scientists refer to as the greenhouse effect is actually essential to human existence. The sun warms the earth, and certain gases (including carbon dioxide and water vapor) act like the glass of a greenhouse, trapping heat and keeping the planet's surface warm enough to support life. However, measuring humanity's effect on the concentration of greenhouse gases is a key issue in understanding global climate change. Industry and other human activity add carbon dioxide to the atmosphere. This strengthens the greenhouse effect and may cause a significant warming trend. During the 1990s, the issue of global warming received attention worldwide.

For nations to attempt to confront this issue, researchers needed to ascertain a reliable method for tracing temperatures and gas levels in the near and distant past. Ice-core sampling made this possible. In the ices found on the earth's frozen poles, scientists drilled core samples from water that had not been thawed for thousands of years. The ice offered snapshots of the air and water from long ago, long before humans began to burn fossil fuels.

Actually, the ice cores contained remnants from snow from long ago. The snow carried with it compounds, including sulfate, nitrate, and even dust, radioactive fallout, and trace metals. In polar areas, this snow falls on top of the previous year's snow without either melting. As this happens repeatedly over many years, the snow compresses to form ice. This is the ice, then, that scientists can use to provide a record of changes in air composition and temperature. This record has suggested the warming trend in the earth's temperature since humans began burning fossil fuels.

Scientific consideration of global warming and the emissions of greenhouse gasses have fueled international debate over the ratification of an international policy on the issue, the Kyoto Protocol. Such legislation would place caps on emissions from all nations. The Kyoto Protocol is a document signed by about 141 countries at Kyoto, Japan, in December 1997. The protocol commits thirty-eight industrialized countries to cut their emissions of greenhouse gases between 2008 and 2012 to levels that are 5.2 percent below 1990 levels. According to one estimate, global warming could cost the world about $5 trillion. Developing countries are expected to be the hardest hit.

Advocates see it as a baby step along the necessary road to reducing human impact on climate before the oceans rise and prairie songbirds emigrate to the Arctic. Some opponents of the Kyoto Protocol refuse to believe the scientific findings. Armed with contrary findings, other scientists have claimed to disprove global warming altogether. Other opponents admit that the science proves global warming, but the Kyoto Protocol is simply too costly to the standing of developed nations such as the United States. The reality, however, is that, since 1990, the United States has increased its release of greenhouse-damaging emissions, including carbon dioxide, methane, nitrous oxide, and other pollutants. Buildup of such gasses contributes to rising temperatures as well as to an increase in skin cancer. It also appears unlikely that American leaders will accept limitations on emissions that may impact American productivity, even in the short term. Philosophically opposed by any effort to limit its own development, the United States under the administration of President George W. Bush has rejected the protocol.

By 2007, political pressure and scientific reality appeared to be bringing the United States into the international discourse on global warming once again.

Sources and Further Reading: Christianson, *Greenhouse*; Hughes, *American Genesis*; McNeil, *Something New Under the Sun: An Environmental History of the Twentieth-Century World*; Pollan, *Omnivore's Dilemma*.

NORTH AMERICAN FREE TRADE AGREEMENT, HUMAN RIGHTS, AND ENVIRONMENTAL CONSEQUENCES

Time Period: 1994 to the present
In This Corner: United States government, Mexican government, industries
In the Other Corner: Some labor groups, some environmental groups, some human rights groups (mainly in Mexico)
Other Interested Parties: Agriculture, indigenous
General Environmental Issue(s): Environmental policies, pollution, sustainability

The NAFTA is an economic pact creating a free trade bloc among Canada, the United States, and Mexico. Signed in 1992, the treaty was ratified in 1993 and came into effect on January 1, 1994, creating the world's largest free trade area. NAFTA called for the elimination of all trade barriers among member countries over a fifteen-year period. Additionally, it incorporated side agreements on labor and the environment in an attempt to balance economic growth with sustainable development.

Inspired by the European Community's success in reducing trade barriers among its member countries, Canada, the United States, and Mexico, agreed to work together to form a North American free trade pact. Using the terms of the Canada–United States Free Trade Agreement of 1988, the three countries extended the provisions of this agreement to Mexico and included terms for the possible expansion of the treaty. As designed, NAFTA does not create a set of supranational governmental bodies, nor does it create a body of law superior to national law. The three countries wanted to retain their sovereignty and a semblance of control over their economies. To reemphasize this supremacy of domestic law, under United States law, NAFTA is classed as a congressional-executive agreement rather than a treaty.

The NAFTA document outlines a fifteen-year phasing out of investment restrictions among Canada, the United States, and Mexico. Duties on the majority of goods shipped between member countries were to be eliminated immediately while the remaining tariffs would be gradually phased out. However, the treaty does specifically protect intellectual property rights, including patents, copyrights, and trademarks. The stipulations of NAFTA apply equally to all three countries, except in the area of agriculture. Agricultural tariff reductions and the protection of select agricultural industries were negotiated bilaterally. In addition, provisions for labor and environmental protection were added as supplemental "side agreements," the North American Agreement on Environmental Cooperation (NAAEC) and the North American Agreement on Labor Cooperation (NAALC).

The environmental and labor side agreements were signed in 1993, before the implementation of the agreement. The NAAEC was a response to environmentalists' concerns that North American companies would relocate to Mexico to take advantage of lower environmental standards and/or that the United States would lower environmental standards if the three countries did not agree on environmental regulations. Thus, the NAAEC was designed to oblige the member countries to enforce their own environmental regulations. The NAAEC did this with the establishment of one trilateral institution, the North American Commission for Environmental Cooperation (CEC), and two additional bilateral (United States and Mexico) institutions, the Border Environmental Cooperation Commission (BECC) and the North American Development Bank (NADBank). The first of these, the CEC, is an organization that addresses environmental concerns within Canada, the United States, and Mexico. It helps prevent potential trade and environmental conflicts and promotes the effective enforcement of environmental law (Commission for Environmental Cooperation 2007). The second institution, the BECC, is a binational institution that works to protect, preserve, and enhance human health and the environment of the United States–Mexico border region (Border Environmental Cooperation Commission 2007). This is achieved by strengthening cooperation among interested parties and supporting sustainable projects. Finally, the NADBank is a binational financial institution managed equally by the United States and Mexico for the purpose of financing BECC-certified environmental projects (North American Development Bank 2007). The NADBank helps to finance infrastructure necessary for a clean and healthy environment along the United States–Mexico border region.

The second side agreement, the NAALC was designed to create a foundation for cooperation in dealing with labor issues among Canada, the United States, and Mexico. In addition, the labor agreement was set up to promote greater coordination among trade unions and social organizations that fight for improved labor conditions. This goal is achieved by encouraging the publication and exchange of information, data development, and studies in addition to fostering transparency in the administration of labor law (Commission for Labor Cooperation 2007).

The effects of NAFTA are varied. Trade has dramatically increased among the three nations, although there are discussions over whether NAFTA caused the trade increase or whether it simply reflects a more wide-ranging increase in world trade (Lederman, Maloney, and Serven 2005). Most economists agree that NAFTA has had some beneficial effects for Mexico. However, they are also quick to point out that NAFTA has not produced any economic convergence among the member nations, nor has it substantially reduced poverty rates (Floudas and Rojas 2000). Furthermore, there is evidence that NAFTA has eliminated hundreds of thousands of agricultural jobs in Mexico because of an influx in cheaper imported agricultural products (Oxfam 2003). There have also been concerns that NAFTA is pulling manufacturing jobs from the United States into Mexico, although most economics contend that NAFTA has not had a large effect on the United States manufacturing economy (Griswold 2002). Of the numerous polls conducted over the past few years, a plurality has expressed the view that, on balance, NAFTA is something positive (Weber 2006).

Groups that consider themselves supporters of NAFTA have changed over time. When the idea was first discussed, support came mainly from free trade economists and industries that believed they would benefit from the agreement. However over time, with the addition of provisions and the two side agreements, the list of supporters has grown to include not only industry but also some agricultural entities and some environmental groups. Supporters note the economic benefits that have come from NAFTA, benefits for not only the United States but also for Canada and Mexico. In addition, environmental groups suggest that, without the side agreements, environmental concerns, particularly in Mexico, would be more difficult to address.

Like supporters, critics of NAFTA have also changed over time. Before the agreement was clearly delineated, many groups put forward concerns: manufacturing, agriculture, labor, environmental organizations, and human rights groups among others. With the defining of the agreement and the addition of the two side agreements, some critics were swayed. However, many others were not. Today, the critics of NAFTA include many who have felt negative impacts from the changes and those who worry about long-term effects. These groups include major industries such as automotive, agriculture, textile and computer manufacturing, telecommunications, financial services, energy, and trucking, in addition to some environmental groups, labor, and human rights groups.

Critics of NAFTA suggest a loss in manufacturing jobs in the United States and a reduction in the market price of many of Mexico's agricultural products, which depresses the incomes of those in the lower income bracket. They also point to the case in which the Canadian government removed a ban on methylcyclopentadienyl manganese tricarbonyl (MMT), a gasoline additive that some studies have linked to nerve damage, to settle a suit by an American company. The American company has brought suit under NAFTA's Chapter 11, which allows a corporation or individuals to sue Canada, the United States, or Mexico when the government's actions have adversely affected their investments.

Finally, critics of NAFTA suggest a proliferation of human rights abuses in the "maquiladoras," industries that import materials and equipment on a duty-free basis for assembly or manufacturing and then re-export the assembled product back to the originating country. Maquiladoras quickly populated the United States–Mexico border region after the ratification of NAFTA. The human rights abuses in question deal with working conditions, pollution, and health-related concerns.

As NAFTA is modified with time, and as critics bring their concerns to the CEC and NAALC, changes will continue to take place in the development and implementation of NAFTA. Additionally, since the ratification of NAFTA, Chile has joined with the agreement. The addition of other countries along with periodic assessment of the agreement and each country's situation will also likely make changes to the agreement, again creating a shift of supporters and critics.

Sources and Further Reading: Commission for Environmental Cooperation, *Who We Are*; Commission for Labor Cooperation, *Objectives, Obligations, and Principles*; Floudas and Rojas, "Some Thoughts on NAFTA and Trade Integration in the American Continent"; Johnson and Beaulieu, *The Environment and NAFTA: Understanding and Implementing the New Continental Law*; Lederman, Maloney, and Serven, *Lessons from NAFTA for Latin America and the Caribbean*; MacArthur, *The Selling of "Free Trade": NAFTA, Washington, and the Subversion of American Democracy*; Markell and Knox, eds., *Greening NAFTA: The North American Commission for Environmental Cooperation*; Mayer, *Interpreting NAFTA: The Science and Art of Political Analysis*; Oxfam, *Dumping without Borders: How U.S. Agricultural Policies Are Destroying the Livelihoods of Mexican Corn Farmers*.

FACING UP TO AMERICAN ENERGY DEPENDENCE

Time Period: 1970s
In This Corner: Carter administration, energy conservationists
In the Other Corner: Energy and petroleum interests, defenders of the status quo
Other Interested Parties: OPEC, American consumers
General Environmental Issue(s): Energy

For the United States, a century of energy decadence came to a screeching halt during the 1970s. Politicians could no longer appear to be take their responsibilities seriously without at least proposing ideas on the subject of energy use and conservation. However, few individuals agreed with each other on how to proceed. In the end, this active disagreement might be part of the reason that many of the 1970s energy reforms disappeared as quickly as they came.

OPEC Emerges as a New Kind of World Power

For Americans, lines at local gas stations served as one example of massive changes in global affairs after World War II. History teachers usually use the term "decolonization" to refer to this era when many additional nations became autonomous, responsible for their own development and governance. Certainly, the Cold War added a new version of quasi-colonial authority; however, overall, nations in Africa and particularly the Middle East could begin to pursue their own futures. The use and management of every resource took on strategic importance, and, therefore, it follows that the administration of the world's most sought after commodity reflected these changes most acutely.

From the dominant stranglehold of Western powers and the large petroleum corporations that ruled them, oil morphed into a tradeable, ultra-volatile commodity. Yergin wrote that this new era in world oil demonstrated that "… oil was now clearly too important to be left to the oil men" (Yergin 1993). As political leaders in each oil nation assessed how best to leverage power for his nation from his supply of crude, it took little time for them to also realize the merit of joining forces with similarly endowed nations.

OPEC was created at the Baghdad Conference in Iraq in September 1960. Its formation was precipitated by changes in the oil market after World War II. Lacking exploration skills, production technology, refining capacity, and distribution networks, oil-producing countries were unable to challenge the dominance of the oil companies before World War II. Although Mexico wrestled control of its oil industry from foreigners in 1938, it quickly receded from the lucrative international market because of insufficient capital for investment.

Other nations, such as Venezuela, sought to establish their own agreements that would allow them to keep more of the revenue from sales. In 1943, Venezuela signed the first "fifty-fifty principle" agreement, which provided oil producers with a lump sum royalty plus a fifty-fifty split of profits. Oil companies, of course, disliked such arrangements. Soon, nations with such agreements were shunned by oil companies. Other nations, however, such as Saudi Arabia, began pursuing their own fifty-fifty contracts.

In the midst of this tumultuous period in oil administration, Iran passed a law demanding the termination of previous agreements with Anglo-Iran (referred to as Anglo-Persian before 1935 and British Petroleum after 1954). When no agreement came, the Iranian Prime Minister Mossadegh nationalized oil operations in May 1951. (A new British-Iranian agreement was signed the following year. The newly restored Shah of Iran became a pillar of American Middle East policy until the Iranian Revolution in 1979.)

The chaos created by the pursuit of separate agendas fueled oil producers to construct a cartel. OPEC's founding members in 1960 were Iran, Iraq, Kuwait, Saudi Arabia, and Venezuela. Eight other countries joined later, including Qatar (1961), Ecuador (1973), and Gabon (1975) (Ecuador and Gabon withdrew from the organization in 1992 and 1994, respectively). What these nations had in common was oil. To varying degrees, however, they also shared small size and political influence. Together, however, OPEC's purpose was obvious: to limit supplies in the hope of keeping prices high.

It seems ironic, today, to talk about oil producers needing to manipulate markets to keep the price of petroleum up; yet, major oil companies colluded from the 1920s to the 1960s to prevent prices (and profits) from falling. As their influence waned, other methods were used. One of the biggest difficulties was that, as prices fell, domestic producers simply could not compete. Moreover, the Eisenhower administration concluded (as the Japanese had before World War II) that dependence on foreign oil placed the country's national security in jeopardy. As a result, the United States implemented import quotas on petroleum. Although meant to stimulate domestic production, the quota kept domestic prices artificially high. In the end, the result was simply a net transfer of wealth from American oil consumers to American oil producers. By 1970, the world price of oil per barrel was $1.30, and the domestic price was $3.18 (Danielsen 1982, 150).

OPEC's ability to manipulate prices did not become a reality until Egyptian leader Anwar Sadat urged his fellow members to "unsheath the oil weapon" in early 1973. The primary rationale for this action was politics. Israel's military aggression outraged its Arab neighbors throughout the late 1960s. Israel's attack on Egypt in 1967 resulted in an earlier embargo, which proved unsuccessful because of an oversupply of crude on the world market. In October 1973, U.S. President Richard Nixon agreed to provide more military jets to Israel after a surprise attack on the nation by Egypt and Syria. On October 19, the Arab states in OPEC elected to cut off oil exports to the United States and to the Netherlands.

In petroleum circles, the embargo is often referred to as the "First Oil Shock." As such, it combines new market features of the early 1970s: first, production restraints that were ultimately supplemented by an additional 5 percent cutback each month; and, second, a total ban on oil exports to the United States and the Netherlands and eventually also to Portugal, South Africa, and Rhodesia. Factoring in production increases elsewhere, the net loss of supplies in December 1973 was 4.4 million barrels per day, which accounts for approximately 9 percent of the total oil available previously. Although these numbers told of a genuine shortfall, the fickle petroleum market accentuated the embargo by inserting a good bit of uncertainty and panic. It was the American consumers who felt the impact most.

To provide oil to consumers, brokers began bidding for existing stores of petroleum. In November 1973, per barrel prices had risen from around $5 to more than $16. Consuming nations bid against each other to ensure sufficient petroleum supplies. For American consumers, retail gasoline prices spiked by more than 40 percent. Although high costs were extremely disconcerting, scarcity also took the form of temporary outages of supply.

The front on this new resource war could be found on the home front: the American gas station.

Previously, many Americans were content to drive their cars until gas gauges neared empty. Now, prudent consumers topped-off their tanks because they were uncertain about gasoline's future availability. One journalist described the scene near New York City in this manner:

> Anxious motorists overwhelmed gasoline stations in the metropolitan [New York] area yesterday, with many stations running out of supplies early in the day, while dealers hoped incoming deliveries under February allocations would restore calm by midweek.
>
> In Brooklyn, Murray Cohen, an owner of the AYS Service Station at Avenue Z and East 17th Street, said he had imposed a $3 maximum for each car's purchases, only to find that most people needed only 75 cents' worth to fill up. One man, he said, waited in line for an hour and could use only 35 cents' worth.
>
> In Washington, William E. Simon, director of the Federal Energy Office, who had asked drivers not to buy more than 10 gallons at a time, yesterday issued an appeal to them to stay away from stations unless they bought at least $3 worth.... "Panic buying isn't helping the situation ..." (Merrill 2007)

Many states implemented odd-even gas purchasing based on the car's license plate number. Regardless of how they arrived at the station, motorists throughout 1973–1974 needed to wait in line for one to two hours or more, ironically, of course, with their engines running the entire time. In other regions, the worst harbinger became signs that read, "Sorry, No Gas Today." Expressway speeds were cut from sixty to seventy miles per hour to fifty. Many communities, as well as the White House, forwent lighting public Christmas trees. Some tolls were suspended for drivers who carpooled in urban areas. Rationing plans were leaked to the public, even if they were not implemented. For instance, in the New York City region, the Federal Energy Office estimated that residents eighteen years of age and older could expect to receive books of vouchers for thirty-seven gallons per month (*New York Times* 1974).

By the end of 1973, in fact, gas lines were plentiful throughout the nation. Supplies of petroleum were least disturbed on the West Coast, but, by February, even California had adopted odd-even day rationing. Gas station operators were subjected to mistreatment, violence, and even death. Drivers also reacted with venom to other drivers attempting to cut into gas lines.

These petroleum shortages extended into 1974; the implications of them, however, extended through the rest of the decade. The shock was an abrupt lesson. Although few Americans understood why the price fluctuated so wildly, for the first time most Americans learned three valuable lessons: petroleum was a finite resource; the United States imported the bulk of its petroleum supply; and the United States was entirely reliant on this commodity.

Although the embargo had economic implications, it had begun as a political act by OPEC. Therefore, the Nixon administration determined that it needed to be dealt with on a variety of fronts, including, of course, political negotiation. These negotiations, which actually had little to do with petroleum trade, needed to occur between Israel and its Arab neighbors, between the United States and its allies, and between the oil-consuming nations and the Arab oil exporters. Convincing the Arab exporters that negotiations would not begin while the embargo was still in effect, the Nixon administration leveraged the restoration of production in March 1974. Although the political contentions grew more complex in the ensuing decade, the primary impact of the embargo came through the residual effects it had on American ideas of petroleum supply.

Oil Embargo Spurs Policy Shift

As energy supplies became a more significant topic after the 1970s Arab oil embargo, a panicked public expected action. Richard M. Nixon, by this point embattled with the growing problem of Watergate but reelected in 1972, appeared before Americans on November 7, 1973 to declare an "energy emergency." He spoke of temporary supply problems:

> We are heading toward the most acute shortages of energy since World War II.... In the short run, this course means that we must use less energy—that means less heat, less electricity, less gasoline. In the long run, it means that we must develop new sources of energy which will give us the capacity to meet our needs without relying on any foreign nation.

> The immediate shortage will affect the lives of each and every one of us. In our factories, our cars, our homes, our offices, we will have to use less fuel than we are accustomed to using....

> This does not mean that we are going to run out of gasoline or that air travel will stop or that we will freeze in our homes or offices anyplace in America. The fuel crisis need not mean genuine suffering for any Americans. But it will require some sacrifice by all Americans.

In Nixon's speech, he went on to introduce "Project Independence," which he viewed "in the spirit of Apollo, with the determination of the Manhattan Project, [would] ... by the end of this decade" help the nation to develop "the potential to meet our own energy needs without depending on any foreign energy source."

In reality, Nixon's energy czar, William Simon, took only restrained action. Rationing was repeatedly debated, but Nixon resisted taking this drastic step on the federal level. Although he had rationing stamps printed, they were kept in reserve. In one memo, Nixon's aide Roy Ash speculated that, "In a few months, I suspect, we will look back on the energy crisis somewhat like we now view beef prices—a continuing and routine governmental problem—but not a Presidential crisis." Nixon's notes on the document read "absolutely right" and, overall, his actions bore out this approach.

As energy supplies became a more significant topic after the 1970s Arab oil embargo, each side of the environmental argument staked out its claim on the issue. Environmentalists used the 1973 oil shortage to argue that Americans needed to learn "living within limits." This lesson would be demonstrated again in 1990 when the nation went to war against Iraq largely to maintain control of oil supplies. Additional concerns came with an increasing awareness of the effects of air pollution and particularly auto emissions' relationship to global warming. The Clean Air Act of 1991 began a process requiring automobile makers to prioritize increased mileage and also to investigate alternative fuels.

Of course, the argument for a conservation ethic to govern American consumers' use of energy was a radical departure from the post-war American urge to resist limits and to flaunt the nation's decadent standard of living. Although this ethical shift did not take over the minds of all Americans in the 1970s, a large segment of the population began to consider a new paradigm of energy accounting. They became interested in energy-saving technologies, such as insulation materials and low wattage light bulbs. As a product of the 1970s, some Americans were ready and willing to consider less convenient ideas of power generation such as alternative fuels.

Amory Lovins' "Soft Energy Paths"

In a 1976 *Foreign Affairs* article entitled "Soft Energy Paths," Amory Lovins became spokesman for the small but vocal American movement for alternative fuels. In his subsequent book, Lovins contrasted the "hard energy path," as forecast at that time by most electrical utilities, and the "soft energy path," as advocated by Lovins and other utility critics:

> The energy problem, according to conventional wisdom, is how to increase energy supplies ... to meet projected demands.... But how much energy we use to accomplish our social goals could instead be considered a measure less of our success than of our failure.... [A] soft [energy] path simultaneously offers jobs for the unemployed, capital for businesspeople, environmental protection for conservationists, enhanced national security for the military, opportunities for small business to innovate and for big business to recycle itself, exciting technologies for the secular, a rebirth of spiritual values for the religious, traditional virtues for the old, radical reforms for the young, world order and equity for globalists, energy independence for isolationists.... Thus, though present policy is consistent with the perceived short-term interests of a few powerful

institutions, a soft path is consistent with far more strands of convergent social change at the grass roots. (Lovins 1979, 121–22)

Carter Considers Alternatives and Conservation

Lovins presented Americans with a radical alteration to their basic approach to energy. How did Americans react to the energy crisis of the 1970s? With heightened environmental awareness after the first Earth Day in 1970, public calls for change found many receptive ears. This especially became the case after the oil embargo of the early 1970s. One set of receptive ears resided in the White House after the 1976 election (Horowitz 2005, 20–25).

President Jimmy Carter's administration would be remembered for events such as the Iranian hostage crisis; however, when he controlled the agenda, he steered American discourse to issues of energy. In a 1977 speech, Carter urged the nation as follows:

Tonight I want to have an unpleasant talk with you about a problem unprecedented in our history. With the exception of preventing war, this is the greatest challenge our country will face during our lifetimes. The energy crisis has not yet overwhelmed us, but it will if we do not act quickly.

It is a problem we will not solve in the next few years, and it is likely to get progressively worse through the rest of this century.

We must not be selfish or timid if we hope to have a decent world for our children and grandchildren.

We simply must balance our demand for energy with our rapidly shrinking resources. By acting now, we can control our future instead of letting the future control us....

Our decision about energy will test the character of the American people and the ability of the President and the Congress to govern. This difficult effort will be the "moral equivalent of war"—except that we will be uniting our efforts to build and not destroy. (Horowitz 2005)

Conclusion: 1970s Energy in Hindsight

Carter would introduce wide-reaching policy initiatives mainly aimed at energy conservation. Although he offered a clear vision of our limited future based on extracted energy resources, by the 1980s, many Americans were returning to business as usual (Horowitz 2005, 43–46).

From energy conservation measures in American homes to vehicle size and weight restrictions, the reactions to the 1970s energy shortage were largely disavowed. Consumers, in the end, clearly remained willing to consume at their pre-1970s level (or more) and to ignore the outcomes forecasted by Carter, Lovins, and others.

Sources and Further Reading: Horowitz, *Jimmy Carter and the Energy Crisis of the 1970s*; Yergin, *The Prize: The Epic Quest for Oil, Money & Power*.

BOSTON HARBOR AS POLITICAL FOOTBALL

Time Period: 1988
In This Corner: Republican National Party and candidate George H. W. Bush

In the Other Corner: The Democratic National Party and presidential candidate Massachu-
setts Governor Michael Dukakis
Other Interested Parties: Boston residents, American voters
General Environmental Issue(s): City planning, watershed development, coastal management

By the end of the 1980s, environmental topics had become an important issue in many elections. However, public reaction was not always entirely predictable. In one of the most revolutionary uses of political advertisements in modern history, the campaign of Vice President
George Bush recast the policies of Massachusetts Governor Michael Dukakis and the environmental commitment of the entire Democratic Party. Most remarkable, they did so without calling their own environmental policies into question.

During the election of 1988, television ads sponsored by the Republican Party accused
the Dukakis administration in Massachusetts of dumping "500 million gallons of barely
treated sewage" into the harbor each day. Bush referenced the ad in one of the candidates'
debates when he quipped in reference to a Dukakis response, "That answer was about as
clear as Boston Harbor."

In actuality, coastal scientists agree that the harbor's treatment and situation was worse
than the ad stated. Dukakis, however, bore little personal responsibility for the utilitarian
ethic with which Bostonians had viewed their harbor for generations. In fact, the harbor had
been used as a dump since the first era of European settlement in early 1600s. The waters
functioned as a trade corridor that was necessarily allowed to become an ecological wasteland
for the next two centuries.

By the late 1980s, Americans had become much more conscious of pollution. The modern environmental movement of the 1970s had brought solid waste and many other forms
of pollution to the attention of grassroots activists. However, in areas such as Boston, ethics
of previous eras guided the management of less noticeable forms of pollution. In the case of
Boston Harbor, sewage treatment plants built in the 1960s continued to release millions of
gallons of sewage into each outgoing tide. These wastes, 25 percent of which was untreated,
it was hoped, could be easily dealt with by the greatest "sink" on Earth: the wide ocean
(bodies of water that naturally break down pollutants are referred to as a sink). Even if the
sink failed to process the wastes, however, the tide would carry the problems away from
the harbor. This laissez-faire system for dealing with pollutants was seen in any coastal
city.

By the 1980s, Boston's use of nature's processing "system" had begun to fail. Many
beaches were closed because of pollution for more than half of every summer. Fish, lobster,
and other inhabitants of the harbor area each showed results of the extended pollution of
the waters. For instance, studies in the mid-1980s showed that harbor fish suffered from the
highest level of liver cancer ever recorded anywhere in the world.

By 1985, two federal agencies, the EPA and the Department of Justice, filed suit against
the Commonwealth of Massachusetts for not complying with the 1972 Clean Water Act.
Meeting these requirements then became the responsibility of the Massachusetts Water
Resource Authority (MWRA). New sewage treatment equipment was a primary part of one
of the largest public works projects ever. Today, the Deer Island Sewage Plant is the second
largest in the United States. Each day, it processes 380 million gallons of sewage. In the past,
the sludge (which are suspended solids, pathogens, and toxic contaminants) would have been

dumped into the harbor. Since 1991, however, the sludge has passed into a recycling plant that turns it into fertilizer.

By 1997, Massachusetts had come into compliance with the twenty-five-year-old Clean Water Act. More innovations came in 2000 when the world's longest sewage tunnel began to dump 350 million gallons of wastewater per day well outside of Boston Harbor. Some critics complained that this wastewater was dumped dangerously near Stellwagon Bank. This fragile ecosystem serves as a refuge for populations of whales, sea birds, turtles, fish, lobster, and scallops, but scientists continue to monitor the safety of the water in the area.

The project to save the Boston Harbor cost approximately $5 billion, but today it is viewed as a success. With increased coastal health, tourism has returned to the rejuvenated Harbor, and the MWRA now spends most of its time above the harbor, monitoring upstream in the Charles, Mystic, and Neponset Rivers. Ironically, much of this improvement was spurred by the inaccurate use of the harbor in political ads.

The 1988 campaign provided a flashpoint to spur action and reflection on the steward-ship of Boston Harbor. More than anything, however, it reflected a growing willingness of American voters to consider environmental issues among their voting considerations.

Sources and Further Reading: Conservation Law Foundation, *Early History of CLF's Fight to Cleanup Boston Harbor 1983–1986*, http://www.clf.org/programs/cases.asp?id=188; Dolin, *Political Waters: The Long, Dirty, Contentious, Incredibly Expensive but Eventually Triumphant History of Boston Harbor—A Unique Environmental Success Story*.

CAN YOU DIG THE BIG DIG?

Time Period: Early twenty-first century
In This Corner: Some residents and politicians, Boston road builders and city planners
In the Other Corner: Traditional-minded residents and politicians, many environmentalists
Other Interested Parties: Other cities considering such a radical plan
General Environmental Issue(s): City planning

One of the grandest urban highway projects ever undertaken in the United States, Boston's "Big Dig" is so new that its success has yet to be fully judged by engineers and historians. Although construction was largely completed by 2006, technological problems, including the fatal accident caused by the fall of one of the massive steel plates used to line the underground and water tunnels, stirred additional criticism of the project just when it should have been permanently altering life in the city. Although everyone agreed that Boston's eighteenth- and nineteenth-century infrastructure created significant traffic and planning challenges, the scale of the Big Dig intimidated engineers and outraged some environmentalists.

The planning of older cities such as Boston is one generation's ideas on top of another's. Each layer represents the accepted ideas of the time period, which might be replaced and outmoded by preferences of the next. More than any other era, urban renewal of the 1950s and 1960s radically altered the fabric of many cities, including Boston. The very concept of "renewal" was organized by planners' trust that they had devised for better ways to solve city problems such as transportation, overcrowding, and pollution. During the mid-twentieth cen-tury, entire sections of existing cities were handed over to planners to execute large-scale renewal efforts.

In Boston, transportation was a primary criticism of the mid-twentieth century. To alleviate traffic problems, planners constructed a massive, elevated six-lane highway, known as the Central Artery, directly through the center of downtown Boston. With an eye toward efficient travel, planners overlooked many things. To clear the way, homes and entire neighborhoods were demolished, displacing an estimated 20,000 Bostonians. In addition to these difficulties, the highway's poor design actually did not increase the efficiency of travel. Critics complained of flaws including no breakdown lanes, sharp curves, and a dangerous abundance of exits and entrances. The most serious criticism, however, concerned what the Central Artery did to its surroundings and nearby residents. Clearly, the highway permanently altered the fabric of the city, creating a forty-foot-high structure that separated Boston's North End from commercial and financial districts downtown. Instead of solving the city's traffic problems, the Central Artery contributed to new ones. Planners soon realized that the Central Artery was going to require renewal or improvement of its own.

Planning for what became known as the Big Dig began in 1969 when Fred Salvucci, the transportation commissioner to the mayor of Boston, set about to solve the city's never-ending traffic woes. He proposed to wave a magic wand and complete the impossible: literally to make Boston's automobile traffic disappear. The magic wand was technology that would allow roadways to be constructed underground. Literally, the traffic would disappear.

Environmental impact reviews of the plan for extensive underground tunnels began around 1983. The technological breakthrough to enable the plan to move forward was slurry wall construction. Based on Italian practices, slurrying allowed engineers to build tunnels without having to first excavate. This was crucial to any construction plan because it would allow Boston's surface life to proceed almost undisturbed while work took place underground.

Construction began in 1991 and lasted for more than a decade. Environmental considerations influenced the tunnel construction as well as other aspects of the process. Although the displacement of so much fill creates some environmental problems, it also helped to solve some others. For instance, from 1993 to 1998, more than 4,400 bargeloads of dirt were taken to Spectacle Island in Boston Harbor. This fill was put on top of an old dump and then used to construct a new public park. Throughout the region, the massive amounts of Boston Blue Clay removed from the tunnel areas (approximately three million cubic yards) have been used to cap and, therefore, close landfills.

Lasting eighteen years and costing $11 billion, the Big Dig is the most complex and expensive highway project ever undertaken in the United States. By 2006, the magic could be glimpsed: much of Boston's auto and truck traffic now passes underground.

Source: Massachusetts Turnpike Authority, *The Big Dig*, http://www.masspike.com/bigdig/background/index.html.

IMMIGRATION ALONG THE UNITED STATES–MEXICO BORDER

Time Period: Throughout history, but particularly since the nineteenth century
In This Corner: Immigrants, human rights activists
In the Other Corner: Nation states
Other Interested Parties: Retirees, conservationists
General Environmental Issue(s): Human rights, habitat destruction

Immigration is the movement of people from one country to another. This issue is not only social but also political. It is directly related to the concept of nation state and the idea of a homeland country for individuals and thus has existed throughout history, but particularly since the creation of many nation states in the nineteenth century. Most immigrants move from low- and lower-middle-income countries to high-income developed countries, often in an attempt to escape poverty or to find a job or gain educational opportunities for themselves or their children. Natural disasters and civil unrest and/or persecution are also reasons for immigration. Occasionally, people immigrate in an attempt to find a more suiting climate or less expensive place to retire. Overall, people immigrate in a search for a better life. According to the U.N., in 2005, about 3 percent of the global population, or 190 million people, immigrated.

Within the post–9-11 United States, immigration has drawn much attention as one of the most controversial issues in the country. This is the most recent reexamination of an issue that throughout history has proven to draw great support and criticism.

The United States is an immigrant country, accepting, at different times in history, people from around the world. Since World War II, the United States has taken in more refugees than any other country, and, as of 2006, the United States accepted more legal immigrants as permanent residents than any other country. By 2006, the number of immigrants, legal and illegal, in the United States reached 37.5 million (Ohlemacher 2007). Mexico is the leading source of this immigration, both legal and illegal. The U.S. Department of Labor has reported that fifty-three of all agricultural workers in the United States are undocumented immigrants; growers and labor unions say the number of younger field hands who are undocumented is closer to 70 percent (Preston 2007).

Arguments in support of immigration generally focus on economic and human rights issues. Economically, the argument usually relates to labor supply and productivity. According to agricultural producers, medium and small businesses, and free marketeers, agriculture is just one of the areas in which jobs will not be filled if illegal immigrants are not available; additionally, a free global labor market, including free immigration, would improve global productivity. Other supporters of immigration focus on human rights issues, noting that the efforts to stop illegal immigration have only pushed the illegal immigrants into ecologically sensitive, dangerous regions of the United States–Mexico border (Massey 2001). The result is an increasing rate of mortality of illegal immigrants. These deaths are from exposure, drowning, killings by vigilantes or "coyotes," and accidents involving the U.S. Border Patrol. Many of these supporters also suggest that people from poor countries should be allowed to enter rich countries to improve their standard of living.

Opponents of immigration also use economic and human rights arguments. The economic arguments include the idea of economic nationalism, the idea that a country's jobs should go to citizens of that country. Additionally, some argue that immigration is a form of corporate welfare, in which corporations benefit and taxpayers end up with the bill for increased services in their communities. Finally, a common argument is that the immigrants are paid wages that are lower than the federally mandated minimum wage, decreasing wages for citizens.

The human rights arguments include the suggestion that some immigrants have brought diseases such as tuberculosis and hepatitis into areas of low incidence. Also, opponents argue that there are not enough school or other social services for citizens and that is it not acceptable for immigrants to not assimilate into the culture of the United States.

Environmental arguments about immigration can be found supporting both sides of the debate. When looking at groups such as the Sierra Club, the controversy over the topic is apparent. Some environmentalists believe population increases in the United States, which is greatly boosted by immigration, are environmentally unsustainable. Others see overpopulation as a global issue and something that needs to be addressed by all countries. Finally, both groups mention the issue of illegal immigrants crossing through the Sonoran and Chihuahuan deserts, environmentally sensitive areas. Those opposed to immigration focus on the destruction created by the crossers. Those in support of immigration suggest that, if the immigrants were allowed to cross legally, the environment would not sustain some level of destruction.

Currently, the discussion surrounding this issue focuses on President George W. Bush's proposal to reinstate the Guest Worker Program or to expand the H-2B program to fill the areas where labor is needed. Although in June 2007 the U.S. Senate failed to bring the immigration bill to cloture and toward passage, supporters of immigration, including President George W. Bush, do not believe the issue is closed. Concerns that the price of failure will be hundreds of more people dying in the desert, increasing shortages of workers as workplace raids and deportation takes those working in many agriculture and service sector jobs, and environmental damage caused by transporting illegal workers into the United States, promises the issues will yet again be reexamined in the near future.

Sources and Further Reading: Cornelius, *Death at the Border: The Efficacy and the "Unintended" Consequences of U.S. Immigration Control Policy 1993–2000*; CNN, "Senate Immigration Bill Suffers Crushing Defeat"; Gelletly, *Mexican Immigration: The Changing Face of North America*; Legrain, *Immigrants: Your Country Needs Them*; Massey, *Smoke and Mirrors: U.S. Immigration Policy in the Age of Globalization*; Meilander, *Towards a Theory of Immigration*; Preston, "U.S. Farmers Go Where Workers Are: Mexico"; U.S. GAO, "Illegal Immigration: Border-Crossing Deaths Have Doubled Since 1995"; Zuniga and Hernandez-Leon, eds., *New Destinations: Mexican Immigration in the United States.*

MOUNTAINTOP REMOVAL IN APPALACHIA

Time Period: 1970s to the present
In This Corner: Coal mining interests, many politicians, many miners
In the Other Corner: Local residents, some state governments, environmentalists
Other Interested Parties: Health officials
General Environmental Issue(s): Mining, coal, industrial waste

In recent decades, coal companies have matched their extraction practices with the growing consumption of energy by American consumers. In the United States, one hundred tons of coal are extracted every two seconds. Around 70 percent of that coal comes from strip mines, which tear away the surface to access layers of the earth's crust, and, over the past twenty years, an increasing amount of this coal comes from sites in which entire mountains have been lowered. These are the coal mines known as mountaintop-removal sites.

Erik Reece wrote, "Not since the glaciers pushed toward these ridgelines a million years ago have the Appalachian Mountains been as threatened as they are today. But the coal-extraction process decimating this landscape, known as mountaintop removal, has generated

Mountains near Kayford, West Virginia, seen in January 2000, show how mountaintop removal mining has flattened many mountain peaks throughout Appalachia. AP Photo/Bob Bird, File.

little press beyond the region." When viewed from the air, this mining technique is described by Reece as "ecological violence." He describes the landscape this way:

> Near Pine Mountain, Kentucky, you'd see an unfolding series of staggered green hills quickly give way to a wide expanse of gray plateaus pocked with dark craters and huge black ponds filled with a toxic byproduct called coal slurry. The desolation stretches like a long scar up the Kentucky-Virginia line, before eating its way across southern West Virginia. Central Appalachia provides much of the country's coal, second only to Wyoming's Powder River Basin.
>
> After decades of prosperity from harvesting coal, much of Appalachia is now fed up with the impacts of modern-day mining. They are protesting the industry on which it relied for a generation. (Reece 2006)

Coal's Legacy on the Land

When coal became an important energy source for American industry over a century ago, concern for the environment was not at the forefront of public attention. For years, smokestacks from electrical and industrial plants emitted pollution into the air but were viewed as a sign of economic progress. In remote areas with very little economic opportunity, coal offered the entire community a livelihood.

Early on, all coal mining required tunneling into the earth to extract the strips of coal. Although inefficient, cheap labor made this mining method cost effective for more than a century. As new technologies became available, large companies that owned far-flung sites began using machines that could transform the coal-bearing mountains into a massive quarry. Not only is mountaintop mining less labor intensive than underground mining, it is also more efficient and profitable than the older form of surface mining: strip mining, in which a mountainside was gradually pealed away. Over the past few decades, the new practice is thought to have impacted more than 400,000 acres in this four-state Appalachian region, including more than 1,200 miles of streambeds.

West Virginia is one of the nation's most productive coal landscapes. Since the evolution of mountaintop removal, more and more of West Virginia's total production of coal, some 154 million tons in 2004, has come with the ancillary costs of this technique. Relative to western coal (Wyoming is the nation's top coal producer), second-ranked West Virginia's low-sulfur bituminous burns with a cleaner, hotter efficiency in the electric power plants of America. This, of course, means that it also experiences more of the hazardous fallout. In addition, it means that the industrial transitions are particularly problematic there.

For instance, in 1948, some 125,000 men worked in the mines of West Virginia. By 2005, there were fewer than 19,000. Most of them were employed in remaining underground mines and were responsible for a small fraction of the state's production. In addition, residents have begun to realize that mountaintop removal has permanently altered the state's ecology.

The Process of Mountaintop Removal

Mountaintop removal is currently being used to mine for coal in West Virginia, Virginia, Kentucky, and Tennessee. In the case of mountaintop removal for coal, the name of the process creates a very clear vision of how the work actually takes place. Computer modeling works with seismographic sensors to create a portrait of the geology beneath the surface of the area. This image guides the placement of the new mining efforts and even allows companies to gauge the size and potential of each layer of coal.

Once a site has been selected, as a first step, hundreds of feet of forest, topsoil, and sandstone (each referred to as "overburden") are unearthed. Thus, the process is initiated with bulldozers and front-end loaders clear-cutting the land and removing the topsoil. Clearing the surface means that the miners can more easily access the thin seams of rich bituminous coal that stretch in horizontal layers throughout these mountains. Almost everything that is not coal is pushed down into the valleys below. "Valley fills" must be approved by the EPA. In central Appalachia between 1985 and 2001, 6,700 valley fills were approved. The U.S. EPA estimates that more than 700 miles of healthy streams have been completely buried by mountaintop removal and thousands more have been damaged.

Advocates of mountaintop removal point to the amount of energy demanded by the United States. The process of mining coal is comparatively inexpensive and relies on domestic resources rather than foreign reserves of resources such as petroleum. In addition, supporters note that, once the coal is removed, the companies restore the mined areas, planting trees and grasses. Furthermore, the completed reseeding has opened the landscape, creating meadows that are beneficial to game animals. Finally, advocates of mountaintop removal note that this coal-mining procedure is safer than sending miners into underground mines.

Once the surface habitat is scraped away, however, explosives are used to break open the rock strata holding the coal. Once the vein is opened, a dragline digs into the rock to expose the coal. These machines can weigh up to eight million pounds with a base as big as a gymnasium and as tall as a twenty-story building. These machines allow coal companies to hire fewer workers. A small crew can tear apart a mountain in less than a year, working night and day. Some of the largest machines on earth, then, are used to scoop out the layers of coal, dumping millions of tons of "overburden," the former mountaintops, into the narrow adjacent valleys, thereby creating valley fills.

After extracting the coal, companies are supposed to reclaim land, but all too often mine sites are left stripped and bare. Even where attempts to replant vegetation have been made, the mountain is never again returned to its previous elevation nor appearance. This change in the landscape, argue critics of the process, is merely the most obvious visual record of other long-term implications.

"What the coal companies are doing to us and our mountains," commented one observer, "is the best kept dirty little secret in America." In recent years, however, critics have made certain that the impacts of such mining practices will no longer remain a secret. They are openly decrying the environmental and social fallout of what they call "strip mining on steroids."

Wendell Berry Comments on Mining's Impact on the Appalachian Landscape

Throughout his writing career, essayist Wendell Berry found inspiration in the rural Kentucky environment in which he grew up and lived. He became an outspoken critic of strip mining and other practices that damaged the natural environment to extract the desired coal. His calls for action in the 1970s helped to fuel the outcry for policies regulating coal industry practices and demanding remediation once coal had been extracted. This passage comes from "Mayhem in the Industrial Paradise," an essay contained in *A Continuous Harmony* (1972):

> I have just spent two days flying over the coal fields of both eastern and western Kentucky, looking at the works of the strip miners.... In scale and desolation—and, I am afraid, in duration—this industrial vandalism can be compared only with the desert badlands of the West. The damage has no human scale. It is a geologic upheaval. In some eastern Kentucky counties, for mile after mile after mile, the land has been literally hacked to pieces. Whole mountain tops have been torn off and cast into the valleys. And the ruin of human life and the possibility is commensurate with the ruin of the land. It is a scene from the Book of Revelation. It is a domestic Vietnam...."
> (Berry 1972)

Long-Term Impacts for the Region

Critics of the mountaintop removal/valley fill process cite a number of concerns: environmental, social, and economic affects. Environmental impacts from mountaintop removal include destruction of parts of the large contiguous temperate forests, destruction of native plants and of wildlife habitat, and vegetation loss, which leads to landslides. Although mining companies do revegetate the mining sites, activists claim that the companies plant non-native grasses rather then native tree species. Environmentalists and activists also believe some mountaintop removal mining violates the Clean Water Act by discharging pollutants into streams.

Critics also argue that mountaintop removal affects communities that neighbor the mining regions. Among the concerns are that dynamite blasts crack foundations and walls of houses and other structures, and the mining itself dries up wells and contaminates water, leaving communities with questions about the purity of their drinking water. Critics often cite a well-known case of water contamination that occurred near Buffalo Creek, West Virginia, in 1972. An earthen dam holding back contaminated water from coal washing broke when heavy rains fell in the region. More than 132 million gallons of black wastewater flowed down the valley and into the town of Buffalo Creek, killing 125 people and injuring another 1,100. Regulations and procedures have improved since the 1970s, but critics of coal mining suggest that the possibility for future problems with water contamination still exists.

Finally, critics also discuss economic impacts. Mountaintop mining uses fewer miners to extract the mineral than would the traditional coal mining procedures. This is mainly attributable to the fact that explosives and large machinery reduce the need for miners. The result is a limited number of job possibilities for people living in traditional mining communities. The UMW representatives suggest that, with fewer miners, the mining companies are able to lessen the impact of the union and call for increasing legal measures to protect both the miners and communities effected by mountaintop removal. The second negative economic impact of the mountaintop removal/valley fill procedures comes from an impact to waterways. Waterways are often one of the Appalachian states' greatest tourist attractions. Mountaintop removal can contaminate the waterways, and the valley fills can bury streams and carry sediment into previously clear fishing streams, all of which negatively impact tourism in the region.

The main environmental problems associated with land disturbance include erosion of soil, dust pollution, and impact on biodiversity. Today, minimizing these impacts and rehabilitating land are legal requirements that are placed on mining companies and enforced by federal regulation; however, for much of coal mining's history, no such requirements existed. Therefore, a great deal of the current work in former coal mining areas focuses on fixing the impacts of abandoned mines or alerting companies and regulators about what has to be done.

Overburden is the industrial term for the earth that must be moved to clear the way for mining. In the past, the overburden removed was usually dumped into low-lying areas, often filling wetlands or other water sources. In addition, the mineral pyrite, which is commonly found in rocks containing coal seams, was often exposed or relocated to clear the way for mining. The exposed pyrite results in the formation of sulfuric acid and iron hydroxide. When rainwater washes over these rocks, the water runoff becomes acidified, which can affect local soils and rivers and streams. This phenomenon is called acid mine drainage. Similar to black lung, this impact of mining has only been realized years after mining has taken place.

Activists, such as an organization known as Mountain Justice, work hard to expose the injustice spreading across the coal fields of Appalachia. Their primary emphasis is the social costs of these new technologies that require fewer workers. In addition, these activists strain to show the public the efforts of coal companies to circumvent or even break existing laws.

One emphasis of these actions is to try to improve the living conditions for residents of the coal regions. For generations, families of coal miners have borne the brunt of the pollution created by various phases of coal's acquisition. For instance, coal washing often results in thousands of gallons of contaminated water that looks like black sludge and contains toxic chemicals and heavy metals. The sludge, or slurry, is often contained behind earthen dams in huge sludge ponds. In Appalachia today, it is estimated that there are 500 similar impoundment pools. Variously referred to as slurry ponds, sludge lagoons, or waste basins, they impound hundreds of billions of gallons of toxic black water and sticky black goo, byproducts of cleaning coal, mostly from underground mines but also from surface mines.

One of the impoundments broke on February 26, 1972 above the community of Buffalo Creek in southern West Virginia. Records showed that the Pittston Coal Company had been warned that the dam was dangerous but had chosen to do nothing. Heavy rain caused the pond to fill up and it breached the dam, sending a wall of black water into the valley below. More than 132 million gallons of black wastewater raged through the valley, killing approximately 125 people and creating millions of dollars worth of damage (Reese 2006).

Even if such tragic failure does not occur, the slurry ponds are simply the most obvious example of the toxic residue that mining leaves behind for members of these communities. Mountaintop removal generates huge amounts of waste. Whereas the solid waste becomes valley fills, liquid waste is stored in massive, dangerous coal slurry impoundments, often built in the headwaters of a watershed. Not only the human residents suffer. The U.S. Fish and Wildlife Service has written that mountaintop removal's destruction of West Virginia's vast contiguous forests destroys key nesting habitat for neo-tropical migrant bird populations and thereby decreases the migratory bird populations throughout the northeast United States.

Forest loss and water despoliation are two of the primary impacts of mountaintop removal. For decades, environmentalists and legislators have attempted to construct mechanisms that would encourage coal companies to begin to consider the long-term impacts of this mining technique that offers such considerable short-term gains. Although most legislation has shied away from telling coal companies how they can or cannot mine, federal regulations have placed severe expectations on how the land must be left after mining is completed.

Regulating Reclamation

Mountaintop removal/valley fill were first used as a mining practice in the 1960s. The success of obtaining the coal from this process, along with the United States energy crisis, led to an increase in the use of this form of mining in the 1970s. Mountaintop removal was further expanded in the 1990s to meet the increasing demand for low-sulfur coal, an event that occurred as the Clean Air Act tightened emissions limits on high-sulfur coal processing. By the mid-1990s, mountaintop removal had became the dominant coal-mining technique; it is practiced in West Virginia, Kentucky, Tennessee, and Virginia.

Among the actions taken within the United States to help mitigate effects of the mountaintop removal/valley fill procedures was the Surface Mining Control and Reclamation Act (SMCRA) of 1977. The objectives of this act were to prevent water contamination and household damage attributable to explosions associated with forms of strip mining and to ensure land restoration. The act does require mining companies to restore land to its previous use and contours, but waivers are often given. At the time the SMCRA was written, some West Virginians attempted to get mountaintop removal prohibited as a form of surface mining. Their argument was that it would destroy their landscape. Despite their efforts, the government did not prohibit mountaintop removal. Since then, help has come through the U.S. Environmental Protection Agency and U.S. Fish and Wildlife Service cooperating with the State of West Virginia to minimize adverse environmental effects. EISs are prepared to help achieve this goal, and recently, the majority of work in opposition of mountaintop removal/valley fill has come from local communities and NGOs.

In 1977, the SMCRA was passed to help former mining areas begin to confront this daunting problem. The act requires that mining sites be restored to their original contours, particularly the sites of surface mining that have seen radical alteration. The legislation also requires that a mining operator submit a plan for restoring the land and for mitigating acid mine drainage before being granted a permit to begin mining operations. The law also required the formation of a fund that could be drawn from to finance the restoration of old abandoned mines by imposing a tax on current coal production.

Here is an excerpt from the text:

Surface Effects of Underground Coal Mining Operations

[30 U.S.C. 1266]

SEC. 516. (a) The Secretary shall promulgate rules and regulations directed toward the surface effects of underground coal mining operations, embodying the following requirements and in accordance with the procedures established under section 501 of this Act: Provided, however, That in adopting any rules and regulations the Secretary shall consider the distinct difference between surface coal mining and underground coal mining. Such rules and regulations shall not conflict with nor supersede any provision of the Federal Coal Mine Health and Safety Act of 1969 nor any regulation issued pursuant thereto, and shall not be promulgated until the Secretary has obtained the written concurrence of the head of the department which administers such Act.

(b) Each permit issued under any approved State or Federal program pursuant to this Act and relating to underground coal mining shall require the operator to—

(1) adopt measures consistent with known technology to prevent subsidence causing material damage to the extent technologically and economically feasible, maximize mine stability, and maintain the value and reasonably foreseeable use of such surface lands, except in those instances where the mining technology used requires planned subsidence in a predictable and controlled manner: Provided, That nothing in this subsection shall be construed to prohibit the standard method of room and pillar mining;

(2) seal all portals, entryways, drifts, shafts, or other openings between the surface and underground mine working when no longer needed for the conduct of the mining operations;

(3) fill or seal exploratory holes no longer necessary for mining, maximizing to the extent technologically and economically feasible return of mine and processing waste, tailings, and any other waste incident to the mining operation, to the mine workings or excavations;

(4) with respect to surface disposal of mine wastes, tailings, coal processing wastes, and other wastes in areas other than the mine workings or excavations, stabilize all waste piles created by the permittee from current operations through construction in compacted layers including the use of incombustible and impervious materials if necessary and assure that the leachate will not degrade below water quality standards established pursuant to applicable Federal and State law surface or ground waters and that the final contour of the waste accumulation will be compatible with natural surroundings and that the site is stabilized and revegetated according to the provisions of this section;

(5) design, locate, construct, operate, maintain, enlarge, modify, and remove, or abandon, in accordance with the standards and criteria developed pursuant to section 515(f), all existing and new coal mine waste piles consisting of mine wastes, tailings, coal processing wastes, or other liquid and solid wastes and used either temporarily or permanently as dams or embankments;

(6) establish on regarded areas and all other lands affected, a diverse and permanent vegetative cover capable of self-regeneration and plant succession and at least equal in extent of cover to the natural vegetation of the area;

(7) protect offsite areas from damages which may result from such mining operations;

(8) eliminate fire hazards and otherwise eliminate conditions which constitute a hazard to health and safety of the public;

(9) minimize the disturbances of the prevailing hydrologic balance at the mine site and in associated offsite areas and to the quantity of water in surface ground water systems both during and after coal mining operations and during reclamation by—

    (A) avoiding acid or other toxic mine drainage by such measures as, but not limited to—

        i. preventing or removing water from contact with toxic producing deposits;

        ii. treating drainage to reduce toxic content which adversely affects downstream water upon being released to water courses;

        iii. casing, sealing, or otherwise managing boreholes, shafts, and wells to keep acid or other toxic drainage from entering ground and surface waters; and

    (B) conducting surface coal mining operations so as to prevent, to the extent possible using the best technology currently available, additional contributions of suspended solids to streamflow or runoff outside the permit area (but in no event shall such contributions be in excess of requirements set by applicable State or Federal law), and avoiding channel deepening or enlargement in operations requiring the discharge of water from mines;

(10) with respect to other surface impacts not specified in this subsection including the construction of new roads or the improvement or use of existing roads to gain access to the site of such activities and for haulage, repair areas, storage areas, processing areas, shipping areas, and other areas upon which are sited structures, facilities, or other property or materials on the surface, resulting from or incident to such activities, operate in accordance with the standards established under section 515 of this title for such effects which result from surface coal mining operations: Provided, That the Secretary shall make such modifications in the requirements imposed by this subparagraph as are necessary to accommodate the distinct difference between surface and underground coal mining;

(11) to the extent possible using the best technology currently available, minimize disturbances and adverse impacts of the operation on fish, wildlife, and related environmental values, and achieve enhancement of such resources where practicable;

(12) locate openings for all new drift mines working acid-producing or iron-producing coal seams in such a manner as to prevent a gravity discharge of water from the mine.

(c) In Order to protect the stability of the land, the regulatory authority shall suspend underground coal mining under urbanized areas, cities, towns, and communities and adjacent to industrial or commercial buildings, major impoundments, or permanent streams if he finds imminent danger to inhabitants of the urbanized areas, cities, towns, and communities.

(d) The provisions of title V of this Act relating to State and Federal programs, permits, bonds, inspections and enforcement, public review, and administrative and judicial review shall be applicable to surface operations and surface impacts incident to an underground coal mine with such modifications to the permit application requirements, permit approval or denial procedures, and bond requirements as are necessary to accommodate the distinct difference between surface and underground coal mining. The Secretary shall promulgate such modifications in accordance with the rulemaking procedure established in section 501 of this Act.

In addition to the SMCRA, efforts to regulate this style of mining are tied to the Clean Water Act. Environmentalists contend that the U.S. Office of Surface Mining should enforce a buffer-zone rule prohibiting any mining activity within one hundred feet of a stream. Under the Clean Water Act, the agency charged with regulating the streambed was the U.S. Army Corps of Engineers. A lack of enforcement has led to many court cases in Appalachia in recent years, and concerns don't stop with active mines.

The SMRCA legislation did not intend to allow coal companies to walk away from their surface mines and leave them denuded. Stripped mountainsides, the law declared, must be restored to their "approximate original contour" and stabilized with grasses and shrubs and, if possible, trees. Therefore, the legislation did not order companies to attempt to perform the expensive work of putting the mountain together again. Most often, companies were allowed to turn the land over to the state for public use as pasture or as fish and wildlife habitat.

Abandoned mines and their environmental problems can be found in over twenty-nine states and tribal lands. The U.S. General Accounting Office estimates that there are 560,000 abandoned mines on federal lands in the United States. In the state of Pennsylvania alone, 5,600 of the 9,000 abandoned mines are considered threats to human health or to environmental quality. More than $1.5 billion has been spent since 1977 on restoring abandoned mines; two-thirds of the funds have been spent in four states: Pennsylvania, Kentucky, West Virginia, and Wyoming (Opie 1998).

In the case of acid mine drainage, the acid runoff can dissolve heavy metals such as copper, lead, and mercury that then leach into ground and surface water. Environmental effects of acid mine drainage include contamination of drinking water and disruption of growth and reproduction in aquatic plants and animals. Treating the problem is time-consuming and expensive (Montrie 2003).

Conclusion: Mountaintop Removal in the Twenty-First Century

In recent years, some federal courts have ruled that the practice of valley fills violates the Clean Water Act. For the most part, the Bush administration has interpreted the law to allow coal companies the opportunity to conduct valley filling and other practices. In August

2007, the Bush administration passed a rule that allowed mine operators to expand the practice as long as they attempted to limit debris and made efforts to cause the least possible environmental harm. "This is a parting gift to the coal industry from this administration," said Joe Lovett, executive director of the Appalachian Center for the Economy and the Environment. "What is at stake is the future of Appalachia. This is an attempt to make legal what has long been illegal."

Mountaintop removal continues today despite environmental, social, and economic concerns. Demand for electricity in the United States has increased by 70 percent in the past twenty years, and coal-fired power plants are a major source of electricity in the United States today, statistics that will likely support the continuation of mountaintop removal/valley fill for years to come.

Sources and Further Reading: Banerjee, "Tanking on a Coal Mining Practice as a Matter of Faith"; Broder, "Rule to Expand Mountaintop Coal Mining"; Gardner and Sainato, "Mountaintop Mining and Sustainable Development in Appalachia"; Loeb, *Moving Mountains: How One Woman and Her Community Won Justice from Big Coal*; Miller, *Coal Energy Systems*; Reese, E *Lost Mountain: A Year in the Vanishing Wilderness: Radical Strip Mining and the Devastation of Appalachia*.

THE WESTERN UNITED STATES, URBAN GROWTH, AND THE NEW WEST

Time Period: 1800s to the present
In This Corner: Ranching and natural resource extraction
In the Other Corner: Growth and development
Other Interested Parties: Environmentalism and tourism
General Environmental Issue(s): Urban sprawl, open space, and habitat protection

As the United States grows, one region that is feeling the push is the West, the region today called the "New West." This region, once renown for ranching and agriculture, is now watching as the former agricultural land is developed into housing and commercial areas. The change is impacting both the natural and cultural environments of the region.

The New West is generally described as Montana, Wyoming, Colorado, New Mexico, Arizona, Utah, Idaho, Nevada, Oregon, and parts of California and Washington. This region was once considered a land of cowboys and Indians, wide-open majestic space, mountains and deserts. With these images came an aura, if not a culture, of individualism, opportunity, and small-town life.

Over time, the activities of the West changed. The region went from the Wild West to communities supported by ranching and natural resources extraction, mainly logging and mining. In addition, large portions of the western landscape were acquired by the federal government and are managed by the NPS, the U.S. Forest Service, the U.S. BLM, and the U.S. Fish and Wildlife Service. The federal government set aside these lands for a variety of reasons. Some of the lands are considered national treasures (Yellowstone National Park and The Canyonlands) and are protected for people's enjoyment, others were known to be important to watersheds (Roosevelt National Forest and Comanche National Grasslands) and are managed to maintain clean water, and yet others were seen as "wastelands" (many of the U.S. BLM areas) and are used for their natural resources and scenic beauty. In each of these cases, the federal government believed it was the best steward to manage the land. In addition, the U.S. Department of Defense owns and uses vast areas within the West.

Today there is another transition taking place. The stereotypes of open spaces, small-town life, and a culture of independence no longer hold as true as they once did; that is a picture of the "Old West." Over the past forty years, changes have displaced many of the traditional elements of the West. This is attributable, in large part, to growth and development: growing urban centers, increasing job possibilities, and developments on former ranch lands. The new picture is a landscape characterized by a patchwork of land uses and conflict over the use of resources.

This new image of the region, both the good and the bad, is what many refer to as the New West. The New West is a place of expanding tourism and recreation activities, adding to the amenities that are attracting migrants faster than any other region of the United States, high-tech industry, creating jobs for the new migrants, and shopping malls and cookie-cutter houses in constantly expanding neighborhoods (Robb and Riebsame 1997). However, between the urban centers of Portland, Denver, Phoenix, Albuquerque, Las Vegas, and Salt Lake City, and alongside the tourist destinations of Aspen, Jackson Hole, Moab, and Lake Powell, are vast areas that are still thinly populated and that maintain their traditional culture and lifestyles.

The two diverse groups attempting to live together in the New West are those who are creating the New West and those who are trying to maintain elements of the traditional West. Among the conflicts arising from this situation are concerns about urban sprawl, questions of land use, and problems with water scarcity.

As America's most rapidly growing area, sprawl is not only noticeable but, in many locations of the New West, is the most notable element of the region. The paradox of people wanting to live in a wide open area and then moving into neighborhoods that are quickly

Houses sit one on top of the other in a subdivision in the east Denver suburb of Aurora, Colorado, as the mountains rise in the background, May 2001. State lawmakers continue to discuss bills that would take a more directed approach to future growth in a state that is struggling to deal with a population explosion in the past decade. AP Photo/David Zalubowski.

consuming the wildness of the area is characteristic of sprawl. It also creates tension among long-time residents who traditionally have worked with the land and the wealthier new residents who are buying up the once open land. It also brings conflict among developers and conservationists. The developers note the economic benefits that come with development, including some sprawl, and the conservationists reference the loss of open space and habitat that comes with development. The ongoing fight against unregulated development and sprawl is uniting two traditional opponents: ranchers and environmentalists. In this case, both parties are working to save the open spaces of the West. The groups understand that development is inevitable; what they want to see is smart growth, regulated development that preserves the value of the West, including habitat and open space, in addition to the traditional cultures of ranching.

The question of land use is another area of conflict in the New West. Unlike past "boom" periods in the West, the current population explosion is not dependent on the extraction of one resource such as oil or gold but rather on a sustained passion for living in the specific landscape. Thus, the citizens of the West are no longer focused on similar lifestyles and type of land use, creating a situation in which many newcomers do not blend into the traditional culture. Although they want the beauty of the open west, they want it without the smell of cattle or the inconvenience of having to drive behind a tractor. This, at times, creates tension between the two groups of people. Many ranchers find that their new neighbors do not appreciate living downwind from a cattle operation and find that, as the land around them is developed, they can no longer afford to pay their property taxes. This creates a situation in which the ranchers feel pressure to sell their operations, often to developers who will pay well but who in many ways represent the change from the Old West they knew to New West.

A final area of conflict in the New West concerns the use of water. Much of the West is considered desert or high plains desert, areas with minimal water resources. With increasing population, municipalities are demanding more water for human use. To add to the pressure, a western drought that began in 1999 has continued after a respite of a couple of wet years. These conditions have pitted urban water needs, including human consumption, landscape and golf course irrigation, and industrial uses, against rural water needs, mainly irrigation. Further complicating these debates is the way water is allocated in the West. Unlike the eastern United States, where riparian water rights are used, in the western United States previous appropriation water rights are used. Thus, water is allocated under the "first in time, first in line" idea rather than as a resource related to land ownership. This means that many of the "first in time" water rights (once human consumption needs are met) are owned by families who have ranched the region for generations, not the municipalities or developers. This creates tension as municipalities, along with developers and others, argue that they bring more economic vitality to the region than do the ranches and thus should have access to the water. This type of conflict is ongoing in many regions of the New West: Las Vegas vs. Northern Nevada, Tucson vs. Southern Arizona, Denver vs. Colorado's Western slope.

As the western United States continues to attract newcomers, the region will continue to change. Conflicts will continue as land uses are disputed and limited resources, such as water, are distributed to limited numbers of stakeholders. Yet also as growth continues, more municipalities are beginning to pursue regulated growth patterns, attempting to accurately value their natural and aesthetic resources so they can wisely use what they have.

Sources and Further Reading: Carleson Ringholz, *Paradise Paved: The Challenge of Growth in the New West*; Egan, "New Fight in Old West: Farmers vs. Condo City"; Flores, "The West That Was, and the West That Can Be"; Hine and Faragher, *The American West: A New Interpretive History*; Klinkenborg, *The New Range Wars*; Martin, "'New Urban' Islands Dot the West"; Robb and Riebsame, *Atlas of the New West: Portrait of a Changing Region*; Stiles, *Brave New West: Morphing Moab at the Speed of Greed*; Travis, *New Geographies of the American West: Land Use and the Changing Patterns of Place*; Wiley and Gottlieb, *Empires in the Sun: The Rise of the New American West*.

WORLD NETWORK OF BIOSPHERE RESERVES

Time Period: 1968 to the present
In This Corner: UNESCO, environmental groups, local communities, scientists, indigenous groups
Other Interested Parties: More than one hundred countries around the world, conservationists
General Environmental Issue(s): Sustainable development, habitat protection, capacity building

The origins of the U.N. Man and the Biosphere (MAB) program date back to 1968. It was in this year that the U.N. Educational Scientific and Cultural Organization (UNESCO) organized the first intergovernmental conference to reconcile conservation and use of the world's natural resources. Although the MAB program was not started until 1970, the foundations for the concepts came from this conference: to establish "terrestrial and coastal areas representing the main ecosystems of the planet in which genetic resources would be protected, and where research on ecosystems as well as monitoring and training would could be carried out" (U.N. Educational Scientific and Cultural Organization 2007b). The U.N. and many of its member countries recognized the need to address environmental issues on a binational and/or international scale, with a primary focus on ecological, social, and economic dimensions of biodiversity loss. The support for and growth of this program are substantial. Starting with fourteen programs in December 2006, there were 527 biosphere reserves in 111 countries (U.N. Educational Scientific and Cultural Organization 2007b).

When UNESCO launched the MAB program in 1970, it focused on fourteen project areas covering different ecosystems. The program was substantially modified in 1974 and 1995 to first develop and then strengthen the concept of the biosphere reserve. The concept of biosphere reserves was strengthened when the UNESCO General Conference adopted the Statutory Framework and the Seville Strategy for Biosphere Reserves. Since that time, the work of the MAB has concentrated on combining scientific and political aspects in the World Network of Biosphere Reserves locations. This is done to "reduce biodiversity loss, improve livelihoods, and enhance social, economic, and cultural conditions for environmental sustainability" (UNESCO MAB web page). More specifically, the MAB will have three main lines of action. The first of these is minimizing biodiversity loss "through the use of ecological and biodiversity science in policy and decision-making" (UNESCO web page). This will include interdisciplinary research, capacity building, and creation of centers for integrated ecosystem management among developing nations. The second main line of action will be promoting "environmental sustainability through the World Network of biosphere Reserves" (UNESCO web page). Actions taken to achieve this will include creating linkages between biodiversity conservation and social-economic development around biosphere reserves, working to prevent conflict around the biosphere reserves, sharing knowledge concerning scientific, educational, and decision-making for the biosphere reserves, and ultimately establishing new biosphere reserves and transboundary biosphere reserves. Finally, the third main line of action for the MAB is the enhancing of "linkages between cultural and biological diversity" (UNESCO web page). Actions to achieve this include working to understand local and indigenous knowledge of the areas under study, fostering local-level sustainable use of the biosphere reserves, and recognizing the role of sacred sites and cultural landscapes in the management and sustainability of ecosystems.

Today UNESCO and the MAB program use the World Network of Biosphere Reserves as vehicles for education, research, and participatory decision making. New biosphere reserves that are added are selected on their ability to provide these functions, along with being representative of a major biogeographic region where ecosystems or animal and plant species or variety need to be conserved. The MAB biosphere reserves contain a core protected area, a buffer zone where nonconservation activities are prohibited, and a transition region where some nonconservation practices are permitted. With appropriate zoning and management and the conservation of the inclusive biodiversity, sustainability of ecosystems can be maintained.

Conservation of the MAB reserves depends on support at the local, national, and international level. The management practices bring together many scientists, local officials, local inhabitants, national representatives, and international government and nongovernmental organizations, including the World Ban, the World Conservation Union, Conservation International, and the World Wide Fund for Nature. Coordination of these participants begins with the MAB governing body. This governing body, the Internal Coordinating Council of the Man and the Biosphere Programme, also referred to as the "MAB Council" or the ICC, comprises thirty-four member states. The member states are elected at UNESCO's biennial General Conference. The MAB council delegates authority and work to its Bureau, a group of representatives nominated from each of UNESCO's geopolitical regions. In this manner, the MAB works to include all sectors and regions of members within the decision-making and management processes, and, by doing so, hopes that biosphere reserve projects will be successful.

Since the creation of the MAB in the 1970s, many individuals have benefited from UNESCO's biosphere reserves. Local communities, including indigenous communities and rural societies, have found their water resources protected and a more stable and diverse economic base. Farmers, foresters, and fishermen have gained access to training, scientists have new research sites and security for permanent plots and monitoring activities, and the global citizen has locations to view the planet's main ecosystems in their natural state.

Sources and Further Reading: German MAB National Committee, *Full of Life: UNESCO Biosphere Reserves—Model Regions for Sustainable Development. The German Contribution to the UNESCO Programme Man and the Biosphere (MAB)*; Price, *Mountain Research in Europe: Overview of MAB Research from the Pyrenees to Siberia. United Nations Educational, Scientific and Cultural Organization*; Risser and Cornelison, *Man and the Biosphere*; Soule and Terborgh, *Continental Conservation: Scientific Foundations of Regional Reserve Networks*; Stokes, *Man and the Biosphere: Toward a Coevolutionary Political Economy*; UNESCO, "The World Network of Biosphere Reserves"; UNESCO, "UNESCO's Man and the Biosphere Programme (MAB)," www.unesco.org/mab/mabProg.shtml; UNESCO, "UNESCO 'Man and Biosphere' Programme FAQs," http://www.gov.mb.ca/conservation/wno/status-report/fa-8.19.pdf.

MONTREAL PROTOCOL ON SUBSTANCES THAT DEPLETE THE OZONE LAYER

Time Period: 1973 to the present

In This Corner: One hundred ninety-one countries around the world

In the Other Corner: CFCs and halons

Other Interested Parties: Scientists, environmentalists

General Environmental Issue(s): Depletion of the ozone layer, human and environmental health

In 1973, two chemists at the University of California, Irvine, Frank Sherwood Rowland and Mario Molina, began studying chlorofluorocarbons (CFCs) and their effect on the earth's atmosphere. The work of these chemists sparked increased funding for additional studies in the area. Ultimately, data from the combined CFC and related studies resulted in world leaders developing the Montreal Protocol, an international treaty designed to protect the ozone layer.

The research done by chemists Rowland and Molina found that CFC molecules in the atmosphere were sufficiently stable to travel to the stratosphere, where they were broken down by ultraviolet radiation, and thus released chlorine atoms. The chemists proposed that this process caused the breakdown of large amounts of ozone in the stratosphere. This breakdown is noteworthy because it is the ozone layer that absorbs most of the ultraviolet-B radiation that reaches the earth's atmosphere. Therefore, depleting substantial parts of the ozone layer would allow more ultraviolet-B radiation to reach the surface of the earth and consequently result in changes such as an increase in skin cancer, damage to crops, and harm to marine phytoplankton.

Rowland and Molina published their research in 1974 and shortly thereafter testified before the U.S. House of Representatives explaining their findings and the hypothesized repercussions. These efforts led to greater recognition of the problems associated with CFCs and consequently to an increase in funding for additional work on CFCs and the Earth's atmosphere. In 1976, the U.S. National Academy of Sciences published a report validating the scientific credibility of the ozone depletion hypothesis. Work on the topic continued for the next decade, leading to the discovery of the Antarctic ozone hole in late 1985. After the discovery of the Antarctic ozone hole, governments recognized the need for "stronger measures to reduce the production and consumption of a number of CFCs ... and several Halons" (UNEP 2004a). On September 16, 1987, the Montreal Protocol on Substances that Deplete the Ozone Layer, or the "Montreal Protocol," was adopted.

The Montreal Protocol is an international treaty to protect the ozone layer from continued exposure to a number of CFCs and several halons. Protection comes through phasing out the production of CFCs 11, 12, 113, 114, and 115 and halons 1211, 1301, and 2402. The Montreal Protocol came into force on January 1, 1989 when twenty-nine countries and the European Economic Community ratified the Protocol. As of August 2007, 191 countries have adopted the Montreal Protocol, making it one of the most widely implemented treaties and a remarkable example of international cooperation (UNEP 2007).

Countries that adopt the Montreal Protocol recognize "that the world-wide emissions of certain substances can significantly deplete and otherwise modify the ozone layer in a manner that is likely to result in adverse effect on human health and the environment" (UNEP 2004b). For the signatory countries, the ultimate objective is the elimination of the global emissions of substances that deplete the ozone, as determined by scientific data. To achieve this objective, phase-out schedules were created for the signatory countries to follow. These phase-out schedules were designed so they could be revised through periodic scientific and technological assessments. Consequently, there have been a number of adjustments to accelerate these phase-out schedules.

In an attempt to maintain support from the signatories for the objectives of the protocol, there are a few exceptions to the phasing out of the named CFCs and halons. Exceptions are granted for developed nations but only for "essential uses" in which no acceptable substitutes can be found. In the case of developing nations, there are a number of special provisions that are given. Because of the difficulties developing countries would face in industrial and social realms if they attempted to eliminate the controlled substances, the Montreal Protocol was designed to provide provisions for developing countries. These provisions make it easier for these countries to implement the protocol; help comes in the form of funds to assist with the phasing out of substances determined to deplete the ozone, the purchasing of rights to use new technologies, and the establishment of national Ozone Offices. Donors, namely developed countries that are party to the Montreal Protocol, support these programs through monetary contributions to the Multilateral Fund for the Implementation of the Montreal Protocol.

Since the signing of the Montreal Protocol, the treaty has undergone five amendments: the London Amendment in 1990, the Copenhagen Amendment in 1992, a revision in Vienna in 1995, the Montreal Amendment in 1999, and the Beijing Amendment in 1999. Each of these introduces other control measures and adds new controlled substances to the phase-out lists (UNEP 2004a). The London Amendment added additional halogenated CFCs to the list: carbon tetrachloride and methyl chloroform (UNEP 2004a). Additionally, it established a mechanism to provide financial and technical assistance to developing countries to ease their compliance with the protocol. The Copenhagen Amendment introduced measures to control both the production and consumption of hydrochlorofluorocarbons (HCFCs) and methyl bromide. The Montreal Amendment established a requirement for licensing systems to allow control and monitoring of trade in the protocol's controlled substances; it did not add new substances to the protocol. Finally, the Beijing Amendment initiated control measures for production of HCFCs, restrictions on trade with non-parties for controlled HCFCs, and added bromochloromethane to the list of controlled substances.

Most countries around the world have adopted the Montreal Protocol. This astounding response to the protocol has led to a leveling off or decreasing of the most important ozone-depleting substances within the atmosphere. Unfortunately, halon concentrations have continued to rise, although their rate of increase has slowed. Likewise, the level of HCFCs has also increased, likely, at least in part, as a result of the substitution of CFCs by HCFCs. Despite some drawbacks, the level of compliance has been high and the number of signatories continues to grow. Overall, there is widespread support for the Montreal Protocol, and it is generally considered one of the most successful international environmental agreements in force today.

Sources and Further Reading: Benedict, *Ozone Diplomacy*; Brack, *International Trade and the Montreal Protocol*; Liftin, *Ozone Discourses*; UNEP, Ozone Secretariat, "The 1987 Montreal Protocol on Substances that Deplete the Ozone Layer (as agreed in 1987)," http://ozone.unep.org/

Ratification_status/montreal_protocol.shtml; UNEP, Ozone Secretariat, "Evolution of the Montreal Protocol," http://ozone.unep.org/Ratification_status/index.shtml.

ECOTOURISM

Time Period: 1950s to the present
In This Corner: Environmental organizations
In the Other Corner: Tourism industry, some governments, some environmentalists
Other Interested Parties: Local communities
General Environmental Issues: Sustainability, local socioeconomic concerns

Ecotourism is a broad term with many definitions. However, generally it can be defined as the practice of responsible travel to natural areas that conserves the environment and improves the welfare of local people (Honey 1999). Ideally this travel is low-impact, educational, ecologically and culturally sensitive, and benefits local communities. The emergence of this form of tourism began in Africa in the 1950s with the legalization of hunting (Kamuaro 1996). The creation of African recreational hunting zones initiated the movement to create protected areas and game reserves. Over time, it became clear that these areas were important revenue-earning venues for the countries where the areas were located. This economic incentive further pushed the idea of conserving areas of environmental interest in a wider range of countries. The concept of ecotourism became prevalent in the late 1980s when a change in tourist perceptions, increased environmental awareness, and a desire to explore natural environments coalesced, making it advantageous to preserve areas deemed to contain "natural wonders."

Within the past decade, ecotourism has arguably experienced the fastest growth of all subsectors in the tourism industry, providing both economic and environmental benefits to regions where ecotourism has taken foot. Among the benefits of ecotourism is the inclusion of local communities through informed consent and participation. Increasing the environmental knowledge of locals, along with showing them the possibilities to gain economic benefits while minimizing environmental impacts, contributes to a long-term support of ecotourism programs and thus desire to protect the natural wonders that bring these benefits.

Conversely, ecotourism is not without its critics. One set of critics suggests that the concept of ecotourism is widely misunderstood and thus is used as a marketing tool rather than criteria for responsible tourism. The lack of a clear definition and delineation of what ecotourism is has led to this disconnect between theory and practice. Ideally, ecotourism helps conserve the biological and cultural diversity of a region by protecting and promoting sustainable use of the ecosystem, thus allowing local peoples to share in the socioeconomic benefits while preserving the distinct characteristics that led to the region's biological and cultural development. In practice, it often consists of developing tourism in a natural environment, with detriment to the ecosystem and without efforts to educate tourists to the fragility of the natural world.

Another set of critics points to the fact that ecotourism can cause conflict concerning issues of land-use rights and the failure to deliver community-level benefits, among other social impacts. At the local level, those in power conflict over control of land, resources, and the distribution of the ecotourism profits. Furthermore, often it is foreign investors and corporations that own most ecotourism, providing minimal benefits to local communities

(Higham 2007). The major concern surrounding this arrangement is that, without helping sustain the local community, local peoples may degrade the environment as a means of sustenance. For example, within many Caribbean Islands, the sale of coral trinkets to affluent ecotourists provides a living for local people. However, this means of survival encouraged the development of a destructive market in wildlife souvenirs. Overall, critics of ecotourism argue that the practice is neither environmentally sustainable nor socially beneficial and thus should not be considered a strategy for conservation and development.

A third criticism of ecotourism concerns the idea that ecotourism leads to the commodification of both natural resources and cultures. When ecotourists are attracted to goods and experiences, the value of these commodities can change. For example, pieces of petrified wood are not viewed as a type of "rock" but rather as a collectable when visiting petrified forests. Likewise, images of the local culture and ethnic groups can also be commodified (Higham 2007). The viewing of local culture as a part of the scenery, along with the wildlife, often leads to an increased struggle for cultural survival. This is particularly true if the local groups' traditional interaction with the natural world is simplified to entertain tourists. To further this problem, at times governments seize on these simplified images rather than taking the time to understand the complicated interactions that groups have with the natural world. The government officials then use the simplified "understandings" to make the appearance of including the local culture in environmental policy making, although in reality, the government focus is not on environmental or cultural preservation but rather on the economic gains that go with development of greater ecotourism. Examples of this type of interaction include many situations in which a government will claim to be preserving land for ethnic groups, although in reality, the land is taken from the local people to develop preserves for ecotourism (Kamauro 1996).

A final criticism of ecotourism deals with the increase in the number of people within a region developing ecotourism. With the growth of an ecotourism industry comes an increase in population to construct and then work for and support the ecotourism industries and activities. This increase puts extra pressure on the local environment and infrastructure, often requiring the construction of additional roads, sanitation facilities, and lodges. This requires the use of non-renewable energy sources and the use of already limited local resources. Where these facilities are not built, such as in many African preserves, garbage and the disposal of campsite sewage into rivers pollutes the region. In many of the sensitive natural environments where ecotourism exists, the tourists, despite their claim to be educated and environmentally concerned, rarely understand the ecological consequences of their visits and the effect of their daily activities on the environment.

How is it possible to maintain the benefits yet quell the concerns of the critics? In part, the answer to this question rests with the issue of developing a more comprehensive definition of ecotourism. On the spectrum of tourism activities from conventional tourism to strict ecotourism, there is contention as to where "ecotourism" starts; at which limit of biodiversity preservation, local socioeconomic benefits, and environmental impact does "ecotourism" begin? This confusion has led environmentalists, special interest groups, and governments to define ecotourism differently. Environmental groups generally suggest that ecotourism includes a sustainably managed, conservation- and education-directed nature-based experience. As a part of this, environmentalists have argued for global standards of accreditation, differentiating ecotourism based on the level of environmental and cultural sustainability and commitment. Conversely, the tourism industry and many governments perceive ecotourism more widely as tourism based in nature. They would like to maintain the economic benefit, which often means either limiting the controls on what can be classified as ecotourism or using terms that sound like ecotourism, terms such as "green tourism" and "bio-tourism," but which do not hold to the same standards. For many supporters of regulating ecotourism, these terms are greenwashing at best and are often environmentally destructive, economically exploitive, and culturally insensitive at worst (McLaren 1998).

Both sides of the ecotourism issue are aware of the need to protect the resources, at least to the point where tourism can be maintained. Thus, in an attempt to achieve this goal, governments and environmentalists have begun working together to set standards for regulating ecotourism. Unfortunately, in many areas, the regulations remain poorly implemented, mislead tourists and manipulate tourists' concerns for the environment, and thus have not worked to create a united, beneficial concept of ecotourism.

Sources and Further Reading: Hardin, "The Tragedy of the Commons"; Higham, *Critical Issues in Ecotourism: Understanding a Complex Tourism Phenomenon*; Honey, *Ecotourism and Sustainable Development: Who Owns Paradise?*; Isaacs, "The Limited Potential of Ecotourism to Contribute to Wildlife Conservation"; Kamauro, "Ecotourism: Suicide or Development?"; McLaren, *Rethinking Tourism and Ecotravel: The Paving of Paradise and What You Can Do to Stop It*; Vivanco, "Ecotourism, Paradise Lost—A Thai Case Study"; Weaver, *Ecotourism*.

ANTARCTICA

Time Period: 1820 to the present
In This Corner: Forty-five signatory countries of the Antarctic Treaty System
In the Other Corner: Resource developers, nations interested in developing new resources
Other Interested Parties: Penguins and seals
General Environmental Issue(s): Scientific research, ecosystem sustainability

The continent of Antarctica is largely south of the Antarctic Circle and overlies the South Pole. The continent comprises 14,000,000 square kilometers, making it larger than both Europe and Australia. Nearly 98 percent of Antarctica is covered in ice, and, aside from penguins and seals, there is no permanent population on the continent. However, at any one time, there are anywhere from 400 to 1,000 scientists and researchers living temporarily in research stations on the continent. Antarctica is the only "uninhabited" continent and belongs to no country; rather it is collectively managed by a number of countries.

Antarctica was "discovered" by a Russian expedition led by Mikhail Lazarev and Fabian Gottlieb von Bellingshausen. This group made the first confirmed sighting of Antarctica in January of 1820, shortly before the British Navy sighted the land and a few months before the United States sealer Nathaniel Palmer sighted Antarctica. Despite the three sightings in 1820, the continent remained for the most part untouched because of the continent's hostile environment and isolation for the next 100 years.

By the late 1950s, seven countries had claims on Antarctica: Argentina, Australia, Chile, France, New Zealand, Norway, and the United Kingdom. In addition, two countries, Russia and the United States, reserved the right to make future claims on the continent. In part to deal with the diverse claims to the land, in 1959, Antarctica was declared to be a politically neutral scientific preserve. At this time, twelve countries signed the Antarctic Treaty and related agreements, the "Antarctic Treaty System." The signatories included the United Kingdom, South Africa, Belgium, Japan, the United States, Norway, France, New Zealand, the Soviet Union (and later Russia), Poland, and Argentina. The Antarctic Treaty System not only created the neutral scientific preserve, it also created the position of the Antarctica Treaty Secretariat to help govern Antarctica. It also banned military activities and mineral mining on the continent and promoted scientific research and protection of the region's ecosystem. This agreement was important not only because of its uniting of these diverse countries, but also because it is considered the first arms control agreement established during the Cold War. Today, an additional thirty-three countries, for a total of forty-five, have signed on to the Antarctic Treaty System.

In an attempt to preserve the environment of Antarctica, the United States passed the Antarctic Conservation Act in 1978. This act provides for the conservation and protection of native species on Antarctica. Protection of the ecosystem is provided through restrictions on the transportation of non-native plants and animals to the continent and extraction of any indigenous species from the continent. This act was amended in 1996 to allow for greater leeway for scientists.

Other legal documents that affect the ecosystem on Antarctica include the Convention for the Conservation of Antarctic Marine Living Resources (CCAMLR) of 1980 and the 1991 Protocol on Environmental Protection of the Antarctic Treaty. The first of these, the CCAMLR, mandates that Southern Ocean fisheries consider their effects on the Antarctic ecosystem as a whole. The second restricts development and exploitation of Antarctica.

Despite these international documents supporting Antarctica, there are still a number of areas of ecological concerns, including illegal fishing in the region. Another area of growing concern is tourism. Today, Antarctica sees limited but increasing tourism. Those wanting to preserve Antarctica's ecosystem fear that, with too much growth of tourism, the ecosystem will be negatively affected.

Today, the main activity on the continent of Antarctica is scientific in nature. Each year, twenty-seven countries send scientists to perform scientific experiments at research stations on Antarctica. The studies include research on biology, glaciers, plate tectonics, and the ozone layer among many other issues. Although the region is isolated, the importance of doing research in this area is well established. For example, one of the most popularly recognized studies done in Antarctica was the 1985 British discovery of the existence of a hole in the ozone layer. The additional research on this topic showed that the destruction of the ozone was caused by CFCs. This ultimately let to the creation of the Montreal Protocol of 1989. More recently, Antarctica has become a region for increasing studies concerning global warming.

Antarctica is a symbol of cooperation among nations. It is also an area of rich scientific research benefiting people around the globe.

Sources and Further Reading: Antarctic Conservation Act of 1978, www.scar.org/treaty/; Heacox, "Antarctica. The Last Continent"; Moss and deLeiris, *Natural History of the Antarctic Peninsula*; Rosove, *Let Heroes Speak: Antarctic Explorers, 1772–1922*; Stonehouse, *The Last Continent: Discovering Antarctica*; U.S. Antarctic Program External Panel of the National Science Foundation, *Antarctica—Past and Present*, www.nsf.gov/pubs/1997/antpanel/antpan05.pdf; U.S. Central Intelligence Agency, "Antarctica," https://www.cia.gov/library/publications/the-world-factbook/geos/ay.html.

THE GREEN PARTY IN THE UNITED STATES

Time Period: 1995 to the present
In This Corner: Environmentalists, Green Party proponents
In the Other Corner: Democrats
Other Interested Parties: Republicans
General Environmental Issue(s): Politics

The political impact of environmentalism became very great by the end of the twentieth century. Most often, members of each major American political party might have an environmental commitment. By the 1980s, however, a growing number of Americans continued an international trend and initiated a party organized around environmental and other humanitarian principles: the Green Party. The party continues to have some influence in American politics, which makes many observers argue whether or not such a third party helps the causes it supports.

The U.S. Green Party traces itself to the European Greens, who first organized as an anti-nuclear, pro-peace movement at the height of the Cold War. It was the German Greens, organized by Petra Kelly, who were most specifically influenced by the U.S. environmental movement of the early 1970s.

The U.S. Green Party convened in 1984 in meetings that focused on local elections. By organizing itself on the local level, the Greens first influenced policy and government in towns, cities, and villages. The first Green Party candidate appeared on the ballot in 1986. In 1990, Alaska became the first state to grant the Greens full ballot status. California followed in 1992.

During the 1990s, the Green Party went through internal changes. Some Greens grew impatient with the strategy of radical, slow, long-term organizing. These Greens called for the immediate creation of state Green parties and some for a national Green party. Their calls created internal division, and two distinct factions took shape within the party. The first division believed the idea that the Greens were a social movement that should work up from the grassroots; the second urged the party to pursue national power and to challenge the American two-party system. In 1991, the groups changed their names, respectively, to The Greens and Green Party USA. The Greens would remained an activist party based on dues-paying members, whereas the Green Party USA sought a national stature. The emergence to the national scene came quickly when, in November 1995, well-known activist Ralph Nader initiated the Green Party's first presidential campaign by officially entering the California Green primary. Nader's unconventional campaign aroused Green Party activity in states that up this point had little activity. By election eve, the Greens had placed Ralph Nader on twenty-two ballots nationwide, with another twenty-three states qualifying him as a write-in candidate. In August 1996, state Green parties held their first national Nominating Convention in Los Angeles, California. The party chose a ticket of Nader and Native American Winona LaDuke, who is a well-known activist for causes related to environmental and indigenous women's issues.

The Nader–LaDuke campaign challenged the candidates and platforms of the Democrats and Republicans and forced the parties to discuss issues that were important to the Green Party. The success of the campaign spurred many additional Green political efforts throughout the United States. In the end, the Nader-LaDuke campaign came in fourth place after that of Ross Perot. They gathered more than 700,000 votes, which accounts for approximately 1 percent of the vote nationwide. The single best state performance came in Oregon, where the Green Party earned 4 percent of the vote. Although the Greens focused on local and state politics, some Greens continued to believe in a presidential run. Reagan and Carter each demonstrated the ways that each president possesses the ability to alter the intensity with which federal regulation of environmental factors is carried out. The chief executive's position became to look most attractive for a Green candidate. In the 2000 presidential election, Nader ran for president again as a member of the Green Party with a platform against development, large corporations, and in support of environmental causes. Although 70 percent of Americans call themselves environmentalists, Nader's campaign garnered less than 5 percent of the national vote. However, genuine pockets of such sentiments were seen in states such as Wisconsin, Florida, Oregon, and California. It seems a growing percentage of Americans are willing to entertain radical political change in the best interests of environmental causes. Today, the Green Party is said to be organized around "Ten Key Values": ecological wisdom, grassroots democracy, decentralization, community-based economics, feminism, respect for diversity, personal and global responsibility, and future focus/sustainability.

By forcing politicians to consider their stands on these issues, members of the Green Party have altered the national landscape of American politics. Throughout the nation, however, the Green Party members elected to state and local office have initiated basic changes to the planning and environmental issues that concern Greens.

Source: *Green Politics.*

LYNDON AND LADY BIRD JOHNSON TAKE FEDERAL ACTION AGAINST BLIGHT

Time Period: 1960s
In This Corner: Lady Bird and Lyndon Johnson, landscape designers
In the Other Corner: Engineers, developers
Other Interested Parties: American drivers
General Environmental Issue(s): Planning, scenery, highways, policy

As Americans committed as never before to automobile travel in the 1950s, the nation began a binge of highway building that rivals the greatest construction feats in human history. Marvels of modern engineering, the sweeping new highwayscapes became part of the everyday lives of many Americans. However, they were built as entirely utilitarian structures.

"Ugliness is so grim," Lady Bird Johnson, First Lady to President Lyndon Johnson in the mid-1960s, once said. "A little beauty, something that is lovely, I think, can help create harmony which will lessen tensions." Through her influence, President Johnson applied his view of federal action and big government to the problem of blight.

On February 8, 1965, Johnson issued a call for a White House Conference on Natural Beauty to be chaired by Laurance Rockefeller, a wealthy businessman and philanthropist who took a special interest in matters involving environmental conservation. In his call for the conference to take place, Johnson urged thoughtful action on issues and spaces completely overlooked by most Americans:

> I hope that, at all levels of government, our planners and builders will remember that highway beautification is more than a matter of planting trees and setting aside scenic areas. The roads themselves must reflect, in location and design, increased respect for the natural and social integrity and unity of the landscape and communities through which they pass.

Through the influence of his wife, Johnson was particularly sensitive to the increasing urbanization of the American landscape. The rapid expansion of highways and the evolution of what we now call sprawl radically altered the American scene. Sounding a bit like Rachel Carson in *Silent Spring*, Johnson critiqued such changes as a product of industrial development and uncontrolled waste. Addressing such problems, stated Johnson, required a "new conservation" that went beyond protecting the landscape from destruction. In addition, he urged federal involvement in the restoration of natural beauty as something essential to "the dignity of the man's spirit." On May 25–26, 1965, panels of conservationists, industrialists, government officials, and private citizens met at the request of President Johnson for a conference on the issue. The conference followed Johnson's list of issues to be discussed, including the following: a solution to the problems of automobile junkyards; the possibility of underground installation of utility transmission lines; policies of taxation that would not penalize or discourage conservation and the preservation of beauty; areas in which the federal government could help communities develop their own programs of natural beauty; and the possibility of a tree-planting program.

Conference goers focused on the idea of positive reinforcement with the White House offering awards or citations for the "President's Highway of the Year". In announcing the Beautification Act on October 22, 1965, President Johnson sought to share his inspiration for the bill. He explained as follows:

> … America likes to think of itself as a strong and stalwart and expanding Nation. It identifies itself gladly with the products of its own hands. We frequently point with pride and with confidence to the products of our great free enterprise system— management and labor.

> These are and these should be a source of pride to every American.

> They are certainly the source of American strength. They are truly the fountainhead of American wealth. They are actually a part of America's soul.

> But there is more to America than raw industrial might. And when you go through what I have gone through the last 2 weeks you constantly think of things like that. You no longer get your computers in and try to count your riches.

> There is a part of America which was here long before we arrived, and will be here, if we preserve it, long after we depart: the forests and the flowers, the open prairies and the slope of the hills, the tall mountains, the granite, the limestone, the caliche, the unmarked trails, the winding little streams—well, this is the America that no amount of science or skill can ever recreate or actually ever duplicate.

This America is the source of America's greatness. It is another part of America's soul as well.

When I was growing up, the land itself was life. And when the day seemed particularly harsh and bitter, the land was always there just as nature had left it—wild, rugged, beautiful, and changing, always changing....

Well, in recent years I think America has sadly neglected this part of America's national heritage. We have placed a wall of civilization between us and between the beauty of our land and of our countryside. In our eagerness to expand and to improve, we have relegated nature to a weekend role, and we have banished it from our daily lives.

Well, I think that we are a poorer Nation because of it, and it is something I am not proud of. And it is something I am going to do something about. Because as long as I am your President, by choice of your people, I do not choose to preside over the destiny of this country and to hide from view what God has gladly given it....

Now, this bill does more than control advertising and junkyards along the billions of dollars of highways that the people have built with their money—public money, not private money. It does more than give us the tools just to landscape some of those highways. This bill will bring the wonders of nature back into our daily lives.

This bill will enrich our spirits and restore a small measure of our national greatness.

As I rode the George Washington Memorial Parkway back to the White House only yesterday afternoon, I saw nature at its purest. And I thought of the honor roll of names—a good many of you are sitting here in the front row today—that made this possible.

And as I thought of you who had helped and stood up against private greed for public good, I looked at those dogwoods that had turned red, and the maple trees that were scarlet and gold. In a pattern of brown and yellow, God's finery was at its finest. And not one single foot of it was marred by a single, unsightly, man-made construction or obstruction—no advertising signs, no old, dilapidated trucks, no junkyards. Well, doctors could prescribe no better medicine for me, and that is what I said to my surgeon as we drove along....

And this administration has no desire to punish or to penalize any private industry, or any private company, or any group, or any organization of complex associations in this Nation. But we are not going to allow them to intrude their own specialized private objective on the larger public trust. Beauty belongs to all the people. And so long as I am President, what has been divinely given to nature will not be taken recklessly away by man....

And unless I miss my guess, history will remember on its honor roll those of you whom the camera brings into focus in this room today, who stood up and were counted when that roll was called that said we are going to preserve at least a part of what God gave us. (Johnson)

As Johnson's presidential legacy became one of the most negative in modern history, Lady Bird, who outlived him by decades, emphasized the beautification initiative that had begun with the 1965 conference and act. Sounding much like early planners such as Olmsted, she wrote in a diary entry on January 27, 1965:

> Getting on the subject of beautification is like picking up a tangled skein of wool. All the threads are interwoven—recreation and pollution and mental health, and the crime rate, and rapid transit, and highway beautification, and the war on poverty, and parks—national, state and local. It is hard to hitch the conversation into one straight line, because everything leads to something else. (Johnson)

During the 1960s, Lady Bird focused much of her efforts on cleaning up Washington, DC, believing that the model of a beautified capital city could become an example to other cities across the country. Her efforts also focused on freeing highway corridors of billboards and junkyards. Instead, she encouraged them to be used as demonstration areas for green landscaping and wildflowers. Back in Texas, Lady Bird spent the last decades of her life establishing the National Wildflower Research Center. In 1998, the center was rechristened the Lady Bird Johnson Wildflower Center in her honor.

Sources and Further Reading: Johnson: http://www.pbs.org/ladybird/shattereddreams/shattereddreams_report.html; Opie, *Nature's Nation*; Rothman, *The Greening of a Nation*; Steinberg, *Down to Earth*.

THE VALDEZ OIL SPILL AND ITS AFTERMATH

Time Period: 1989 to the present
In This Corner: Environmentalists, marine biologists, fishermen
In the Other Corner: ExxonMobil, other petroleum companies
Other Interested Parties: Alaskans, petroleum companies, federal regulators
General Environmental Issue(s): Pollution, petroleum, oil spills, marine

On March 24, 1989, shortly after midnight, the oil tanker *Exxon Valdez* struck Bligh Reef in Prince William Sound, Alaska, spilling more than eleven million gallons of crude oil. The largest spill in U.S. history, the *Exxon Valdez* tested the reaction and cleanup abilities of local and national responders. In addition, however, the ensuing legal fight with ExxonMobil demonstrated the limits of legal efforts to enforce corporate responsibility and also ask how one puts a price on portions of the natural environment.

Valdez is the center of oil transshipment for all of Alaska. Located toward the southern end of the state, the shipping port marked the point where oil, brought south from the Northern Slope of Alaska via the Alaskan Pipeline, would be put on tanker ships to be carried to other areas in the United States. Many of the ships were designed with a double hull to make spills less likely. The *Exxon Valdez* was single hulled; however, the primary reason for its spill is credited to operational error.

Initially, the U.S. Coast Guard's On-Scene Coordinator had authority for all activities related to the cleanup effort because it had occurred in a primary shipping channel. His first action was to immediately close the Port of Valdez to all traffic. Next, a investigator from the Coast Guard along with a representative from the Alaska Department of Environmental Conservation arrived to assess the damage. By noon on Friday, March 25, the Alaska Regional Response Team was gathered and a National Response Team was also activated.

Bill Scheer of Valdez, Alaska, is covered in crude oil while working on a beach fouled by the spill of the tanker *Exxon Valdez* at Prince William Sound, April 13, 1989. Experts continue to debate the extent of the ecological impact while the courts still are considering the extent of ExxonMobil's liability. AP Photo/John Gaps III.

Alaskan oil development combines seven oil companies, including a variety of native groups, into a corporate body known as Alyeska. This group, which included Exxon, first assumed responsibility for the cleanup and established an emergency communications center in Valdez. However, the entire response was a learning experience for everyone involved. Who was responsible for the emergency? For the cleanup? Soon, federal agencies, particularly representatives of EPA and the National Oceanic and Atmospheric Administration, arrived.

In addition, specialists from the Hubbs Marine Institute in San Diego, California, set up a facility to clean oil from otters, and the International Bird Research Center of Berkeley, California, established a center to clean and rehabilitate oiled waterfowl.

A variety of approaches were considered for treating the spill. Burning the oil was one method of removing it. To perform a trial burn, a fire-resistant boom was placed on tow lines, and two ends of the boom were each attached to a ship. The two ships with the boom between them moved slowly throughout the main portion of the slick until the boom was full of oil. The two ships then towed the boom away from the slick and the oil was ignited. Difficult weather, however, precluded the large-scale use of this method of cleaning up.

Chemical dispersants were also used, but many people criticized this method. Alyeska had less than 4,000 gallons of dispersant available in its terminal in Valdez and no application equipment or aircraft. After the application on March 24 by a private contractor, a lack of wave action caused the chemical to improperly mix with the oil. This method also was discontinued.

The primary method used in Prince William Sound was a time-consuming, mechanical cleanup that used booms and skimmers. This method proved most difficult because thick oil and heavy kelp clogged the equipment and caused damage to the skimmers, which then required repair. In addition, the cleanup efforts were complicated by the size of the spill and its remote location, accessible only by helicopter and boat. In Prince William Sound, the spill threatened a delicate food chain that supported the region's commercial fishing industry and approximately ten million migratory shore birds and waterfowl, hundreds of sea otters, dozens of other species, such as harbor porpoises and sea lions, and several varieties of whales.

Although the spilled oil presented an emergency, the situation was most acute because of the pristine wilderness in which the spill had occurred. With limited time in which to act, rapid decisions were required to determine the best course of action. Early in the cleanup process, personnel attempted to identify sensitive environments and then to rank their priority for cleanup. For instance, seal pupping locations and fish hatcheries were given the highest importance.

In the aftermath of the *Exxon Valdez* incident and as a result of many of these lessons, Congress passed the Oil Pollution Act of 1990, which required the Coast Guard to strengthen its regulations on oil tank vessels and oil tank owners and operators. Today, tank hulls provide better protection against spills resulting from a similar accident, and communications between vessel captains and vessel traffic centers have improved to make for safer sailing.

Controversy, however, continued to emanate from the incident, primarily as interested parties filed lawsuits against Exxon for its negligence in operating the tanker and in enforcing regulations over the behavior of tanker pilots and captains. The spill, of course, was a public relations nightmare for Exxon. Therefore, in 1991, ExxonMobil pleaded guilty to breaking several environmental laws and settled criminal and civil lawsuits of more than $1 billion. This was the most extensive attempt in human history to mitigate the environmental damage caused by an industrial disaster.

When an additional $5 billion was added in punitive damages over the next years, Exxon began to more aggressively limit its legal responsibilities for the incident. In the early 1990s, ExxonMobil funded research that claimed the Sound was well on its way to recovery, but new scientific research, conducted over the past fourteen years, states the opposite. Many

Alaskans and members of the environmental movement have accused Exxon of paying scientists to create misinformation about the impact of the spill.

Some observers have argued that petroleum is a naturally occurring substance, and, thus, it would have only a small amount of impact on the natural surroundings (Wheelright 1996). In the last word on this spill, however, scientists have ranked the impact on specific species, listed here: (1) recovery unknown: cutthroat trout, dolly varden, Kittlitz's murrelet, rockfish; (2) not recovering: common loon, cormorants, harbor seal, harlequin duck, Pacific herring, Pigeon Guillemot; (3) recovering: clams, killer whales, marbled murrelet, mussels, sea otter, sediments, tidal communities, wilderness areas; (4) recovered: archaeological sites, bald eagle, black oystercatcher, common murre, pink salmon, river otter, sockeye salmon (http://library.thinkquest.org/10867/results/status/index.shtml).

Sources and Further Reading: Coates, *Trans-Alaskan Pipeline Controversy: Technology, Conservation, and the Frontier*; Strohmeyer, *Extreme Conditions: Big Oil and the Transformation of Alaska*; Wheelwright, *Degrees of Disaster: Prince William Sound, How Nature Reels and Rebounds*; Yergin, *The Prize: The Epic Quest for Oil, Money & Power*; http://library.thinkquest.org/10867/results/status/index.shtml; http://www.epa.gov/oilspill/exxon.htm; http://www.evostc.state.ak.us/index.cfm.

WHO IS RESPONSIBLE FOR FIXING POLYCHLORINATED BIPHENYLS IN THE HUDSON RIVER?

Time Period: 1977 to the present
In This Corner: Residents of Hudson River Valley, environmental organizations, litigators
In the Other Corner: General Electric Corporation, litigators
Other Interested Parties: NIMBY environmentalists, federal regulators, policy makers
General Environmental Issue(s): Industrial waste, rivers, litigation

Although many waterways suffer the effects of industrial pollution, the Hudson River's story is unique. In fact, the experience of the Hudson region in the 1970s helped to show the way for others. This became one of the seminal events of the modern environmental movement, especially because it involved actions by one of the largest companies in the United States, GE.

The use of polychlorinated biphenyls (PCBs), which are substances used for a wide range of industrial purposes, began for GE in the late 1940s. Without knowing their danger to the surrounding community, 1.3 million pounds of PCBs were released by GE into the upper Hudson around Albany, New York, from 1947 to 1977. The primary contributors of PCBs to the Hudson River were GE's two electrical capacitor manufacturing plants located at Hudson Falls and Fort Edward, New York. Both plants discharged manufacturing process wastewater containing PCBs directly into the Hudson River until 1977. Investigations of plant discharges by Department of Environmental Conservation staff in 1975 also revealed PCB discharges from the Hudson Falls plant to the sanitary sewer system leading to the Hudson Falls Village Sewage Treatment Plant and PCB-contaminated storm water discharges to the Hudson River from both plants.

Therefore, the Hudson River PCB site is a complicated one, encompassing a nearly 200-mile stretch of the Hudson River in eastern New York State from Hudson Falls, New York, to the Battery in New York City and includes communities in fourteen New York counties and two counties in New Jersey. The site is divided into the Upper Hudson River, which runs from Hudson Falls to the Federal Dam at Troy (a distance of approximately forty miles), and the Lower Hudson River, which runs from the Federal Dam at Troy to the southern tip of Manhattan at the Battery in New York City.

Scientific evidence about the dangers of PCBs grew during the early 1970s. By the mid-1970s, a growing number of studies found links to premature births and developmental disorders and had shown that PCBs caused cancer in laboratory animals. The primary health risk associated with the site came from the accumulation of PCBs in the human body through eating contaminated fish. Since 1976, high levels of PCBs in fish have led New York State to close various recreational and commercial fisheries and to issue advisories restricting the consumption of fish caught in the Hudson River. PCBs are considered probable human carcinogens and are linked to other adverse health effects, such as low birth weight, thyroid disease, and learning, memory, and immune system disorders. PCBs in the river sediment also affect fish and wildlife.

Environmentalists and health advocates initiated their calls for attention in the early 1970s. In fact, one of the first organizations of its kind, Riverkeepers, had been established to defend the Hudson River in 1966. Originally established by fishermen, Riverkeepers was most concerned with the rumors about PCBs. Because of the danger, recreational fishing in the forty-mile reach of the upper Hudson between Hudson Falls and the Troy Dam was prohibited from 1976 until 1995. In 1975, New York State successfully sued GE to stop dumping PCBs into the Hudson. Two years later, the federal government banned PCBs. The difficulty then became cleaning up the mess that the Hudson had become. This process was helped by the 1980 passage of the Superfund law, which requires the use of federal funds to clean up sites contaminated with toxic waste. Furthermore, in 1983, with strong support from Hudson River advocates, the U.S. EPA declared a 200-mile swath of the river, from Hudson Falls to New York City, a Superfund site.

The debate was not over, however. Dredging, GE said, would stir up PCBs. The company claimed that sedimentation was cleaning the river naturally by burying the PCBs under silt. They argued that the river should be left alone so that it might heal itself. Using a public relations blitz throughout the region, GE told the public that dredging would destabilize the lives and economies of Hudson Valley towns.

Environmentalists argued that GE's claims were simply not scientifically well founded. For instance, the Natural Resources Defense Commission argued that the idea that the river is somehow cleaning itself has been refuted by numerous EPA studies, which have been extensively peer-reviewed by independent scientists. A study by the New York State Department of Environmental Conservation even found that PCBs are spreading beyond the river, contaminating soil and wildlife along its banks (Natural Resources Defense Commission, http://www.nrdc.org/water/pollution/hhudson.asp).

Additionally, critics argued that GE's attempts to sow fears about the economic impact of dredging were also deceptive. The company implied that a huge area would be targeted instead of the actual forty-mile stretch where dredging will be concentrated.

In early 2001, when George W. Bush became president, his new EPA administrator, Christine Todd Whitman, called for removing more than 100,000 pounds of PCBs from targeted hotspots in the upper Hudson, at a projected cost of $460 million. The project curbed by 40 percent the amount of PCBs washed downstream in the ten years after the project's completion. The text of EPA's February 2002 Record of Decision (ROD) for the Hudson River PCBs Superfund Site reads as follows:

> The targeting of Hot Spots 36, 37 and the southern portion of 39 is based on cur-rently available data showing that those areas have high PCB concentrations, and potential for loss to the water column or uptake by biota. Additional sampling will be conducted during remedial design to determine whether other areas in River Section 3 have these characteristics and therefore need to be remediated as part of the selected remedy.
>
> Remedial dredging will be conducted in two phases. The first phase will be the first construction season of remedial dredging. The dredging during that year will be imple-mented initially at less than full-scale operation. It will include an extensive monitoring program of all operations. An independent external peer review of the dredging resus-pension, PCB residuals, and production rate performance standards will be conducted during design. Monitoring data will be compared to performance standards identified in this ROD or developed during the remedial design with input from the public and in consultation with the State and federal natural resource trustees. The second phase will be the remainder of the dredging operation, which will be conducted at full-scale. During the full-scale remedial dredging, EPA will continue to monitor, evaluate per-formance data and make necessary adjustments. But the debate has not entirely abated. Faced with massive cleanup costs, GE has sought to lessen its responsibility while not losing face publicly. Riverkeepers is having none of it. In their official state-ment, the environmental organization reports: GE's response to the EPA's February 2002 directive to clean up its Hudson River PCB mess presented itself as a concession but was instead a continuing effort to evade responsibility. Riverkeeper and its alliance of environmental groups are determined to continue to apply pressure on GE until the company commits itself to financing and expeditiously executing on the full EPA cleanup plan....

GE's response contains a commitment only to conduct the design of the cleanup, estimated at approximately $30 million. However, it does not contain assurance that it will conduct or pay for the estimated $460 million cleanup itself. GE instead specifically leaves this cleanup and the vast majority of costs to future negotiation and makes those negotiations contingent on conditions GE alone determines. Additionally, GE has refused to pay its outstanding $37 million debt to the government. It instead defers payment of what it owes until the EPA accepts its new terms.

For residents of the Hudson River region, as well as those in similarly infected areas, the saga continues.

Sources and Further Reading: Clearwater, *Hudson River PCB Pollution Timeline*, http://www.clear water.org/news/timeline.html; National Resources Defense Council, *Historic Hudson River Cleanup to Begin After Years of Delay*, But Will General Electric Finish the Job?, http://www. nrdc.org/water/pollution/hhudson.asp; Opie, *Nature's Nation*; Price, *Flight Maps*; Riverkeeper, http://riverkeeper.org/resources_legal.php; Rothman, *The Greening of a Nation*; Steinberg, *Down to Earth*; U.S. Environmental Protection Agency, *Record of Decision*, http://www.epa.gov/hudson/d_rod.htm#record.

YELLOWSTONE TO YUKON CONSERVATION INITIATIVE

Time Period: 1993 to the present
In This Corner: Around 250 organizations of United States and Canadian origin
In the Other Corner: Some companies in the timber and coal mining industries
General Environmental Issue(s): Establishing a wildlife/ecosystem corridor

In 1993 a group of scientists and conservationists met in Alberta, Canada, to discuss establishing an interconnected wildlife corridor from Yellowstone National Park to the Yukon region in Alaska. From that initial meeting spawned a growing interest, within both the United States and Canada, to put the idea of "Yellowstone to Yukon" into practice. In 1997, the Yellowstone to Yukon Conservation Initiative (Y2Y) was formally established.

Y2Y includes around 250 organizations from both the United States and Canada (Wilderness Society 2007). These organizations include a wide variety of environmental nonprofit groups, institutions, and foundations, in addition to ecologists, conservationists, and other interested individuals. These organizations and individuals share the vision of creating a wildlife corridor covering approximately 2,100 miles, mainly encompassing the Rocky Mountains. Starting in west central Wyoming, Y2Y stretches northwest to the Peel River in the northern Yukon, an area only thirty-seven miles south of the Artic Circle. The Y2Y region will range from 125 to 500 miles wide and will encompass grasslands in the plains, cedar-hemlock forests in the mountains and wild rivers in the western inland-coastal watersheds. The ultimate goal is to establish more than 460,000 square miles of connected wildlife habitat by connecting existing protected areas.

As of 2007, the proposed Y2Y area consists of both public and private lands. The objective is to develop a long-term strategy incorporating science and stewardship in an effort to establish a connected wildlife linkage to protect world-renowned wilderness, wildlife, and native plants. The so-called "movement corridor" would function as a bridge, providing the needed territory to sustain large carnivores such as wolves and grizzlies, species that need a large habitat for both a healthy prey base and breeding grounds. Within the Y2Y, these species could move freely and naturally within a protected, fully functioning ecosystem. In addition to maintaining a healthy ecosystem of clean air and water, it would help conserve the natural beauty of the areas, the wilderness, and the indigenous plant and animal species.

Opposition to the Y2Y focuses mainly on the natural resources in the region. Some of the timber, coal mining, and agricultural industries as well as local ranchers are concerned that the corridor will create an area devoid of resource extraction and use. In addition, they worry that restrictions on resource extraction will have a negative impact on local economies dependent on those industries.

Supporters of the Y2Y counter that the one goal of the initiative is to involve affected communities and industries to promote consensus building and stronger supportive communities. By working together with communities such as Native Americans and First Nations, resource-dependent towns, and ranchers, the hope is to find common ground and economic incentives that would support working toward land conservation and stewardship. Supporters realize this will be challenging in a region once so dependent on resource extraction but believe it is possible to create thriving communities based on new economic industries such as tourism and outdoor recreation, activities that could ecologically help conserve the areas, while simultaneously generating local income and interest.

The Y2Y has gained a great deal of support since its inception. This is attributable not only to the fact that the region in question is one of the last remaining areas that have a natural breeding population of grizzly bears and wolves and that it includes many diverse and unique ecosystems, but also because of the growing trend of retirees coming to the region specifically for its pristine, unaffected wilderness. In addition, traditional outdoor enthusiasts like hunters and anglers continue to support these types of conservation efforts to maintain a healthy sporting tradition and an appreciation for nature. With the support of these groups and individuals, the Y2Y is working toward creating a wildlife corridor that strikes a balance between needs of the human communities and the wildlife.

Sources and Further Reading: Chadwick, "Yellowstone to Yukon"; Chester, "Landscape Vision and the Yellowstone to Yukon Conservation Initiative"; Finkel, "From Yellowstone to Yukon"; Gadd, "The Yellowstone to Yukon Landscape"; Higgs, *Nature by Design: People, Natural Process, and Ecological Restoration*; Levesque, "The Yellowstone to Yukon Conservation Initiative: Reconstructing Boundaries, Biodiversity, and Beliefs"; Posewitz, "Yellowstone to the Yukon (Y2Y): Enhancing Prospects for a Conservation Initiative"; Wilderness Society, "Yellowstone to Yukon"; http://www.wilderness.org/WhereWeWork/Montana/y2y.cfm?TopLevel=Y2Y.

YUCCA MOUNTAIN REPOSITORY

Time Period: 1982 to the present
In One Corner: U.S. Department of Energy
In the Other Corner: State of Nevada
Other Interested Parties: Environmentalists
General Environmental Issue(s): Spent nuclear reactor fuel and other radioactive waste

Yucca Mountain, a ridgeline in the desert of south-central Nevada, is the site of the proposed Yucca Mountain Repository, a U.S. Department of Energy (DOE) terminal storage facility for radioactive waste, including spent nuclear reactor fuel. The placement of the terminal storage facility has caused controversy concerning the issue of a state's right to determine its economic and environmental future.

The NWPA of 1982 mandated that the DOE find a site for operating an underground disposal facility for spent nuclear fuel and other radioactive waste. In an attempt to fulfill this obligation, the DOE began studying Yucca Mountain to determine whether it would be an acceptable location to store the spent nuclear fuel and other radioactive waste currently stored at 126 sites across the United States. In 1984, the DOE selected Yucca Mountain as one of ten locations deemed suitable as a potential repository site. Then, in 1985, Yucca Mountain, along with Hanford, Washington, and Deaf Smith County, Texas, were chosen for intensive scientific study for possible selection as the future site of the repository.

Two years after choosing these three sites, the U.S. Congress amended the NWPA, instructing the DOE to study only Yucca Mountain for possible selection as the future site of the repository. The rationale behind this change was the fact that Yucca Mountain is already located within a former nuclear test site and is located in a remote area in a sparsely populated state. The amended act did state that, if Yucca Mountain is found unsuitable, the studies would be stopped immediately and the U.S. Congress would determine the next step for the DOE.

In 2002, President George W. Bush signed House Joint Resolution 87, permitting the DOE to continue with their work to establish a safe repository for the country's nuclear waste. Shortly thereafter, the U.S. secretary of energy, Spencer Abraham, declared Yucca Mountain a suitable location for the nation's nuclear and radioactive waste repository. Although the governor of Nevada objected to this declaration, the U.S. Congress overrode the objection, allowing the project to move forward. After additional studies, in mid-2006, the DOE agreed on March 31, 2017 as the date to open the repository. As an additional assurance, in March 2006, the DOE announced that the Oak Ridge Associated Universities/Oak Ridge Institute for Science and Education (a nonprofit consortium that includes ninety-six doctoral-degree-granting institutions and eleven associate member universities) would provide independent expert reviews of scientific and technical work for the Yucca Mountain Repository project. The Oak Ridge Associated Universities/Oak Ridge Institute for Science and Education will join the EPA, Department of Homeland Security, and DOE's Office of Science to ensure that the project uses sound science.

In November 2006, the Democratic Party won the mid-term elections, positioning Nevada's Senator Harry Reid, a strong opponent to Yucca Mountain Repository, as Senate majority leader. This change in political control strengthened the opponents to the Yucca Mountain Repository Project. As of July 2007, the DOE studies continue, but Nevada's leaders, including Senator Reid, are working to stop the Yucca Mountain project.

Opposition to the Yucca Mountain Repository includes not only leaders in Nevada but also the majority of citizens in Nevada; most Nevadans believe it is unwarranted for a state with no nuclear power plants to store the nuclear waste for the entire country. Furthermore, leaders and citizens in a number of other states are also opposed, particularly those in the states that will make up the transportation corridor for the radioactive waste. Although not finalized, the likely northern transportation path would heavily affect cities such as Buffalo, Cleveland, Pittsburgh, Chicago, Omaha, Denver, and Salt Lake City, whereas the likely southern route would affect Atlanta, Nashville, St. Louis, Kansas City, and Las Vegas. Opposition also includes a group of Native Americans, including the Western Shoshone, Southern Paiute, and Owens Valley Paiute and Shoshone peoples, who view the region of Yucca Mountain as traditionally important for their cultural ecosystems and histories.

Other issues put forth by those opposed to the Yucca Mountain project include questions about damage to the environment and water sources, along with concerns about the geological stability of the Yucca Mountain region. The area is composed largely of tuff, a volcanic material that some experts suggest has physical, chemical, and thermal characteristics that make it suitable for containing radioactive waste for hundreds of thousands of years. However, opponents to the Yucca Mountain Repository fear that the cracks that naturally occur in tuff could serve as a route for radioactive waste to leak or as a path for groundwater to come into contact with the waste. Additionally, Nevada ranks third in the nation for current seismic activity. Finally, opponents also focus on the March 31, 2017 date that the facility will begin to accept radioactive waste. For the repository to open in 2017, the project needs to be fully funded, progress cannot be slowed with litigation, and the NRC needs to complete the review of the License Application, within three years of submission, and accept the DOE efforts to meet standards for nuclear quality and safety. The opposition is not confident that all of these requirements will be satisfactorily met.

In reaction to these concerns, the EPA requires that the DOE show that the Yucca Mountain site can safely contain wastes, even with the possibility of earthquakes, volcanic activity, climate change, and container corrosion over one million years. Furthermore, supporters of the Yucca Mountain Repository argue that the project needs to move forward to protect U.S. citizens and the environment as a whole. They note that there have been extensive governmental studies that show Yucca Mountain to be a suitable site for nuclear and radioactive waste disposal and that the cost of not moving forward with the project is high and increasing with the growing number of nuclear plants forced to resort to on-site storage. The supporters also note that it is possible for a temporary facility to be opened at the Yucca Mountain site or somewhere else in the American west if opening of the Yucca Mountain underground storage repository is delayed. This would not only do exactly what the opponents do not want to see but would also do it in a manner that is temporary and thus not as secure for people or the environment.

Neither side of the Yucca Mountain repository project appears to be willing to accommodate the other. Although compromise will have to be made, currently the controversy surrounding the Yucca Mountain project remains.

Sources and Further Reading: Eureka Country, Nevada Nuclear Waste Office, http://www.yucca mountain.org/new.htm; Ginsburg, *Nuclear Waste Disposal: Gambling on Yucca Mountain*; Macfarlane and Ewing, eds., *Uncertainty Underground: Yucca Mountain and the Nation's High Level Nuclear Waste*; Norrell, "Yucca Mountain Lawsuit Filed," www.indiancountry.com/content. cfm?id=1096410530; U.S. Department of Energy, "DOE Awards $3 Million Contract to Oak Ridge Associated Universities for Expert Review of Yucca Mountain Work"; Vandenbosch and Vandenbosch, *Nuclear Waste Stalemate: Political and Scientific Controversies*; White House, "President Signs Yucca Mountain Bill."

PEOPLE FOR THE ETHICAL TREATMENT OF ANIMALS (PETA)

Time Period: 1980 to the present
In This Corner: PETA, animal rights activists
In the Other Corner: Fast food restaurants, wool producers
Other Interested Parties: Anti-Defamation League, National Association for the Advancement of Colored People
General Environmental Issue(s): Animal rights

People for the Ethical Treatment of Animals (PETA) claims to be the world's largest animal rights organization. PETA's slogan "animals are not ours to eat, wear, experiment on, or use for entertainment" reflects the primary goal of the organization, to stop the suffering of animals. To meet this goal, PETA focuses on what they consider the areas in which the largest numbers of animals suffer the most intensely for the longest periods of time. Those areas include factory farms, laboratories, the clothing trade, and the entertainment industry. In addition, the organization also works to eliminate cock fighting, the extermination of animals considered pests, chained backyard dogs, and eating meat.

PETA was founded in 1980 in Virginia. By using public education, animal rescue, lobbying, celebrity involvement, and protest campaigns to achieve its goals, the organization has increased its membership and public attention. By 2007, PETA had grown into an international organization with 1.6 million members and supporters.

PETA's actions first drew attention in 1981 for the Silver Spring monkeys case. Alex Pacheco, a cofounder of PETA, went undercover to expose the abuse of monkeys at the Institute of Behavioral Research in Silver Spring, Maryland. The experiment under question was headed by Dr. Edward Taub and consisted of cutting nerves in the limbs of monkeys and then placing the monkeys in stressful situations to see what would lead the monkeys to use their damaged limbs. Pacheco presented photos as evidence to the local police. Taub was convicted of animal cruelty, although it was later overturned on appeal. Although PETA did not get custody of the monkeys, which they had held as one of their goals, their actions did lead to greater publicity and thus an increase in members and donations. Additionally, the case ultimately led to an amendment to strengthen the Animal Welfare Act in 1985.

Since the Silver Spring monkeys case, the organization has become known for protest campaigns that include throwing paint at men and women wearing fur coats to protest the fur industry, using the media and internet activities to speak out against Kentucky Fried Chicken, and designing a billboard campaign picturing naked supermodels with "I'd Rather Go Naked than Wear Fur" printed across their chests. PETA also runs campaigns to change the names of cities and towns whose names suggest animal exploitation. For example, PETA campaigned to change the name of Fishkill, New York, and Rodeo, California. Rodeo did not change its name, even after PETA offered to donate $20,000 worth of veggie burgers to local schools if the town would change their name (*Contra Costa Times* 2003). Criticism of PETA comes from both animal rights activists and opponents. The organization is criticized for a variety of its protest campaigns, some which by their own admission include civil disobedience, and for euthanizing animals. PETA's support of euthanasia consists of offering free euthanasia service to counties that kill unwanted animals by gassing or shooting and for some animals that have been mistreated. PETA is also criticized for functioning as voice for the radical groups ELF and Animal Liberation Front and for objectifying the female body with their "I'd Rather Go Naked than Wear Fur" campaign. Finally, the connection that PETA makes between the killing of animals and the Holocaust, and work animals and slaves has drawn strong opposition from groups such as the Anti-Defamation League and the National Association for the Advancement of Colored People.

Supporters of the organization point to the successes of PETA. The organization has won numerous lawsuits and has brought animal cruelty to public attention. The combination of their education campaigns, their undercover work (which some consider "blackmail"), and public support has achieved changes in laboratory practices and has convinced companies such as Polo Ralph Lauren and J. Crew to no longer use fur in their fashions and Estee Lauder and L'Oreal to halt animal testing.

Although there is substantial controversy surrounding PETA and the group's goals and tactics, the group has made a name for itself. PETA does draw publicity, and their actions, if considered beneficial or blackmail, have drawn attention to animal rights issues achieving change in both public attitudes and public policies.

Sources and Further Reading: Anti-Defamation League, http://www.adl.org/; Associated Press, "PETA Claims Victory as Fashion House Drops Fur"; *Contra Costa Times*, "Rodeo Residents' Beef with PETA Must Wait," http://www.beefusa.org/newsrodeoresidentsbeefwithpetamustwait 13689.aspx; Newkirk, "The ALF: Who, Why, and What?"; People for the Ethical Treatment of Animals, "Researchers study Silver Spring Monkeys under terminal anesthesia prior to euthanasia, and discover new roles ..." www.peta.org/; Singer, *Animal Liberation*; Singer, *In Defense of Animals: The Second Wave*; Wennberg, *God, Humans, and Animals: An Invitation to Enlarge Our Moral Universe*; Workman, *PETA Files: The Dark Side of the Animal Rights Movement*.

TIMBER CERTIFICATION/CERTIFIED WOOD PRODUCTS

Time Period: 1990s

In This Corner: Sustainable-timber industry representatives, environmental groups

In the Other Corner: Timber industry representatives

Other Interested Parties: Consumers, Home Depot, IKEA, and other wood and wood product vendors

General Environmental Issue(s): Sustainable forestry

In the early 1990s, as the American public became more concerned with clear-cutting and deforestation around the world, the idea of sustainable forestry became an issue for consumers. Following these consumer desires, the producers of wood and wood products, along with a number of environmental groups looking to protect forests, sought a way to distinguish wood products that come from sustainable forestry. The solution to this dilemma came in the form of timber certification, a process that results in a written certificate attesting to the origin of wood. The outcome of the timber certification process was the creation, in 1993, of the Forest Stewardship Council (FSC). Today, more than 6,000 companies participate in the FSC system (Forest Stewardship Council 2007).

The sustainable forestry certification process measures the forest management practices against a set of standards. By doing this, the process validates environmentally sustainable practices of producers. Certification of forest management includes management planning, silviculture, harvesting, road construction, and related activities. The process examines how these activities affect environmental, economic, and social aspects of the surrounding region. The certification process links the consumer who desires environmentally sustainable and socially responsible products with the producers of wood that meets those standards. The objective of the certification process is to evaluate the integrity of wood producers' claims of sustainable management and the authenticity of product origin.

The process of wood certification must be seen as objective and impartial, and, although currently imprecise, as the movement matures, certification is likely to develop into a more standardized procedure. Certification programs exist at international, regional, and national levels in both developed and developing countries. However, according to EcoTimber (2007), "the only forest certification system that enjoys the support of environmental groups worldwide is that of the Forest Stewardship Council (FSC), which is independent, nonprofit, and has a mechanism for tracking wood from the forest (green floor) to the consumer." The FSC, based in Oaxaca, Mexico, is supported by the world's top environmental groups. It created a certification system that accredits international certifiers that follows its standards. The standards include looking at alternatives to herbicides and pesticides, as well as maintaining native ecosystems rather than monoculture tree plantations. FSC works with third-party certifiers to conduct two- to three-day on-the-ground inspections of logging sites to ensure that FSC standards are met or exceeded and then to conduct follow-up inspections of the forests to ensure that the loggers continue to meet the criteria.

Within the United States, FSC recognizes two wood certifiers, SmartWood Program, which is related to the Rainforest Alliance and is located in Vermont, and Scientific Certification Systems, based in California. These certifiers make sure there are no clear-cuts and no erosion along rivers and creeks. Conversely, they look for snags (standing dead trees) and other wood debris to benefit bird and prevent soil erosion, ample forest cover, narrow haul roads, and abundant wildlife.

Home Depot, the world's largest lumber and home-improvement retailer, is one company working to address these sustainable forestry issues. The company is striving to stop selling wood from endangered forests and is pressing its suppliers to provide FSC-certified wood. The number of consumers around the world that support sustainable forestry has become so high that Home Depot, IKEA, and other like-minded companies are having problems finding supplies to meet the demands of consumers who want to buy FSC-certified wood and wood products.

Aware of the growing opportunity in sustainable wood, other organizations have begun to create wood certifications. These organizations, such as the American Forest and Paper Association and the Canadian Standards Association, do not have the same rigorous standards as the FSC. Environmental groups suggest that groups such as these allow large-scale clearing of old growth and often fail to protect First Nations (EcoTimber 2007). Furthermore, it is implied that some of these groups operate under the certification label more so that they too may gain a competitive advantage within the niche market rather than for true support of sustainable forestry.

Although the FSC and sustainable forestry are drawing many supporters, there are also critics of the organization and the certification process. Critics of the certification process mention the expenses of the process, including lower yields that producers suggest will be made up in other forests that may have adverse management practices, and higher opportunity costs. In addition, the critics mention the possibility of free-riding, making the claim of sustainable forestry without paying for the process. All of these are considered disincentives for compliance of the certification process. Many within the producer sector have been reluctant to buy into the certification movement, citing unproven market demands and competing certification systems. Other critics express doubt as to whether the certification truly considers the social implications for First Nations and other communities living around the forest.

Supporters of sustainable forestry and the certification process suggest that a loss in profit from lower timber yields can be partially compensated by lower operating costs and increased recovery from better planning of and protection for future timber and nontimber products and services. Furthermore, they suggest that there are areas within the market in which consumers are willing to pay a premium for sustainable wood and wood products and thus there could be an increase in profit for the producers.

Consumer demand for more environmentally sustainable and socially responsible business pushed the emergency of sustainable timber and wood product certification processes. Although there is not a single accepted forest management standard worldwide, there is a growing acceptance for what should be considered certified timber and wood products.

Overall, the certification process is an important tool for those seeking to ensure that the wood and wood products they purchase come from forests that are sustainably managed and legally harvested.

Sources and Further Reading: Aplet et al., *Defining Sustainable Forestry*; Baharuddin, "Timber Certification: An Overview," http://www.fao.org/docrep/v7850e/V7850e04.htm; Cabarle et al., "Certification Accreditation: The Need for Credible Claims"; EcoTimber, "Environmental and Sustainable Flooring: Is it Really Eco? Don't be Fooled!", http://www.ecotimber.com/info/eco.asp; Forest Stewardship Council, http://www.fsc.org/en/; Jenkins and Smith, *The Business of Sustainable Forestry: Strategies for an Industry in Transition*; McEvoy and Jeffords, *Positive Impact Forestry: A Sustainable Approach To Managing Woodlands*; U.S. Environmental Protection Agency, *Status Report on the Use of Environmental Labels Worldwide*; World Wildlife Fund, *Truth or Trickery?: Timber Labeling Past and Future*.

DEBT FOR NATURE AND DEVELOPMENT SWAPS

Time Period: 1990 to the present
In This Corner: Environmental groups (World Wide Fund for Nature, Conservation International, The Nature Conservancy, among others), aid agencies, developing countries' governments
In the Other Corner: Smaller conservation groups and some analysts
Others Interested Parties: Local communities
General Environmental Issue(s): Conservation, developing countries' debt

In 1984, the World Wildlife Fund (now World Wide Fund for Nature) initiated the debt for nature swap as a means to lessen financial strain and to increase conservation efforts in developing countries. Thomas Lovejoy, the deputy chairperson for the World Wildlife Fund, recognized that many of the countries that held great debt also held much of the world's biological diversity. Debt for nature and development swaps "leverage funds for use in local conservation efforts, based on the model of debt-equity swaps" but without generating profits (Resor 1997). Countries in Africa, Asia, and Latin America have participated in debt for nature and development swaps.

Many of the world's most threatened tropical forests are in some of the poorest and most indebted countries. These countries face numerous economic problems, including underdevelopment and immense debt interest payments. Traditionally, many of these countries could not meet their financial obligations and thus were forced to cut back on government spending and encourage logging of their forests to pay at least the interest on the debt. Furthermore, people desperate for land on which to scratch a living would clear forest for grazing pastures. These countries were sacrificing their forests and other pristine resources in an attempt to deal with their economic problems.

The idea behind debt for nature and development swaps involves purchasing foreign debt at a discount, converting the debt into local currency, and using the resources for local conservation projects. The swaps do not seek profit for the countries or private organizations eliminating the debt; rather, they provide additional resources for the debtor country to put toward the conservation of a valuable resource.

Debt for nature and development swaps take two different forms: commercial and bilateral debt swaps. A commercial debt for nature and development swap involves a NGO soliciting debt donations or purchasing debt from an international commercial bank at a discount. The NGO then negotiates with the debtor government by offering to cancel the debt in exchange for conservation activities or funding. The second form, bilateral debt for nature and development swaps, involves the cancellation of a given amount of debt owed to a creditor government in exchange for a predetermined amount of local currency put into conservation programs in the indebted country.

The first successful swap was implemented in 1987 between Conservation International and Bolivia. Conservation International paid $100,000 to acquire $650,000 of Bolivia's debt. The government of Bolivia took actions to provide the Beni Biosphere Reserve with maximum legal protection and to create three adjacent protected areas in exchange for Conservation International forgiving the debt. Despite some controversy and delays, in the end the debt for nature and development swap drew attention to the conservation challenges and helped the debt for nature and development swaps become a creative tool for financing conservation.

For supporters, debt for nature and development swaps are viewed as a clever way to expand contributions to conservation in developing countries. Donor NGOs view the swaps as a method to increase funding for conservation activities in the debtor countries, and debtor country governments embrace it as a method to help manage their foreign debt situation as well as to preserve resources within their county.

Critics point to the limitations of the debt for nature and development swaps. One clear drawback is lack of implementation. In many of the swaps, the debtor countries lack incentives to follow through with their conservation activities. In other debtor countries, the lack of supporting institutional infrastructure has hindered progress of environmental protection projects. Furthermore, who is responsible to make sure the debtor countries follow through long-term with their agreements? Recently, NGOs began assessing the likelihood of debtor countries living up to their part of the bargain. Critics suggest that, because the NGOs' cost-benefit analyses show diminishing returns, they too will be less likely to become involved in additional debt for nature and development swaps.

Another drawback to the debt for nature and development swap is that the swaps are largely dependent on donor grants and rely on foreign assistance budget allocations from developed counties. At times when foreign assistance and national budgets are contracting, the possibilities for debt for nature and development swaps are lessened. In addition, since the 1990s, many developing countries have undergone structural adjustment, including a restructuring of their external debt. These adjustments have decreased the money owed, improving the situation for developing countries but decreasing the possibilities for debt for nature and development swaps.

Debt for nature and development swaps have created a new way of thinking about conservation in developing nations. They also support local institutions that otherwise would have limited resources for their conservation efforts. The hope is that, over time, the continuation of these swaps will aid in greater conservation and particularly in some of the most biologically diverse areas around the world.

Sources and Further Reading: Asiedu-Akrofi, "Debt-for-nature Swaps: Extending the Frontiers of Innovative Financing in Support of the Global Environment"; Conservation International, *The Debt-for-nature Exchange: A Tool for International Conservation*; Ginn, *Investing in Nature: Case Studies of Land Conservation in Collaboration with Business*; Henry and Price, *Integrating*

Conservation and Development in Papua New Guinea; Lovejoy, "Aid Debtor Nations Ecology"; Morgan, *People and Nature: An Introduction to Human Ecological Relations*; Resor, "Debt-for-nature Swaps: A Decade of Experience and New Directions for the Future"; Sher, "Can Lawyers Save the Rain Forest? Enforcing the Second Generation of Debt-for-nature Swaps"; Sustainability Institute, "Trading Debt for Nature Instead of Nature for Debt," http://sustainer.org/dhm_archive/index.php?display_article=vn221debtfornatureed; World Bank, *The Multilateral Debt Facility for Heavily Indebted Poor Countries.*

THE CONVENTION ON INTERNATIONAL TRADE IN ENDANGERED SPECIES OF WILD FAUNA AND FLORA

Time Period: 1960s to the present
In This Corner: Conservationists
In the Other Corner: Trade organizations
Other Interested Parties: Native groups
General Environmental Issue(s): Conservation, endangered species, threatened species

In the early 1960s, an international discussion began on the issue of international trade in specimens of wild animals and plants. The dialogue focused on the concerns surrounding threats to the survival of a number of species and the possible regulation of wildlife trade for conservation purposes. Of these, at the time of innovative discussions came an international agreement between governments, the Convention on International Trade in Endangered Species of Wild Fauna and Flora (CITES).

CITES was drafted in 1963 as a result of a resolution adopted at a meeting of The World Conservation Union. The text of CITES was finalized among eighty countries in March 1973, written in five languages: Chinese, English, French, Russian, and Spanish. Within the text, the main objective of CITES was to ensure that international trade in wildlife does not threaten the survival of species. The convention was put into effect on July 1, 1975. It has not taken the place of national laws, but instead each country adopts its own domestic legislation to ensure that CITES is implemented within their country. As of July 2007, 172 countries have joined CITES.

CITES was created in the spirit of conservation and cooperation. The signatories recognized that, although the majority of international trade in wildlife does not involve endangered or threatened species, it is important to protect those that are threatened and endangered. Furthermore, the signatories also recognized the difficulties in cross-border and multicountry regulations and thus that cooperation is necessary. To address these and other concerns, the Conference of the Parties (CoP) was created and given supreme decision-making powers. All member countries are represented on the CoP. Funding and support for the CoP and CITES activities come from a trust fund created by contributions from signatory countries. Domestic activities done in support of CITES are not funded by this trust.

The countries that signed on to CITES recognize that international trade in wild animals and plants is worth billions of dollars and that the trade includes not only live animals and plants but also products derived from these species, including food, hides and furs, wood products, and curios and medicines. The signatories recognized that varying degrees of protection are needed for different species depending on the number of products made from each species. To deal with these issues, CITES includes a list of species that are threatened with extinction. The trade of these species is permitted only in exceptional circumstances. There is also a list of species that are not threatened with extinction but which the signatories believe trade needs to be regulated to avoid exploitation.

Since the creation of CITES, the species covered by the convention have benefited from enhanced protection, and the number of species that have become extinct as a result of trade has dropped drastically. This has been achieved mainly through "peer pressure." When a signatory country does not follow the regulations it has set up to support CITES, all countries involved with CITES are notified. Among tactics used to bring the country back into line with CITES are recommendations to all CITES countries to suspend wildlife trade with the offending country, suspension of cooperation from the CoP, and occasional bilateral sanctions, although these are imposed through national legislation and not officially through CITES.

The conflicts that do occur among CITES countries usually occur during discussions of proposed listing of species. Any member country, not only the country where the species is native, can list species. Even if there are objections by the country that holds the traditional range of the species, as long as there is a two-thirds majority supporting the listing, the species will be placed on the CITES list.

Another area of conflict concerns the split listing of species. Some populations of a species, for example African elephant populations in Botswana, Namibia, South Africa, and Zimbabwe, will be listed in the regulated but not threatened listing, whereas all other African elephant populations are considered threatened. Some people argue that this split listing leaves open the possibility of "laundering" specimens as a part of the less-protected populations. This issue has led to great discussion because people who want to avoid specimen laundering also want to advantage countries with good management practices.

A final area of discussion with CITES deals with what some see as the drawbacks of the convention. Some conservationists find fault with the structure and philosophy of CITES. The convention is focused on individual species rather than on ecosystems. Thus, CITES does not address issues such as habitat loss or poverty, issues that directly effect the survival of the listed species. Furthermore, critics believe greater enforcement and better reporting, along with access to more funds, would improve the functioning of the convention.

Sources and Further Reading: CITES, www.cites.org/; Ellis, *Tiger Bone and Rhino Horn: The Destruction of Wildlife for Traditional Chinese Medicine*; Heinen and Reibstein, "Convention on International Trade in Endangered Species of Wild Fauna and Flora (CITES)"; Hill, "The Convention on International Trade in Endangered Species: Fifteen Years Later"; Hutton and Dickenson, *Endangered Species Threatened Convention: The Past, Present and Future of CITES, the Convention on International Trade in Endangered Species of Wild Fauna and Flora*; Reeve, *Policing International Trade in Endangered Species: The CITES Treaty and Compliance*; U.S. Fish and Wildlife Service, *Convention on International Trade in Endangered Species of Wild Flora and Fauna*, http://www.fws.gov/international/laws/citestxt.html; Wijnstekers, *The Evolution of CITES 2003: A Reference to the Convention on International Trade in Endangered Species of Wild Fauna and Flora*.

GENETICALLY MODIFIED ORGANISMS

Time Period: 1970s to the present
In This Corner: Organic farms, family farms
In the Other Corner: Biotech companies

Other Interested Parties: Environmental groups, developing countries
General Environmental Issue(s): Genetically engineered food, biodiversity

Genetically modified organisms (GMOs) are organisms that have been altered with genetic engineering techniques. These engineering procedures add genetic material into an organism's genome. This is done to generate new traits within the organism. GMOs were made possible through scientific advances in the 1970s. By the mid- to late 1980s, GMOs were being tested by numerous companies for use in medicines and vaccines, foods and food ingredients, and feeds and fibers.

The process for GMOs is not "biotechnology," although the two are often used interchangeably. Biotechnology refers to using living organisms or their components to make products such as bean and cheese. In contrast, genetic modification is a process that alters the genetic makeup of living organisms, including animals, plants, and bacteria. Although technically GMOs would include organisms whose genetic makeup has been altered by conventional crossbreeding or mutation, these organisms are not included in the classifications of GMOs. Genetic engineering involves crossing species that could not cross in nature; it is this type of modification that is classified as "genetically modified."

The wide variety of uses of GMOs includes production of pharmaceuticals, human health (gene therapy), and agriculture. Although controversy surrounds many aspects of GMO procedures, transgenic plants within the food supply is likely the most contentious. The controversy surrounding GMO food products has intensified since 1996 with the first commercial cultivation of genetically modified plants.

Supporters of genetically modified food and organisms stress that the plants are developed to withstand harsh environmental conditions, improve shelf life, and increase nutritional value. Additionally, the process enhances the taste and quality of the foods. The GMOs and foods are considered safe; no major health hazards have been discovered. Additionally, the genetically enriched foods, such as Gold rice, have nutritional benefits such as the addition of vitamin A, which potentially can prevent disease. Supporters also note that the nutritional benefits from these foods have improved the health of many people in the developing world. With GMOs, the people can get the vitamins and minerals they need from eating foods, such as rice, that they can afford to buy rather than relying on meat and fruits and vegetables that are more expensive in many cases.

Supporters also purport ecological benefits from using genetically modified plants. It is suggested that the use of genetically modified crops may reduce greenhouse gas emissions and pesticide loads in the environment. This argument is largely made via the assertion that genetically modified crops produce more food from less land, leaving more room for open space. Additionally, advocates note that the majority of genetically modified crops are herbicide- and insect-resistant soybeans, corn, cotton, canola, and alfalfa (U.S. Department of Energy Office of Science 2007). For these crops, there is less of a need for spraying herbicides or pesticides, saving the environment from both the applied chemicals and the greenhouse gases emitted from the vehicles that would apply the chemicals.

Critics of genetically modified foods and organisms cite a number of concerns. First, genetically modified plants are grown on open fields. Thus, they often coexist with conventional and organic crops, causing a fear of environmental risk. Most countries require biosafety study before the approval of genetically modified plants for cultivation, although different countries have different standards. Europe is particularly sensitive to this issue, requiring not only the biosafety studies but also monitoring programs to detect long-term environmental impacts. In addition, many European countries are setting standards for food and feed derived from genetically modified plants because of a high demand from consumers for access to nongenetically modified foods.

A second concern focuses on the possibility of unintended consequences and of long-term health risks that are yet to be discovered. Critics suggest that, in the process of "improving" foods, there is a potential for a transfer of allergens and of antibiotic resistance markers and for there to be an unknown effect on other organisms such as soil microbes. Some allergenic traits have already been unintentionally transferred. The primary example of this is a gene from the Brazil nut genetically engineered into soybeans, although soybeans are used for animal feed. Critics of GMOs cite this example when discussing other possible results of using GMOs.

Many of the controversies surrounding genetically modified plants focus on farmers. Originally, farmers bought into the benefits, using less pesticide, gaining a larger yield. However, over time, as problems have surfaced, some farmers have shied away from planting genetically modified seeds. In 2004, Mendocino County, California, became the first county in the United States to ban the production of genetically modified plants, in part to alleviate concerns that conventional crops can be cross-pollinated with the pollen of genetically modified plants. This is a concern because, if genetically modified plants cross-pollinate conventional plants and patented genes are transferred, the patent holder has the right to investigate the crops and possibly even control the use of the crops.

This has become an issue for the largest gene-manipulation biotech company, Monsanto. A representative example of this controversy occurred in North Dakota. Monsanto sued the Nelson family farm for allegedly saving seeds from one season to plant in the next. If the Nelson family had saved seeds, it would be a violation of the company's patent. The Nelsons insist they did no such thing and that there is increasing potential for contamination at all stages of agricultural production: when buying seeds, from combines, and even within grain elevators. Farmers purport that it is all but impossible to guarantee there will be no trace of genetically modified crops within conventional fields. Although the Nelsons' suit was settled in an out of court settlement, other similar stories have ended in bankruptcy for the farmers. Monsanto and other biotech companies want to maintain their intellectual property rights, but it is not always clear when these rights have been stolen and when the existence of GMOs within crops is a cause of unintended contamination.

Finding solutions to these concerns is not proving to be easy. Domestically, some groups advocate the prohibition of genetically modified foods, and others call for mandatory labeling. Overall, despite benefits and drawbacks, American consumers do not support banning new uses of technology, although they do want assurance the products are safe (Pew Initiative on Food and Biotechnology 2005). Internationally, there are similar desires and suggestions. To date, there is little international consensus regarding genetically modified organisms and the role they should play in society. The complexity of this issue and the fact that it is new technology suggest that supporters and critics alike have further research and debates to tackle.

Sources and Further Reading: Associated Press, "Monsanto Drops Seed Patent Lawsuit Against North Dakota Family"; Cook, *Genetically Modified Language: The Discourse of the GM Debate*; Food and Agriculture Organization of the United Nations, *Weighing the GMO Arguments*, http://www.fao.org/english/newsroom/focus/2003/gmo7.htm; Pew Initiative on Food and Bio-technology, *Public Sentiments About Genetically Modified Food*, http://pewagbiotech.org/research/2005update/; Ruse and Castle, *Genetically Modified Foods: Debating Biotechnology*; Schubert, "Monsanto Sues Nelson Farm: A North Dakota Family's Frustrations with Genetically Engi-neered Soybeans"; Teitel, *Genetically Engineered Food: Changing the Nature of Nature*; U.S. Depart-ment of Energy Office of Science, *Genetically Modified Foods and Organisms: What Are Genetically Modified (GM) Foods?*, http://www.ornl.gov/sci/techresources/Human_Genome/elsi/gmfood.shtml.

DIVERSE IMPACTS OF PETROLEUM AND MINERAL USE AND THE EVOLUTION OF THE "CARBON FOOTPRINT" CONCEPT

Time Period: Twenty-first century
In This Corner: American consumers, petroleum companies, mining companies
In the Other Corner: Biologists, scientists, climatologists
Other Interested Parties: Politicians
General Environmental Issue(s): Energy, pollution

The American standard of living that emerged by the end of the twentieth century is one of the most energy intensive in human history. A century of cheap energy, particularly coal and petroleum, allowed Americans to take for granted the impacts of such energy consump-tion patterns. Computer modeling and scientific inquiry have begun to offer a picture of the complications created by the mining for and burning of fossil fuels. Environmentalists and other critics are demanding a new model of accounting for one's impact on the environment in terms of energy consumption. Although some companies and governments continue to fight this type of thinking, many citizens have begun to think much more wholly in terms of the energy that powers their lifestyle. It has become increasingly difficult to ignore the impact of such energy consumption.

Oil Spills and the Environment

Of course, one of the most troubling aspects of offshore drilling is the increased likelihood of oceanic oil spills. Even without offshore drilling, however, shipping oil throughout the globe also creates the possibility of spills. On average, more than 100 million tons of crude oil are shipped every month, often in huge tankers, a few of which can hold as much as 500,000 tons. With such traffic, there is always a risk of accidental oil spills. Crude oils and petro-leum products are complex substances, and their different chemical compounds can react with sea life in a variety of ways. Some poison and kill plankton, the microscopic plants and animals on which many other sea creatures feed. This causes additional damage to the popu-lations of fish and mammals that rely on these plants for food.

Additionally, some crude oils leave sticky residues as they weather, which may either float on the surface until they reach the shore, smothering animals and seaweeds, or form solid balls of "tar." A coating of thick oil can kill sea birds by either poisoning them or damaging the portion of their nervous system that controls body temperature. Conversely, the recovery potential of many marine species is such that the impacts of an oil spill are often short-lived (Wheelwright 1996).

According to the Alaska Sea Grant, a marine research program at the School of Fisheries and Ocean Science at the University of Alaska Fairbanks, oil spills into water place an enormous variety of animals and plants at severe risk from smothering and poisoning. The group says that the infamous 1989 *Exxon Valdez* disaster off the coast of Alaska, America's largest oil spill to date, directly killed between 300,000 and 645,000 birds, including bald eagles and many types of ducks and other sea birds.

The *Exxon Valdez* spill also wreaked untold harm on the health and reproductive success of surviving birds in the surrounding area. Seals, otters, killer whales, and fish were also killed and injured in alarming numbers. In addition, researchers have found long-term impacts, including damage to beach ecosystems and contaminated sediments. These damages are cited for disrupting local economies dependent on fishing and sightseeing. Unfortunately, according to Judith McDowell of the Woods Hole Oceanographic Institution, less than 10 percent of the oil that makes its way into marine environments is actually attributable to spills like that of the *Exxon Valdez*. Most oil ends up in seawater as a result of business as usual. Typically, oil arrives from a combination of natural seeps from the ocean floor and "run-off" from both offshore drilling facilities and land-based automobiles and machinery. Indeed, a significant amount of oil eventually makes its way into both marine and freshwater environments, including underground aquifers and other sources of drinking water, from the millions of cars and trucks that leak oil onto driveways, parking lots, and roads, which is then washed into water systems (Wheelwright 1996).

Environmental Impacts of Burning Fossil Fuels

Many health and environmental problems that the United States faces today we have learned are a result of the country's fossil fuel dependence. The coal industry's most troublesome problem today is removing organic sulfur, a substance that is chemically bound to coal. All fossil fuels, such as coal, petroleum, and natural gas, contain sulfur. When these fuels are burned, the organic sulfur is released into the air, where it combines with oxygen to form sulfur dioxide. Sulfur dioxide is an invisible gas that has been shown to have adverse effects on the quality of air. It also contributes to acid rain, an environmental problem that many scientists think adversely affects fish, wildlife, and forests.

In an effort to solve the problem, some coal-burning power plants are installing "scrubbers" to remove the sulfur in coal smoke. Scrubbers are installed at coal-fired electric and industrial plants where a water and limestone mixture reacts with sulfur dioxide to form a sludge. Scrubbers eliminate up to 98 percent of the sulfur dioxide, but they are very expensive to build (Gelbspan 1998). Efforts to create regulations to require the addition of scrubbers have been resisted by the Bush administration.

Acid Rain and Ground Level Ozone

The problem of acid rain that was mentioned above deserves specific discussion here. The combustion of fossil fuels, particularly coal, produces gaseous emissions of sulphur dioxide and nitrous oxides. Scientists have shown that these emissions are responsible for the production of acid rain and ground level ozone. Acid rain occurs when these two gases react in the atmosphere with water, oxygen, and other chemicals to form acidic compounds. These compounds form in the air but fall back to earth in the form of toxic rain, which is called "acid rain."

Ground level ozone is mainly responsible for smog that forms a brown haze over many American cities. Ground level ozone is formed when nitrogen oxide gases react with other chemicals in the atmosphere before being intensified by bright sunlight.

These environmental implications are particularly problematic because they are transboundary issues: the air pollution from one area may create acid rain or ground level ozone problems in other geographical areas.

In more developed countries, modern emission control technologies and the greater use of low sulphur coal have greatly reduced acid rain. In the United States, despite continued growth in coal-fired electricity production, emissions of sulfur dioxide from utilities have fallen by around 3 percent a year since 1980 (Gelbspan 1998).

Air Pollution and Global Warming

The related implications of the emissions from burning fossil fuels are just beginning to be realized; however, it is clear that the air pollution created by our commitment to these energy resources has cost the earth and the humans who live on it dearly. Among the gases emitted when fossil fuels are burned, one of the most significant is carbon dioxide, a gas that traps heat in the earth's atmosphere. Over the past 150 years, burning fossil fuels has resulted in more than a 25 percent increase in the amount of carbon dioxide in our atmosphere. Fossil fuels are also implicated in increased levels of atmospheric methane and nitrous oxide, although they are not the major source of these gases.

Finally, many researchers have connected these changes in atmosphere to a global rise in temperature. Since reliable records began in the late 1800s, the global average surface temperature has risen 0.5-1.1 degrees Fahrenheit (0.3–0.6 degrees Celsius). Scientists with the Intergovernmental Panel on Climate Change concluded in a 1995 report that the observed increase in global average temperature over the past century "is unlikely to be entirely natural in origin" and that "the balance of evidence suggests that there is a discernible human influence on global climate." This finding has compelled many climate scientists to predict that, if carbon dioxide levels continue to increase, the planet will become warmer in the next century. Projected temperature increases will most likely result in a variety of impacts, including the following: sea-level rise attributable to the warming of the oceans and the melting of glaciers, which may lead to the inundation of wetlands, river deltas, and even populated areas; altered weather patterns that may result in more extreme weather events; and inland agricultural zones that could suffer an increase in the frequency of droughts (Gelbspan 1998).

Harvesting Federal Lands: The Debate Over ANWR

The reliance on energy supplies, of course, has influenced international affairs and national security. Domestically, however, the perceived need to prioritize energy supply has also required officials to think differently about some of America's most special locales, including national parks and wildlife refuges.

In recent years, an increasingly important issue has linked coal and oil development in the United States. Throughout the late 1900s, politicians have debated a basic question of development: if these energy resources are located on federally controlled land, under what conditions should they be opened up for mining? Sharing a commitment to the need to harvest our existing energy resources, no matter where they are located, James Watt, secretary of the interior under President Ronald Reagan, and President George W. Bush are linked over a generation. Picking up on the ideas of Albert Fall, a group of eleven western states rallied in the 1970s to gain control of the development of the resources on their public lands. In what has become known as the "Sagebrush Rebellion" and the "Wise Use Movement," these groups have gained national attention and often been able to bypass government regulations (Gottleib 1993).

These ideas, however, have found new life in the administration of George W. Bush, who has specifically guided the mindset for efforts to harvest remaining domestic supplies of energy-producing resources. Guided by leaders such as the Assistant Secretary of the Interior J. Steven Griles, who served under Watt and then spent years as a consultant to the energy industry, the Bush administration pushed to open up vast new tracts of federal land to oil, gas, and coal development. This was most pronounced with the administration's view of drilling for oil in ANWR.

In agreeing to the construction of the Trans-Alaska Pipeline in 1977, President Jimmy Carter created ANWR to ensure that oil development in northern Alaska was carried out under strict limits. The 1970s debate over drilling in northern Alaska and constructing a Trans-Alaska Pipeline to Valdez was referred to as "a step toward energy independence." Specifically, the Nixon administration assured Americans that the Trans-Alaska Pipeline would cause oil imports from the Middle East to fall from 27 percent of the U.S. supply to 20 percent by 1990.

President Jimmy Carter's willingness to allow these developments was predicated on his effort to formalize restrictions on additional such activities in undeveloped areas of Alaska. From 1978 to 1980, President Carter established ANWR and set aside 28 percent of Alaska as wilderness. Carter announced this watershed legislation as follows: "We have the imagination and the will as a people to both develop our last great natural frontier and also preserve its priceless beauty for our children...." With the passage of the Alaska National Interest Lands Conservation Act in 1980, the BLM was ordered to oversee 11 million acres of Alaska as wilderness. This included ANWR (Nash 1982).

Debate on further development in these areas of Alaska awaited the administration of George W. Bush. As part of the administration's energy plan, a new priority was placed on developing energy resources on federal lands. It appears that drilling was approved by a Republican-led Congress in Spring 2005.

Conclusion: Carbon Counting and the Hidden Costs of Mined Energy Resources

Although there are environmental impacts to mining for any mineral, it appears that the greatest impact of the fossil fuel era will be the pollution that burning these resources for energy placed in our ecological commons, including the air and ocean that all humans need to survive. These are the costs of cheap energy use that are not contained in consumer utility or gas bills, nor are they paid for by the companies that produce or sell the energy. The related environmental costs, then, include human health problems caused by air pollution from the burning of coal and oil, damage to land from coal mining and to miners from black lung disease, and environmental degradation caused by global warming, acid rain, and water pollution. Finally, it has also become increasingly clear that protecting our access to foreign sources of oil has become the primary threat to national security, even meriting the use of military force. To offset the almost inevitable impacts of travel and even simply living in the twenty-first century, some individuals have begun determining their impact in terms of the amount of carbon it has put into the environment. Environmental organizations have offered numerical formulas that put a price tag on these emissions. The practice of carbon offsetting is when individuals use these calculations to compute the financial impact of their actions and then pay this amount to an environmental organization or fund that contributes to solving the problem.

Wanting to appear environmentally aware, some organizations have also begun to search for ways of offsetting their carbon impact. In September 2007, for instance, the Vatican accepted one such donation: A Hungarian company named Klimata paid for the restoration of an ancient forest in their country. The replacement of the thirty-seven-acre forest will absorb as much carbon dioxide as the Vatican will produce during the year.

Sources and Further Reading: Opie, *Nature's Nation*; Price, *Flight Maps*; Rothman, *The Greening of a Nation*; Steinberg, *Down to Earth*.

OFFSHORE OIL DRILLING IN THE GULF OF MEXICO

Time Period: 1950s to the present
In This Corner: Natural elements and limitations
In the Other Corner: Oil companies, regional developers, offshore companies
Other Interested Parties: Petroleum consumers
General Environmental Issue(s): Petroleum, marine, offshore drilling

When the first offshore wells were drilled in the shallow Oil Creek of Pennsylvania during the world's first oil boom in the 1860s, few could imagine that such a feat could be carried out on the earth's wildest seas. When wells came in successfully on edges of California beaches in the 1890s, oilmen began to seriously ponder the possibility of leaving sturdy ground to search for oil. Of course, this idea had one primary appeal going for it: the ocean was owned by no one.

From negligible production in 1945, offshore drilling today produces about 34 percent of the world's crude oil and about 25 percent of the world's natural gas. Of all offshore provinces in the world, the Gulf of Mexico is the most explored, drilled, and developed. Today, in the continental shelf waters off Louisiana and Texas, there are nearly 4,000 active platforms servicing 35,000 wells, and 29,000 miles of pipelines. Output from the Gulf, providing close to one-third of U.S. oil and gas production, already exceeds Texas's onshore output and will soon surpass that of Alaska (Priest 2007).

Just as no one owns the property used for offshore petroleum development, few agencies monitor or limit development away from land. This makes many critics urge restraint or, in some areas, moratoriums on offshore development. One thing is clear, however: despite its difficulties and debate, offshore production is a critical part of the petroleum industry.

During its half-century of development, the offshore industry has faced one primary difficulty: overcoming the natural limitations of its task with technological innovation. In addition, however, there remain critics of offshore development who debate the threat of oil spills, rusted well heads, or the blight of working rigs (or abandoned rigs) littering coastal areas.

Technical Difficulties

The first experiments with offshore drilling in the ocean occurred in California. After drilling a large number of wells, early California oilmen noticed that those nearest the ocean were the best producers. Eventually, they drilled several wells on the beach itself. Then in 1887, the oilman H. L. Williams began experimenting with building a wharf that would support a drilling rig that could then be used offshore. His first offshore well extended about 300 feet into the ocean. When Williams' well produced, competing wharfs sprung up along shore, some reaching 1,200 feet in the Pacific. Offshore developments in the Gulf began in the early 1900s. In Louisiana, state initiatives spurred the changes. State and local levee districts leased millions of acres to oil interests. Most famously, Governor Huey Long brokered deals, some of which were corrupt, through his Win or Lose Oil Company that had obtained leases on public lands very cheaply. Once they owned the land, they turned around and subleased it to oil companies. Corrupt or not, such leasing practices in Louisiana stimulated oil development in marine locations onshore as well as in the open sea. Texaco, which had emerged as a leader in fields along the coastal plain, used some of its growing capital to experiment with new techniques, particularly the use of a "submergible barge" in 1933–1934 to drill in soft-bottomed wetlands (Priest 2007).

In 1938, the first free-standing structure to produce oil in the Gulf was built by Pure Oil and Superior Oil in the Creole field, a mile and a half from the city of Cameron, Louisiana. After World War II, the submergible barge was used more widely in shallow open water. In 1947, the Kerr-McGee Corporation drilled the first well from a fixed platform offshore out of sight of land in the Gulf of Mexico. Its platform and the barge used to drill the well set the standard for a new era in offshore development. By 1949, the Gulf of Mexico held eleven fields and approximately forty-four exploratory wells (Priest 2007).

Noticing the enormous possibilities of offshore development, the U.S. government in 1953 passed the Submerged Lands Act to give the federal government title and ownership of submerged lands, extending three miles out to sea. Additionally, the Outer Continental Shelf (OCS) Lands Act authorized the secretary of the interior to lease additional submerged lands for mineral development. In short order, the legal framework was set for large-scale American offshore exploration (and for the federal government to profit from it). During the 1970s, investment in the Gulf of Mexico rose to $16 billion and the number of oil platforms rose to approximately 800.

New Technologies and Opportunities

Tied in with massive amounts of capital, offshore technology used in the Gulf and elsewhere improved dramatically after 1950. Although the technology for locating wells changed most dramatically, methods of drilling and other aspects of the process changed as well, including computer modeling of the ocean floor that also allowed designers to factor ocean waves and tides into platforms and the use of robotics and remote controls on production rigs, and Shell Oil Company even began using remotely operated submersibles as self-positioning drill-ships. These innovations continued, and today this unimaginable effort to bring oil up from beneath the ocean is fairly straightforward.

Today, the typical procedure for locating any type of petroleum well begins with geologists referring to surface features on the landscape. This, of course, was of little use on the ocean bottom before methods were found to make the deep ocean floor visible. Satellite images and gravity meters have opened the ocean bottom to the eye of geologists. The magnetometers are used to measure tiny changes in the earth's magnetic field, which is an indicator of flowing oil. Finally, possibly the biggest change has occurred with the use of seismology. By creating shock waves that pass through hidden rock layers, geologists can then interpret the waves that are reflected back to the surface and deduce where a reservoir of oil is most likely to occur. Today, much of the guesswork has been removed from locating wells anywhere on earth; therefore, locating undersea wells is more likely as well. After locating the supply of petroleum, drilling is carried out by one of four different devices: submersibles; jackups, which have legs that can reach to the sea bottom; drill ships, which drill through a hole in their hulls; or semi-submersibles, which are more stable than ships resting on the ocean surface. Once the oil or gas is struck, the drilling rig is replaced by a production platform. These are made of steel and fixed to the seabed with steel piles. The design includes concrete structures, on the seabed, that are big enough to store oil. The world's biggest platforms are bigger than a football field and rise above the water as high as a twenty-five-story office tower. Often, 500 or more workers live aboard the platform, making it a modern-day boomtown for oil development.

Difficulties of Design in the Gulf

From the earliest days of American offshore design, the Gulf of Mexico confronted engineers with novel conditions arising from the force of waves and from frequent, powerful storms and hurricanes. Oil companies developed a particular technological infrastructure for use in the Gulf, including platforms, pipelines, drilling and support vessels, as well as onshore support centers. In most cases, if oil was located in the Gulf, marshes had to be cut and dredged all along the Gulf Coast to allow entry for the thousands of miles of pipelines from the ocean. Entire towns developed to support this undertaking.

Other challenges of the Gulf include widely varying deposits of hydrocarbons. In most petroleum reserves, concentrated supplies allow for rapid extraction. Historian Tyler Priest observed that "… substantial discoveries have been made in the Gulf basin for the past nine decades." In contrast to most other major fields in which reserves are concentrated in a small number of world-class "giant" fields (fields with a known recovery of more than 500 million barrels of oil equivalent [boe]), "the Gulf basin has yielded thousands of smaller fields of less than 50 million boe, as well as 'large' fields of 50 to 500 million boe and giant fields" (Priest 2007).

From the 1950s through the 1970s, production from offshore Gulf of Mexico helped stave off the rapid exhaustion of U.S. oil and gas reserves. Since then, wrote Priest, "even with the new deepwater discoveries of the 1990s and 2000s, the overall trend in the Gulf has been the depletion of reserves and declining overall production." To succeed in the Gulf, however, oil companies needed to develop platforms that could endure frequent hurricanes.

Just when development in the Gulf really began to accelerate, three "100-year" hurricanes in 1964, 1965, and 1969 (Hilda, Betsy, and Camille) damaged many producing platforms and inflicted damage on the complex network of marine pipelines (Priest 2007). On the heels of this difficulty came the public response to the oil spill in Santa Barbara, California, in 1969. In 1970–1971, three Gulf platforms owned by Shell, Chevron, and Amoco experienced similar difficulties. As criticism of offshore development increased, companies reconsidered their safety practices. With more awareness given to safety after 1970, offshore petroleum and gas activity in the Gulf increased through the end of the twentieth century.

Using Federal Regulation to Stimulate Growth

As much as new exploration and drilling technology allowed Gulf development to accelerate, federal government policies also were responsible for the revival of offshore oil in the Gulf. A primary portion of this federal role was to create an easier to use system for land acquisition and dispersal. By the late 1950s, confusion over property ambiguity was slowing down oil development in the Gulf.

After a lengthy political battle over the "tidelands," in which federal vs. state offshore jurisdiction became an important issue in the 1952 presidential election, President Dwight Eisenhower signed legislation that "quitclaimed" submerged lands out to three miles from the coast back to the states and authorized the federal government to lease land beyond three miles. During the mid-1950s, the federal government created Gulf of Mexico lease auctions. Louisiana filed suit to stop the auctions and demanded a more precise legal definition for the term "coastline." A complicated interim agreement between Louisiana and the federal government dividing the Gulf into zones of overlapping jurisdiction was finally worked out in 1956 (Priest 2007). By 1962, the numbers had grown dramatically and the federal government leased nearly two million acres (more than all previous offshore sales combined). Expansion of offshore development continued during the ensuing decades. In the 1980s, oil companies reached depths of 7,500 feet, or nearly 1.5 miles. Platforms became massive, portable entities. In fact, the U.S. National Aeronautics and Space Administration reported that two man-made objects were observable from the moon's surface: one was the Great Wall of China; the second was Shell Oil's platform "Troll," which stood in the 1,000 feet of water in the North Sea, approximately 1,500 feet high. In addition, three-dimensional seismic imaging provided scientists with better information than ever before on the location and size of undersea reserves. Legal changes also arrived during the 1980s. Congress passed the Federal Oil and Gas Royalty Management Act in 1982, which forced all oil and gas facilities under construction to integrate methods for maintaining the environment and conserving resources. To this end, the Department of the Interior created the Minerals Management Service (MMS) whose job it was to manage mineral resources in an environmentally sound and safe manner and to timely collect, verify, and distribute mineral revenues from federal and Indian lands. In 1981, Congress protected America's coasts, beaches, and marine ecosystems from the threats of oil and gas development when they adopted the OCS Moratorium. The Reagan administration seized the sea commons with Proclamation 5030 in 1983, which claimed rights to land up to 200 miles off the U.S. coastline; however, the moratorium has prevented the leasing of coastal waters for the purpose of fossil fuel development. Every year since then, Congress has renewed the moratorium on new oil and gas development off the Atlantic and

Pacific coasts as well as Bristol Bay, Alaska. In 1990, President George H. W. Bush authored an additional level of protection, and in 1998 Bill Clinton extended these protections and set them to expire in 2012.

Conclusion: Favorable Political Winds Belie a Limited Supply

Often, policy initiatives have been behind offshore growth. For instance, the 1980s brought the controversial appointment of James Watt, an antagonist of the environmental movement, as secretary of the interior under President Ronald Reagan. Watt established a new "area-wide leasing" system offshore. Priest explains that this system "put into play entire planning areas (e.g., the central Gulf of Mexico) up to 50 million acres, as opposed to tracts specifically nominated and offered under the tract selection system. In other words, companies could bid on any tract they wanted in the planning area, rather than have to choose from a limited number of carefully selected ones, and they would be more likely to acquire them in bunches, giving them greater control over large prospects."

In recent years, Congress has debated increasing offshore development even further. In particular, some members wish to overcome the limits of the OCS Moratorium. Fueled by the administration of George W. Bush, new initiatives were explored to vastly expand development along the coasts of California and Florida.

In 2006, the MMS released its five-year OCS planning document that details future leasing and development in areas of the Gulf as well as Alaska and the Mid-Atlantic coast. The plan calls for aggressively opening up leasing in vast new areas. The document also calls for the end of the current presidential deferrals protecting Bristol Bay in Alaska. The MMS held public hearings on the proposal, nine hearings in Alaska and one apiece in Alabama, Florida, Louisiana, Virginia, and Texas. Critics contend that such radical changes in offshore development will have implications for the marine environment for generations.

Other groups, of course, oppose offshore wells anywhere. These opponents argue that offshore wells cause a wide range of health and reproductive problems for fish and other marine life, expose wildlife to the threat of oil spills that would devastate their populations, and destroy kelp beds, reefs, and coastal wetlands. They cite statistics that estimate that, during its lifetime, a single oil platform will dump more than 90,000 metric tons of drilling fluid and metal cuttings into the ocean, drill between fifty and one hundred wells, each dumping 25,000 pounds of toxic metals, such as lead, chromium, and mercury, and potent carcinogens such as toluene, benzene, and xylene into the ocean, and pollute the air as much as 7,000 cars driving fifty miles a day. However, as decreasing oil supplies force companies to go after more difficult to reach reserves, more offshore drilling is likely to be needed, and, very likely, the federal government will evolve an even more active role in fostering such development.

Sources and Further Reading: Pratt et al., *Offshore Pioneers*; Priest, *Extraction Not Creation: The History of Offshore Petroleum in the Gulf of Mexico*, http://es.oxfordjournals.org/cgi/content/abstract/khm027v1; Priest, *Offshore Imperative*; Yergin, *The Prize: The Epic Quest for Oil, Money & Power*.

ENRON BANKRUPTCY AND THE NEW ROLE OF ENERGY BROKERS

Time Period: 1990s
In This Corner: Energy traders, politicians

In the Other Corner: Federal regulators, American consumers
Other Interested Parties: Energy producers, mining interests
General Environmental Issue(s): Energy

Seen as one of the nation's most innovative companies, Enron became the leading example of a late 1990s precedent for corporate corruption and greed. Even when it was doing business as usual, however, Enron represented a new development in the commodification of energy. Taking the concepts that were at the root of Rockefeller's Standard Oil Trust, Enron and others created the field of energy trading, which was now possible with electronic stock trading.

A Houston Powerhouse

After its founding in 1985, Enron, which was based in Houston, Texas, became the nation's seventh-biggest company in revenue by emphasizing the transmission and distribution of electricity and gas throughout the United States and the development, construction, and operation of power plants and pipelines worldwide. Following the trend (particularly in states such as California) toward energy deregulation, Enron became the essential middleman between energy producers and consumers. As a result, Enron was named "America's Most Innovative Company" by *Fortune* magazine for five consecutive years, from 1996 to 2000. Viewed as a great technological and business innovator, Enron, and its chairman, Kenneth Lay, were credited with creating the energy markets that it grew to dominate.

For most of the twentieth century, utility companies had generated, transmitted, and sold electricity as state-regulated monopolies. They also built and maintained the "electrical grid," the network of transmission wires that carries electricity to homes and businesses. With deregulation, companies needed to ensure that their supply of power on the grid would remain consistent. This odd role of energy trading, without necessarily owning power plants or supplies of raw material, was Enron's route to success and ultimately to failure. With its transactions taking place entirely on paper, Enron's business was tantalizingly tempting to illegal manipulation. For instance, the company has been accused of manipulating energy prices to create shortages in California in the summer of 1999. These practices, however, made Enron great sums of money, which, of course, made the company even more enticing to investors.

Traders at Enron were among those who took advantage of California's poorly constructed deregulation law and helped to bring about the state's energy crisis of 2000 and 2001. They concocted schemes to manipulate electricity markets and to maximize Enron's profit, using names like Fat Boy and Death Star to describe the strategies. Some bantered casually in 2000 about how they were "stealing" from California and sticking it to "Grandma Millie" by overcharging for power, according to audiotapes of their conversations that have been made public.

The Bottom Falls Out

In a six-week downward spiral during 2000, Enron disclosed a stunning $638 million third-quarter loss, the Securities and Exchange Commission opened an investigation into the partnerships, and the company's main rival backed out of an $8.4 billion merger deal. After a series of scandals involving irregular accounting procedures bordering on fraud involving the company and its accounting firm, Arthur Andersen, Enron filed for protection from creditors on December 2, 2001, in the biggest corporate bankruptcy in U.S. history. Its stock, worth more than $80 about a year before, had tumbled to less than $1 per share. Enron's collapse left investors burned and thousands of employees out of work with lost retirement savings. In addition, Enron, which had 20,000 employees, barred them from selling Enron shares from their retirement accounts as the stock price plunged, saying the accounts were being switched to a new plan administrator. Former Enron Chief Financial Officer Andy Fastow was indicted on November 1, 2002 by a federal grand jury in Houston on seventy-eight counts, including fraud, money laundering, and conspiracy. He is serving a ten-year prison sentence and forfeited $23.8 million. The swift fall of this corporate giant caught investors by surprise. Maybe the best symbol of how quickly corporate entities could come and go was the new baseball stadium opened in Houston in 2000, which was named Enron Field. After the company collapsed, the Houston Astros baseball team needed to pay the company $5 million to have its name removed from the stadium. The team cited the need to do so to avoid the negative publicity associated with the former model of corporate success.

Ultimately, Enron's failure was not tied directly to the actions of traders, who made hundreds of millions of dollars for the company, but company traders were speculating on energy prices and aggressive accounting of risky long-term energy contracts made Enron even more susceptible to a blowup. Therefore, although Enron was a symbol of society's angst about corporate ethics, its business of trading on energy prices was a product of a new era in resource management, and it has outlived the company that perfected it.

Conclusion: Energy Trading on Scarce Resources

Enron, once the country's seventh-largest company, introduced its modern trading floor on national television. Today, companies such as Centaurus have learned from Enron's example. Since its 2002 founding, Centaurus has amassed $1.5 billion in assets under management, but it is doing so with a low profile. Energy trading, given a bad name by Enron, is springing to life again.

Volatile energy markets and record-high commodity prices are prompting renewed interest from investors. That has pushed banks and a growing number of hedge funds to hire more energy traders to maximize their ability to profit from the volatile sector. Whether Americans agree with such ethics, many appreciate the growth that it enables in their investment portfolios.

Sources and Further Reading: McLean, *The Smartest Guys in the Room*; Munson, *From Edison to Enron: The Business of Power and What It Means for the Future of Electricity*; Swartz, *Power Failure*.

DOMINICAN REPUBLIC–CENTRAL AMERICAN FREE TRADE AGREEMENT

Time Period: 2003 to the present

In This Corner: United States industries and Central American and Dominican Republic governments

In the Other Corner: Environmental groups, labor organizations

Other Interested Parties: Agriculture

General Environmental Issue(s): Sustainable development, trade, environment

The Central American Free Trade Agreement (CAFTA) started as a regional free trade agreement among the United States, Guatemala, El Salvador, Honduras, Nicaragua, and Costa Rica. Finalized in December 2003, this agreement created a plan to phase out tariffs within a ten-year period to create a free trade zone. In March of 2004, the member nations added one more country, the Dominican Republic, to the agreement. With this addition, the number of member nations was raised to seven, and the official name of the agreement was changed to the Dominican Republic–Central American Free Trade Agreement (DR-CAFTA).

The DR-CAFTA was created in the name of fostering commerce. The region covered by the DR-CAFTA is the second-largest Latin American export market for the United States (with Mexico being the largest), buying $15 billion of goods a year (Cisneros 2005). It was designed to include the economic sectors of agriculture, manufacturing, and services, in addition to addressing the issues of intellectual property rights, investment, and food safety, among others. Designed to create a free trade zone similar to the NAFTA, DR-CAFTA will immediately eliminate tariffs on about 80 percent of the United States exports to the participating countries. According to the agreement, remaining tariffs will be phased out over the following ten years. The DR-CAFTA does not require reduction in U.S. import duties mainly because of the fact that the majority of these countries' goods already enter the United States duty-free as a result of the U.S. government's Caribbean Basin Initiative.

The U.S. President George W. Bush signed the legislation in August 2005, and the Dominican Republic, El Salvador, Guatemala, Nicaragua, and Honduras all approved the agreement by mid-2006. However, for the agreement to come into effect, it must be approved by all of the member countries. Unlike the other countries, as of July 2007, Costa Rica had not approved the DR-CAFTA. The delay in Costa Rica's approval is rooted in ongoing opposition from a wide range of civil society groups (Washington Office on Latin America 2007). A nationwide coalition, including environmental and labor rights groups, among others, continues to organize frequent protest and marches, maintaining a strong opposition to the DR-CAFTA. Costa Rica held a referendum on DR-CAFTA in October 2007 to vote "yes" or "no" for the DR-CAFTA.

Similar to NAFTA, the DR-CAFTA had a wide variety of opponents from the beginning, but these changed with the development of the DR-CAFTA. In the United States, originally many members of Congress opposed another free trade agreement with Latin American countries. Those members of Congress who most strongly opposed the DR-CAFTA were, not surprisingly, from regions of the United States where there was likely to be competition among local industry and agriculture and the DR-CAFTA products. After NAFTA was passed, U.S. Congressional districts most heavily hit by free trade experienced an "aftershock." This aftershock and the coordinated outrage of constituents who found themselves negatively economically impacted by NAFTA, emboldened representatives and senators of the affected areas against further trade agreements. In addition to these groups, many labor and environmental organizations were also opposed to the DR-CAFTA from the beginning, believing conversations for the development of the agreement did not adequately address environmental and labor concerns. Once the DR-CAFTA was finalized, there was some, albeit minimal, change in the supporters and opponents of the agreement.

Once the DR-CAFTA was finalized, supporters of the agreement included many U.S. corporate backers such as the National Association of Wheat Growers, Microsoft, the National Association of Manufacturers, the Heritage Foundation, and the United States Chamber of Commerce, along with several Central American environmental and development organizations. These groups claim the DR-CAFTA will open new markets to manufacturers, improve the economies of the Central American nations and the Dominican Republic, create worker rights protections, and improve labor laws and environmental standards in the member countries. They note that the DR-CAFTA includes a chapter on the environment, encompassing many important provisions that will help promote improved governance and environmental stewardship in Central America and the Dominican Republic. The member countries are obligated to effectively enforce their own environmental regulations and to strive to improve their domestic laws. Similarly, supporters also emphasize that each member country in the DR-CAFTA pledges to not fail to effectively enforce its own labor laws and to strive to protect internationally recognized worker rights.

In contrast, critics of the DR-CAFTA include many U.S. environmental groups and trade unions, along with small businesses and labor groups in Central America and the Dominican Republic, and some economists. These groups and individuals believe that the agreement is a push toward corporate globalization. They contend the agreement will negatively impact the small Central American countries and the Dominican Republic by lowering living standards and labor protection, eliminating jobs in the region, harming environmental protection, exacerbating unsustainable trade deficits, and promoting privatization and deregulation of fundamental public services in the region. In addition, there is a fear that the U.S. historical dominance in Central America would be expanded, devastating local small business as a result of their inability to compete on economies of scale. Economists' main critique of the DR-CAFTA is that this type of bilateral and regional free trade agreement might undermine the push for a global trade agreement through the WTO. This is of essence because a global trade agreement through the WTO has greater potential for increasing total social welfare because all members of the WTO would be bound by its terms.

The DR-CAFTA is not yet in full effect. If Costa Rica chooses not to approve the agreement, it will need to be modified to deal with the changes. Although many actors are hoping that Costa Rica will not approve the DR-CAFTA, many other powerful actors are hoping it will be approved. Time will tell what changes are needed in the DR-CAFTA and whether the supporters or the opponents were right with their assessments.

Sources and Further Reading: Andrews, "How CAFTA Passed House by 2 Votes"; Cisneros, "DR-CAFTA Holds Promise for San Antonio and Texas," http://ied.utsa.edu/itc/itt2/June 05/feature 2060805.html; Jaramillo and Lederman, *Challenges of CAFTA: Challenges and Opportunities for Central America*; Mazzei, "Two Months of CAFTA," http://americas.irc-online.org/am/3664; Spotts, *CAFTA and Free Trade: What Every American Should Know*; Trejos, "CAFTA in Costa Rica Would Cause Deepening Inequality," http://americas.irc-online.org/am/4575; Tucker, "New Year Sees Delay in CAFTA Implementation," http://americas.irc-online.org/am/3016; Washington Office on Latin America, "Central American–Dominican Republic Free Trade Agreement," http://www.wola.org/cafta.

ROADLESS AREA CONSERVATION RULE ("THE ROADLESS RULE")

Time Period: 1999 to the present

In This Corner: President Clinton, environmental and conservation groups, California, Oregon, New Mexico, Washington, Maine, Montana

In the Other Corner: President George W. Bush, timber industry, minerals industries, Wyoming, Idaho

Other Interested Parties: BlueRibbon Coalition
General Environmental Issue(s): Wilderness protection

The Roadless Area Conservation Rule, otherwise known as "The Roadless Rule," is a piece of legislation that supports the idea of areas of land without roads. Roads mean access. Today, their presence on federal lands has become a key manifestation of the duel between conservation and preservation camps. To preservationists, roads signal the start down a slippery slope of use toward exploitation. For conservationists and developers, roads are the first step toward accessing valuable resources held in many lands owned by the federal government.

Inventoried roadless areas in the United States possess social and ecological values that are important to preserve in an increasingly developed landscape. These areas help to protect air and water quality, biodiversity, wildlife corridors, and opportunities for personal renewal. According to the USGS, conserving inventoried roadless areas allows the present generation to leave a legacy of natural areas for future generations. The Roadless Rule attempts to do just this.

Wilderness Means Roadless

Passed in 1963, the Wilderness Act Established the National Wilderness Preservation System and made roads the enemy of wild areas. The secretary of the interior was directed to review every roadless area of 5,000 acres or more and every roadless island within the national wildlife refuge and national park systems for possible inclusion in the system. The act also included some national forest lands in the system and directed the secretary of agriculture to recommend others. Roadlessness became a key measure for an area's qualification as wilderness. Today, more than one hundred million acres have been included in the National Wilderness Preservation System so far.

The Wilderness Act also required the review of any area within a national forest that was already classified as wilderness, wild, or canoe before the act was adopted was automatically considered a wilderness area and included in the system. With limited exceptions, no commercial enterprise or permanent road is allowed within a wilderness area. Temporary roads, motor vehicles, motorized equipment, landing of aircraft, structures, and installations are only allowed for administration of the area.

Establishing the Roadless Rule

The development of The Roadless Rule began on October 13, 1999, when President Clinton instructed the USFS to develop regulations to provide appropriate long-term protection for USFS inventoried roadless areas. The rulemaking process eventually included more than 180 American Indian and Alaska Native groups and seven federal agencies in addition to the USFS. Additionally, more than 600 public meetings were held nationwide, and seven hearings were held before the U.S. House and Senate (United States Forest Service 2007). In January 2001, President Clinton issued the Roadless Area Conservation policy directive, ending virtually all road construction, logging, and mineral development and leasing on the inventoried roadless lands. The policy was published in the Federal Register on January 12, 2001. The 2001 act states the following:

> As urban areas grow, undeveloped private lands continue to be converted to urban and developed areas, and rural infrastructure (such as roads, airports, and railways).

An average of 3.2 million acres per year of forest, wetland, farmland, and open space were converted to more urban uses between 1992 and 1997. In comparison, 1.4 million acres per year were developed between 1982 and 1992. The rate of land development and urbanization between 1992 and 1997 was more than twice that of the previous decade, while the population growth rate remained fairly constant (FEIS Vol. 1, 3–12). In an increasingly developed landscape, large, unfragmented tracts of land become more important.... Subdivision and other diminishment of tract size of these lands can discourage long-term stewardship and conservation.

Inventoried roadless areas provide clean drinking water and function as biological strongholds for populations of threatened and endangered species. They provide large, relatively undisturbed landscapes that are important to biological diversity and the long-term survival of many at risk species. Inventoried roadless areas provide opportunities for dispersed outdoor recreation, opportunities that diminish as open space and natural settings are developed elsewhere. They also serve as bulwarks against the spread of non-native invasive plant species and provide reference areas for study and research. (Fire Effects Information System, Vol. 1, 1–1 to 1–4)

Generally, The Roadless Rule limits or prohibits activities that would most negatively affect the social and ecological element of areas determined to be without roads. More specifically, the rule prohibits new road construction and reconstruction in inventoried roadless areas on USFS lands, except in cases where the roads are needed to protect health and safety, to conduct environmental cleanup or other activity as required by federal law, to allow access to a mineral lease on lands that are under lease, or to prevent irreversible resource damage. The rule also prohibits cutting, sale, and removal of timber in the inventoried areas, again with a few exceptions. Overall, according to the USFS, the number of acres affected by this piece of legislation totals 58.5 million acres, or 2 percent of the U.S. landbase and 31 percent of the USFS landbase (U.S. Forest Service 2007b).

Supporters of The Roadless Rule note that inventoried roadless areas provide benefits to more than 220 wildlife species listed or proposed as threatened or endangered. The roadless areas also provide large, relatively undisturbed blocks of habitat and solitary areas for hikers and other outdoor recreationalists.

Critics of The Roadless Rule point out that the rule would decrease the amount of timber harvested on USFS lands by 2 percent (U.S. Forest Service 2007). This drop would affect more than 450 timber-related jobs within five years and another estimated 550 jobs related to coal and phosphate commodities (U.S. Forest Service 2007). They also suggest that the forests are owned by all taxpayers and that the rule places unnecessary limitations on these public lands. Finally, critics contend that limiting the construction of roads will increase the risk of wildfires (Brazil 2001).

In late January 2001, with the induction of the George W. Bush administration, the implementation of all policies pushed into place during the final days of the Clinton administration were delayed for sixty days. The Roadless Rule fell into this category. In mid-March, the Bush administration delayed the rule for a second time in response to a court challenge seeking to overturn the rule. The Boise Cascade Company, the State of Idaho, and others filed the challenge. By May, the Bush administration said it would allow The Roadless Rule to go into effect but would move to amend it at a later time. Less than a week later, Federal

Judge Edward Lodge issued a preliminary injunction barring The Roadless Rule from taking effect.

Over the next three years, the Bush administration reversed its opposition to the Roadless Rule, actually defending the rule in a case filed by the State of North Dakota, formally settling a case brought by the State of Alaska temporarily exempting the Tongass National Forest, and ultimately proposing a new rule that would allow a state-by-state petition process. The timber industry praised the administration's new rule, whereas environmental organizations objected to the changes. Formal adoption of the new rule occurred in May 2005 with an announcement in the Federal Register.

During 2005 and 2006, states began to petition for protection of all roadless areas, lining behind either the old or new rule. Finally, in September 2006, a federal district court ordered reinstatement of the old Roadless Rule, as it was proposed under the Clinton administration. The court found that, in the process of repealing The Roadless Rule, the Bush administration had failed to comply with the basic legal requirement of the NEPA and the ESA. The court also ruled that the USFS must stop work on the oil and gas and road projects that were approved during the five years that The Roadless Rule was illegally repealed. In the decision on the case of the *People of California v. the Department of Agriculture et al.*, Judge Patricia LaPorte wrote the following:

> FURTHER FOREST SERVICE ACTIONS As the Court previously ordered, federal defendants are enjoined from taking any further action contrary to the Roadless Rule without first remedying the legal violations identified in the Court's opinion of September 20, 2006. Such further actions by the Forest Service include, but are not limited to, approving or authorizing any management activities in inventoried roadless areas that would be prohibited by the 2001 Roadless Rule (including the Tongass Amendment), and issuing or awarding leases or contracts for projects in inventoried roadless areas that would be prohibited by the 2001 Roadless Rule, including the Tongass Amendment. The effective date of this injunction is September 20, 2006. III. SPECIFIC ACTIVITIES IN ROADLESS AREAS: A. Oil and Gas Leases The 2001 Roadless Rule shall apply to activities commenced hereafter with respect to any and all mineral leases in IRAs in National Forest lands not affected by the Tongass Amendment that issued after January 12, 2001. The Forest Service is enjoined from approving or allowing any surface use of a mineral lease issued after January 12, 2001, that has not already commenced on the ground and which would violate the Roadless Rule (including the Tongass Amendment). This order does not apply to roads that have already been constructed or reconstructed on lease parcels pursuant to approved surface use plans of operation, nor does it apply to leases that include a strict "no surface occupancy" condition that already prohibits road construction that would violate the Roadless Rule. (LaPorte Decision)

To date no court orders address the roadless areas in Alaska's Tongass National Forest that were exempted by separate procedure taken by the Bush administration in 2003.

Sources and Further Reading: Andrews, *Managing the Environment, Managing Ourselves*; Brazil, "Forest Plan Greeted with Skepticism: Critics Claim Bush's Roadless Scheme Deceptive"; Earth-Justice, http://www.earthjustice.org/index.html; Eilperin, "Roadless Rules for Forests Set Aside";

Goodlatte, *Status of the Roadless Area Conservation Rule: Congressional Hearing*; Heritage Forests: http://www.ourforests.org/roadless/environmental_benefits.html; Knickerbocker, "Roadless Areas Get Protection—For Now," http://www.csmonitor.com/2006/1207/p03s03-uspo.html; Longley, "'Roadless Rule' Forest Logging Ban Officially Lifted," http://usgovinfo.about.com/od/thepresi dentandcabinet/a/roadlessgone.htm; Opie, *Nature's Nation*; Price, *Flight Maps*; Rothman, *The Greening of a Nation*; Steinberg, *Down to Earth*; U.S. Forest Service, *Roadless Area Conservation: Rulemaking Facts*, www.roadless.fs.fed.us; U.S. Forest Service, *Roadless Area Conservation: Quick Numbers*, www.roadless.fs.fed.us.

AN INCONVENIENT TRUTH BRINGS CLIMATE CHANGE MAINSTREAM

Time Period: 1990 to the present
In This Corner: Environmentalists, atmospheric scientists
In the Other Corner: Energy companies, many politicians of the status quo
Other Interested Parties: World public
General Environmental Issue(s): Climate, energy, pollution

The debate over global climate change took a dramatic turn in 2006–2007 with the release of and ensuing publicity related to the film *An Inconvenient Truth*. In addition to bringing this environmental issue to an immense audience, the film created the twenty-first century's first environmental celebrity: former Vice President Al Gore. As he helped to accept the Academy Award for best documentary for the film, Gore became the darling for a generation searching for ways to combat this problem. Of course, he also moved into the crosshairs of the interest groups arguing that global warming was a great hoax.

Although the film was categorized as a documentary, it took a form that enabled it to power a movement of interest unlike any film before it. The film by director Davis Guggenheim began its journey as an artistic hit at the well-known Sundance Film Festival. Organized around one man's (Gore) fervent crusade to expose the myths and misconceptions that surround global warming, the film is pitched as a last-ditch, all-out effort to help save the planet from irrevocable change. The bulk of the film records Gore giving a slide lecture that he has offered around the world. The film's success is at least partly attributable to the power of Gore's slideshow and his careful, intricate control of the scientific findings.

David Roberts described the film for *GRIST*, the online magazine, in this way:

> It's something of a miracle that *An Inconvenient Truth*, the chronicle of Al Gore's quest to raise alarm about "climate chaos," exists at all. A movie with a scantily clad Jessica Alba presenting a computer slideshow on climate science is implausible enough. Al Gore doing it, well ... even C-SPAN could be forgiven for having second thoughts. (Roberts 2006)

It must also be noted that only a few years before, a fictional account of possible outcomes of global warming had been released as the feature film *The Day After Tomorrow*. The public response was minimal. *An Inconvenient Truth* at once moved the issue into a progressive, hip, mainstream portion of American popular culture and also resurrected the fortunes of Gore, who had been largely out of the public eye since losing the 2000 election.

After the Oscar ceremony, Gore made his first visit to Congress since appearing there to grudgingly accept the results of the 2000 presidential election. He came to Capitol Hill to educate Congress about the implications of global climate change. During his appearance, Gore appeared intermittently as a nerdy science teacher teaching the older generation about scientific theories and cutting-edge findings and at other times as a preacher attempting to convert his listeners. He forecasted what lawmakers would hear from their grandchildren in a few decades, as they asked, "What in God's name were they doing? Didn't they see the evidence? Didn't they realize that four times in 15 years the entire scientific community of this world issued unanimous reports calling upon them to act?"

At other times, Gore appeared as a historian attempting to put the climate crisis in perspective. He reminded lawmakers of the resolve it took to fight Nazism and Communism. He told them that climate change requires the same kind of commitment: "What we're facing now is a crisis that is by far the most serious we've ever faced," he said.

Accepting questions from lawmakers, Gore faced off with Republican Senator James Inhoffe from Oklahoma. Other exchanges, however, were much more productive, even doting. Consider a few excerpts from the transcript:

Warner: International marketplace. Jobs. China and India. How do we persuade them?

Gore: It's a global problem that requires a global solution. Every international deal has the same structure: rich countries have to go first. On China: first, if we go first, they'll follow. Second, they're saying all the right things. They get it.

Sanders: Thanks not only for focusing the world on this, but for giving the young generation the hope that they can become a great generation. The hearts of many young people are beating a little faster because of you.... If we're aggressive on efficiency and new industries, we can create millions of new, good jobs. Speak on the economic advantages.

Lautenberg: Science is settled. Also, I love my ten grandchildren. How's the morale among scientists facing these attacks?

Gore: Hansen is gutsy, but yeah, a lot of them are stressed.

Lautenberg: Skeptics are hard to fathom. Thanks for keeping at it.

Gore: This interference with science should not be a partisan issue. Stand up for the damn scientists.

Lieberman: This work you're doing will be your greatest service. You have intellectual rigor. I think we've reached the tipping point. Everybody's getting on board. We have a chance to do something about it, sooner than I thought we would.

Alexander: About cap-and-trade. Was the cost of sulfur cap-and-trade prohibitive? And why couldn't we start a climate change effort by putting such a cap on electrical utilities?

Gore: Great question. Cap-and-trade was a Republican idea. I was for it. I had no idea how good it would turn out. It was wildly successful. A new proposal would auction credits off rather than giving them away—that would help with the expense. It's a good idea.

Boxer: This means a lot to all of us. Even to those on the other side of the aisle.... Let me just say: you did good, Mr. VP. You're a role model for us all—as elected leaders and as citizens.

Gore: You don't give out any kind of statue, do you? [laughter]

Boxer: I want to make environmentalism bipartisan again. And I want to focus on global warming, to make up time. We'll be having many more hearings (*An Inconvenient Truth*).

This was hardly the first time Gore tried to motivate Congress to respond to global warming. He first held a hearing on the topic more than twenty-five years ago, not long after he started in the House of Representatives. At that time, he served with another young idealist, Representative Ed Markey (D-MA). Markey told Gore he was ahead of his time on climate change and other issues. "What you were saying about information technologies, what you were saying about environmental issues back then, now retrospectively really do make you look like a prophet," Markey said. "And I think that it would be wise for the Congress to listen to your warnings, because I think that history now has borne you out." As if this recognition was not enough, late in 2007 Gore and his partner, UN's Intergovernmental Panel on Climate Change, were awarded the 2007 Nobel Peace Prize for their efforts to publicize the issue of climate change and to educate humans about what must be done.

Conclusion: A Rock Star for Planet Earth

Although Al Gore's appearance before Congress on behalf of the issue of climate change was remarkable, the oddest byproduct of this initiative was a series of rock concerts called Live Earth. Surrounded by many genuine rock stars, Gore, a policy wonk, became a rock star himself.

Live Earth, which took place on seven continents (in cities Sydney, Tokyo, Shanghai, Johannesburg, Hamburg, London, New York, and Rio de Janeiro) during a span of twenty-four hours, hoped to bring the issue of climate change to a global audience. Although most ticket buyers came for the music, they were also treated to lessons, by rock stars and speakers, about how to reduce their own carbon emissions and campaign for serious political action.

Everyone watching (in person or on live television broadcasts) was urged to sign up by text or online for a sevenfold pledge to plant trees, protect forests, buy from eco-friendly businesses, vote for green-minded politicians, and make "a dramatic increase" in energy savings, but the pledge also involved promising to fight for new laws and policies, to demand that their country sign a new treaty, and the very specific demand that any new coal power station be able to trap and store the carbon dioxide it produces.

The featured speaker was Gore himself. In an era of multimedia information technology, Gore had positioned himself as the Rachel Carson of the twenty-first century.

Sources and Further Reading: *An Inconvenient Truth*: http://www.climatecrisis.net/; Opie, *Nature's Nation*; Price, *Flight Maps*; Roberts, "Gore's New Flick, *An Inconvenient Truth*, Improbably Succeeds" Rothman, *The Greening of a Nation*; Steinberg, *Down to Earth*.

INVASIVE SPECIES

Time Period: Since the 1500s
In This Corner: Native species, environmentalists, economists

In the Other Corner: Invasive species
Other Interested Parties: Ecologists, farmers, U.S. government
General Environmental Issue(s): Habitat loss, biodiversity, elimination of native species

With increasing levels of international trade and travel, the possibility for introducing non-native species is on the rise. The introduction of most non-native species does not prove to be a problem, but if the species is invasive the impact on native species and habitats can be extensive. Invasive species are non-native species that invade habitats and ecosystems, causing adverse economic, ecological, and/or health affects. Although invasive species are a small percentage of non-native species, their numbers are increasing and, along with that, the damage they do.

According to Executive Order 13112 and the National Invasive Species Information Center website (2007), the definition of "invasive species" is twofold. First, invasive species are "non-native to the ecosystem under consideration." Second, the introduction of invasive species must cause or be likely to cause "economic or environmental harm or harm to human health." This is in contrast to most non-native species, which, when introduced to a region, do not cause extensive negative impacts on the economy or ecology (National Invasive Species Information Center 2007). In general, both non-native and invasive species are introduced from other areas around the world; however, some species native to one region of the United States can become invasive in a different region of the country. Thus, as both domestic and international connections expand, care needs to be taken to minimize the introduction of non-native species.

For species to take hold in a new environment, they need to find a niche in which to grow and develop. According to the National Invasive Species Information Center, less than 15 percent of all introduced species will find a niche in the wild that they will be able to exploit. Even of the species that do find a niche, all but 1 percent will grow in harmony with native species. It is the 1 percent of all introduced species that cause harm to the ecosystems they invade; these are the invasive species.

Characteristics of invasive species vary greatly, but there are a few things that many seem to have in common: they do not have natural predators in the invaded area, they favor ecosystems that are generally lacking diversity (it is easier for them to find a niche they can exploit), and they are likely to invade in areas disturbed by fire, construction, agriculture, or other form of ecosystem disruption.

Invasive species in the United States originate from all regions of the world. They began invading the country in the mid-1500s, and their invasions continue. Most of the invasive species reach the United States as a result of human contact, most frequently via "hitch-hiking." Unbeknownst to carriers, invasive species will be transported to the United States in ship ballast water, in shipping crates, or on tourists and travelers. In addition, on occasion, invasive species are intentionally brought into the country. Most frequently, this occurs when individuals want to use the species in the United States for landscaping or as a food source or pest control. Finally, there are a few natural pathways that can also transport invasive species. Ocean currents, wind currents, or migrating animal species can carry invasive species.

The effects of invasive species include economic, environmental, social, recreational, and health costs. Examples of costs include decreased species diversity, deteriorating infrastructure such as clogged water pipes attributable to zebra mussels, loss of sport fishing in regions such as the Great Lakes, illness caused by West Nile virus, and higher prices for food as a result of the expense of increasing use of pesticides against invasive weeds. Each of these costs results in a change in lifestyle for many people and is consequently creating greater levels of support to deal with the problems of invasive species. Furthermore, because of the wide variety of ecosystems being impacted, people in all types of fields and across the United States now recognize the need to minimize further invasions and thus are supportive of steps to stop the transportation of additional non-native species.

With globalization, the United States has seen an increasing number of invasive species. Today, the U.S. federal government views invasive species and habitat destruction as the greatest threats to biodiversity. To protect native ecosystems, avert habitat and crop destruction, and avoid the elimination of native species used as natural resources, as food products, or for recreational activities, billions of dollars are spent in the United States each year to combat invasive species. It appears this money is not being spent in vain. Although damage by invasive species is still pervasive, surveys and educational programs show that the public is not only taking notice but is working to help stop the advancement of invasive species. Efforts made by the public include support, in 1999, of President Bill Clinton's Executive Order 13112. This executive order created the National Invasive Species Council to help coordinate federal agency efforts to eliminate invasive species. Even at the local and state levels, interest from private citizens has increased control and eradication efforts, setting up measures for even greater protection in the future.

Sources and Further Reading: Blumenthal, "In East Texas, Residents Take On a Lake-Eating Monster"; Burdick, "The Truth About Invasive Species: How to Stop Worrying and Learn to Love Ecological Intruders," http://discovermagazine.com/2005/may/cover; Coates, *American Perceptions of Immigrant and Invasive Species: Strangers on the Land*; Elton, *The Ecology of Invasions by Animals and Plants*; Long, "Targeting Ships' Ballast Stowaways," http://online.wsj.com/public/article/SB118540871364378290.html; Mooney and Hobbes, *Invasive Species in a Changing World*; National Invasive Species Information Center, http://www.invasivespeciesinfo.gov/; Ramsey, Kaufman, and Kaufman, *Invasive Plants: A Guide to Identification, Impacts, and Control of Common North American Species*; Sax, Stachowicz, and Gaines, *Species Invasions: Insights into Ecology, Evolution, and Biogeography*.

IVORY-BILLED WOODPECKER

Time Period: 2004 to the present
In This Corner: Ornithologists, the National Audubon Society, the Cornell Laboratory of Ornithology
In the Other Corner: Timber industry, bird collectors
Other Interested Parties: Southern governors in the 1930s
General Environmental Issue(s): Deforestation, habitat loss, species extinction

The ivory-billed woodpecker (*Campephilus principalis*) was first assumed to be extinct in the 1920s. Today, it is officially listed as an endangered species but is considered extinct by a number of researchers.

The ivory-billed woodpecker was described first in 1731; a century later, John James Audubon painted the bird and wrote a detailed description (Audubon Arkansas 2005). The ivory-billed woodpecker is a member of the woodpecker family. At twenty inches tall, twenty ounces in weight, and with a thirty-inch wingspan, the ivory-billed woodpecker is the largest of the woodpecker species in the United States. Distinctive characteristics of the ivory-billed woodpecker include a white bill and dramatic crest along with blue-black with white markings on its neck and back, and white markings on the trailing edge of the upper- and underwings. In addition, the ivory-billed woodpecker's bill is unique, with a tip that is flattened laterally, and these birds have a distinctive drum, often a double knock, and a call that sounds like a toy trumpet.

The natural habitat of the ivory-billed woodpecker is the thick hardwood swamps and pine forests of the southern United States. At one time, the bird ranged as far west as eastern Texas east to North Carolina and included territory north to the southern part of Illinois and as far south as Cuba. After the American Civil War and the ensuing timber boom, much of the traditional habitat was deforested, leaving only isolated tracts of habitat for the ivory-billed woodpecker.

In the 1870s, a lumber boom started in the forest of the south, directly impacting the prime habitat of the ivory-billed woodpecker. This, coupled with collectors hunting the bird, led to a rapid decline in the species. Thus, by the 1920s, the ivory-billed woodpecker was assumed to be extinct. However, in 1924, Arthur A. Allen, founder of the Cornell Laboratory of Ornithology, reported seeing the bird in Florida. Allen's interest in the bird continued, and, in 1935, he confirmed the existence of ivory-billed woodpeckers in the wild, making the only known audio and video recordings of the ivory-billed woodpecker. In the late 1930s, James Tanner estimated that only twenty-five of these birds remained in the wild. Six to eight of the remaining birds were located in the Singer Tract in Louisiana, an area of old-growth forest, and others were located in Florida. After Tanner's assessment of the ivory-billed woodpecker, the southern governors and the National Audubon Society led a movement to publicly purchase the Singer Tract in Louisiana. The movement supported the creation of a reserve for the ivory-billed woodpecker. Unfortunately, before this was achieved, the Chicago Mill and Lumber Company took advantage of their logging rights and clear-cut the Singer Tract forest (Jackson 2004).

The last officially documented sighting of the ivory-billed woodpecker in the United States occurred in 1944. In March 1967, the bird was listed as an endangered species. Finally, the last officially documented international sighting occurred in 1987 in Cuba (Audubon Arkansas 2005). Although in both the United States and Cuba there were numerous unconfirmed sightings of the ivory-billed woodpecker though the 1990s (Mendenhall 2005), in 1994 the International Union for Conservation of Nature and Natural Resources considered the ivory-billed woodpecker to be extinct.

Then in 1999, a credible report of a sighting of the ivory-billed woodpecker occurred in the Pearl River Wildlife Management Area in Louisiana. This report led to a concerted, high-tech and unfortunately unsuccessful search of the area in 2002 (Audubon Arkansas 2005). However, that event led to the assessment of other reported sightings. In April 2005, the journal *Science* published a report about the discovery of an ivory-billed woodpecker in the Big Woods areas of Arkansas. The discovery was reported by a team put together by the Cornell Laboratory of Ornithology and included many experienced ornithologists.

Even with top ornithologists, questions arose about the evidence. For example, the brief video of the birds is grainy, and, even with advanced technology, the Cornell Laboratory could not say with absolute certainly that the sound recordings were that of ivory-billed woodpeckers.

Unfortunately, there were no findings of dead ivory-billed woodpeckers nor of any nests that would positively document the existence of the bird (Cornell Laboratory of Ornithology 2005). In June 2005, ornithologists from Yale University, the University of Kansas, and Florida Gulf Coast University submitted a scientific article skeptical of the Cornell Laboratory's rediscovery of the ivory-billed woodpecker. In August, this group of ornithologists withdrew their paper, stating that additional evidence supplied to them by the Cornell team alleviated their doubts. However, skeptics still remain. Some researchers, including David A. Sibley, a well-known ornithologist, suggest that the Cornell team could have seen the fairly common, smaller relative of the ivory-billed woodpecker, the pileated woodpecker (Sibley et al. 2006).

Since that time, a research group from Auburn University in Alabama and the University of Windsor in Ontario, Canada, presented a collection of evidence that ivory-billed woodpeckers may still exist in Florida. Although their evidence includes fourteen sightings, 300 sound recordings, indications of foraging, and appropriately sized tree nest cavities, they do not have conclusive evidence such as DNA or photographic proof (Hill et al. 2006). Evidence is still being collected by both the Cornell Laboratory team and the researchers working in Florida.

Even if no conclusive evidence is found, the renewed interest in searching for a so-called extinct species has had two definitive impacts. First, it has spurred hope for conservationists. The possibility of taking species off the extinct list is a rare event in today's world. Second, it started a renewed tourism boom in the protected areas of the forest in the south. This has brought economic improvement to surrounding communities, which has increased their level of support for finding the ivory-billed woodpecker, and has created a boom in an already $85 billion industry in U.S. birding (U.S. Fish and Wildlife Service 2003).

Sources and Further Reading: Audubon Arkansas, "Audubon's 'Cache-Lower White Rivers Important Bird Area' Is Home to Ivory-billed Woodpecker," http://www.ar.audubon.org/News_Release_IvoryBilledWoodpecker_April28th2005.html; Cornell Laboratory of Ornithology, "Final Report: Summary and Conclusions of the 2005–06 Ivory-billed Woodpecker Search in Arkansas," http://www.birds.cornell.edu/ivory/current0607/FinalReportIVBWOO.pdf; Fitzpatrick et al., "Ivory-billed Woodpecker (*Campephilus principalis*) Persists in Continental North America"; Fitzpatrick et al., "Response to Comment on 'Ivory-billed Woodpecker (*Campephilus principalis*) Persists in Continental North America'"; Fitzpatrick et al., "Clarification about Current Research on the Status of Ivory-billed Woodpecker (*Campephilus principalis*) in Arkansas"; Gallagher, *The Grail Bird: Hot on the Trail of the Ivory-Billed Woodpecker*; Hill et al., "Evidence Suggesting the Ivory-billed Woodpeckers (*Campephilus principalis*) Exist in Florida"; Jackson, *In Search of the Ivory-Billed Woodpecker*; Jackson, "The Public Perception of Science and Reported Confirmation of the Ivory-billed Woodpecker in Arkansas"; Mendenhall, "Reported Ivory-billed Sightings Since 1944"; Milius, "Comeback Bird," http://www.sciencenews.org/articles/20050611/bob9.asp; Milius, "Alive and Knocking: Glimpses of an Ivory-billed Legend," http://www.sciencenews.org/articles/20050507/fob1.asp; Milius, "Oops. Woodpecker Raps Were Actually Gunshots," http://www.sciencenews.org/articles/20020622/note13.asp; Sibley et al., "Comment on 'Ivory-billed Woodpecker (*Campephilus principalis*) Persists in Continental North America'"; U.S. Fish and Wildlife Service, "U.S. Fish and Wildlife Service Releases 'Birding in the United States,'" News Release, http://www.fws.gov/mountain-prairie/pressrel/03-78.htm.

WIND POWER DEBATE ALONG CAPE COD

Time Period: 2005 to the present
In This Corner: Environmentalists, alternative energy enthusiasts, Cape Cod residents, tourists
In the Other Corner: Environmentalists, power companies

Other Interested Parties: Policy makers, American consumers
General Environmental Issue(s): Energy

Wind power is the fastest-growing commercial energy source in the world. Wind farms already dot the countryside in nearly thirty U.S. states, Europe, and Australia. Denmark, among the biggest European users of wind-generated electricity, has one of the world's most successful offshore wind farms. As other nations turn to alternative sources of energy such as wind power, Americans continue to be reluctant for a number of different reasons. Although initiatives to develop wind power began in earnest throughout the United States by about 2005, one of the most intriguing cases is that of offshore wind farms near Massachusetts.

Wind power became cost competitive in the sites with the most favorable wind regimes around 2000 (Herzog et al. 2001). The obvious wind resources such as those in the Great Plains of the United States, northern Canada, and central Canada are too distant from most cities. Electricity cannot be sent such long distances at this time (Grubb and Meyer 1993). Therefore, wind developers began to actively consider areas with less wind that were more advantageously located. In recent years, three or four manufacturers have developed large wind electric turbines designed to be placed offshore, in waters up to twenty to thirty meters in depth. Europe has put these new turbines to work already. In the United States, some states have implemented new laws that require utilities to increase their use of renewable energy. In Massachusetts, a new law requires that, by 2009, the state must produce enough renewable power to light about 100,000 homes. This has fueled efforts by developers interested in the Cape Cod area off of Massachusetts for the first American offshore turbine.

Cape Wind Associates proposes to install 130 wind turbines over twenty-four square miles of Nantucket Sound, about six miles from Hyannis and nine miles from Martha's Vineyard. Their plan calls for the use of new GE model wind generators that are designed exclusively for offshore use. Each turbine measures forty stories from sea level to the tip of the top blade. Together, the turbine would generate up to 1,491,384 megawatt hours of electricity per year, which is about three-quarters the electrical needs of Cape Cod, or one-tenth of the demand of the entire state of Massachusetts (Cape Wind 2004a). Geographically, the developer states that Nantucket Sound is a highly favorable site for wind development, arguably the best on the East Coast (strong steady winds, close to power lines on shore, shallow water, protected from high waves, and minimal conflicts with transportation systems). Financially, private investors have supplied more than $700 million for the project to date. "We believe the public benefits of this project are going to far outweigh the negatives," said James Gordon who heads Cape Wind Associates, the project's backers.

Although an offshore project has many promising attributes, it also possesses serious concerns, some of which differ from wind development on land. The fight against the Cape Cod wind project is well organized, well financed, and politically potent. "We feel this is going to endanger the environment and hurt both sea birds and mammals," said Isaac Rosen, of the Alliance to Protect Nantucket Sound. "We fear there will be a deleterious effect on tourism, which is the backbone of the cape's economy." The Massachusetts Division of Marine Fisheries anticipates "direct negative impacts to fisheries resources and habitat." The Massachusetts Audubon Society is calling for a three-year study and says it does "not agree … that avian risks are small or that bird use in the area is low." U.S. Fish and Wildlife Service calls the developers' proposal "not sufficient" and recommends at least a three-year study for birds alone, and the Massachusetts Department of Marine Fisheries has "serious concerns centering on the potential risks to migratory birds."

Ironically, many Cape Codders are environmentalists who support wind power, just not here. The development threatens one of the resources held most dear by residents: scenery. Critics see the wind turbines as an ugly intrusion on the seascape. By extension, some residents believe that anything that decreases scenery is a possible threat to tourism and jobs related to it. Other critics emphasize one of the problems with wind power in general: its impact on surrounding species. Birds are a clear problem for the spinning turbines, but, unique to offshore development, some experts believe the turbines would also be a problem for marine species, including whales.

Initially, one might suspect that offshore development is simpler and requires less expense to purchase land, etc. In fact, the portion of the Nantucket Sound under development falls under federal jurisdiction. Therefore, federal agencies have also weighed in with their belief that the proposed Cape Wind power plant has the potential of violating one or more federal laws, including the following: The ESA believes that the power plant may adversely affect several species listed as federally endangered or threatened. The Marine Mammal Protection Act (MMPA) argues that, if the power plant construction or operation results in the killing, harming, or harassment of seals, dolphins, or whales, the project will violate the MMPA. The Migratory Bird Treaty Act (MBTA) states that, if the power plant harms migratory birds, it would be in violation of the MBTA. Finally, the Fisheries Conservation and Management Act (FCMA) contends that the area is a designated Essential Fish Habitat.

This current debate has been fascinating in that environmentalists, who favor the use of alternative fuels to create much-needed energy, have been forced to take an uncomfortable stand. Although they appreciate the energy production, the value of protecting the ocean and keeping it free from human intrusion is one of their highest priorities. In a locale that prioritizes tourism, the sea is a vital portion of history and recreation for Cape Cod. Sailing, fishing, and other traditional activities would be influenced by the wind development.

Source: Kempton et al., "The Offshore Wind Power Debate: Views from Cape Cod," http://www.
    ocean.udel.edu/windpower/docs/KempEtAl-OffshoreWindDebate05.pdf.

SLOW FOOD MOVEMENT

Time Period: 1986 to the present
In This Corner: Slow Food Movement, small farmers, organic growers
In the Other Corner: Agribusiness
Other Interested Parties: Consumers
General Environmental Issue(s): Organics, small family farms, biodiversity

The Slow Food Movement began in Italy when, in 1986, Carlo Petrini protested the opening of a McDonald's in Rome. The movement was first known as "Arcigola" and was designed to resist fast food. Today, the objective has expanded to not only combat fast food but also to help preserve traditional plants, seeds, animals, and farming practices, in addition to the cultural cuisine associated with these traditions. The movement includes more than 80,000 members and is active in more than fifty countries.

As it stands today, each Slow Food chapter is responsible for promoting local farmers, flavors, and culinary artisans through programs such as farmers markets and taste workshops. These programs are designed to meet the objectives of the Slow Food Movement: preserving and celebrating local culinary traditions, forming and maintaining seed banks to preserve local heirloom varieties, supporting small-scale production and processing, and educating people about the qualities of good food and the risks of fast food and agribusiness. The Slow Food Movement also lobbies to include organic farming concerns within agricultural policies and to limit funding of GMOs and agrochemical use. In addition, the movement works with primary and secondary school students creating school gardens, educating the students on nutrition, and perpetuating the skills of farming.

Beyond the local programs, the Slow Food Movement also takes some international actions. The movement opened the University of Gastronomic Sciences in Emilia-Romagna, Italy, in 2004. The goals are to work with students from many locations to not only preserve the artistic elements of traditional cooking but also to promote awareness of good food and nutrition.

Within the United States, the Slow Food Movement membership totals 12,000 and includes more than 140 local chapters. Slow Food U.S.A. is a nonprofit organization dedicated to supporting, celebrating, and educating people about the food traditions in North America. Programs focus on everything from the purity of the organic movement to animal breeds and heirloom varieties of plant species, to handcrafted beer and farmhouse cheeses. The goal is to find pleasure and quality in everyday life by slowing down, respecting traditions of the table, and celebrating the diversity of the earth's bounty.

Critics of the Slow Food Movement claim it is elitist. They also suggest that the movement is disparaging cheaper, alternative methods of food processing and preparing, methods that are important to lower-income people and families. The Slow Food Movement counters these critiques by suggesting that they support local production and preparation because ultimately these processes are less expensive because they do not necessitate long-haul transportation or energy and chemical use. According to the Slow Food Movement, most food travels long distances, using both substantial amounts of fuel and preservatives and other additives. The movement notes that the so-called cheaper alternatives do not accurately reflect their true costs because of government subsidies, which keep transportation and other costs artificially low.

Overall, the Slow Food Movement believes they are protecting not only cultural heritage but also regional diversity and biodiversity. By supporting local, small farms and organics, the movement works to preserve each regional ecosystem and the culture that has developed over generations of living within each ecosystem.

Sources and Further Reading: Hopkins, "'Slow Food' Movement Gathers Momentum"; Nabhan, *Coming Home to Eat: The Pleasures and Politics of Local Foods*; Petrini and Padovani, *Slow Food Revolution: A New Culture for Eating and Living*; Slow Food International, http://www.slowfood.com/; Slow Food USA, http://www.slowfoodusa.org/.

SUSTAINABLE COFFEE

Time Period: 1980 to the present
In This Corner: Environmental groups
In the Other Corner: Super farms

Other Interested Parties: Small coffee growers
General Environmental Issue(s): Sustainability, habitat loss, migratory birds, organics

For centuries, coffee was grown under a canopy of shade trees, a practice that provided a refuge for migratory birds. With the development of high-yield, sun-resistant coffees in the 1970s, substantial acreage was clear-cut for greater production of coffee. This change has led to the creation of monoculture and an increase in the use of pesticides, herbicides, and fertilizers, all changes that have adversely affected the surrounding ecosystems and wildlife. To combat this change and to support local communities, there has been a movement in support of sustainable coffee. Sustainable coffee is grown on farms with high biological diversity and low chemical inputs and with a concern for the issues of deforestation and watershed protection.

Sustainable coffee farms conserve resources, protect the environment, and enhance the quality of life for farmers, society, and wildlife. By some estimates, sun-grown coffee farms shelter only one-tenth as many bird species as shade-grown coffee farms (Rain Forest Alliance 2001). This difference is attributable at least in part to habitat loss surrounding the monoculture, sun-grown farms. The clear-cutting eliminates not only the vegetation that is physically removed but also the associated birds. This loss of nitrogen-fixing canopy trees and the insect-hunting birds that would otherwise reduce the pest population demand that sun-grown coffee farms use both pesticides and fertilizers.

Proponents of shade-grown coffee argue that the often organic, shade-grown coffee not only eliminates chemicals but also produces a richer coffee. Shade-grown coffee develops more slowly than sun-grown coffee, creating a higher sugar content that results in a richer, fuller flavor in the roasted bean. Furthermore, proponents of shade-grown coffee emphasize that it is not uncommon for sustainable coffee farms to be family-owned or cooperatively-run enterprises. In many of these family-owned or cooperatively-run enterprises, greater emphasis is placed on the workers: their education, livelihood, and well-being.

By the late 1990s, these issues began to draw increased interest from consumers and environmental groups. To support this increased interest, in May 2001, three of the United States' leading conservation groups, Conservation International, the Rainforest Alliance, and the Smithsonian Migratory Bird Center, released guidelines for growing earth-friendly coffee. The ideas included within the document and the groups' concept of sustainable coffee were developed by conservation groups in consultation with coffee stakeholders around the world. The ideas include fundamental characteristics that coffee farms and processing facilities must protect ecological health in coffee-growing regions.

To ensure both economic and biodiversity benefits, an international group of coffee stakeholders, including growers, importers, roasters, retailers, environmentalists, and consumers, provided an advisory role. Those involved included Amcafe, Brazilian Specialty Coffee Association, Colombian Coffee Federation, Coordinadora Estatal de Productores de Café de Oaxaca, Deutsche Gesellschaft fur Technische Zusammenarbeit, Dunkin' Donuts Inc., East African Fine Coffees Association, Fair Trade Labeling Organizations, Falls Brook Centre, Global Environment Facility, Indian Institute of Plantation Management, International Federation of Organic Agricultural Movements, Royal Coffee, Specialty Coffee Association of America, Starbucks Coffee Company, US Agency for International Development, and the World Bank.

These players wanted to support not only shade-grown coffee but also the increasing powerful movement for organic and environmentally responsible agriculture, along with the "fair trade" interests that advocate for the creation of sustainable livelihoods for coffee farmers. The pressure from the public supporting these movements, in addition to a shared concern at all levels of the coffee industry for ensuring the long-term health of the ecosystems that support coffee production, led to changes. Attention has focused on minimizing the clearing of rainforest, losses of bird and wildlife habitat, declines in biodiversity, and increased dependency on chemical and pesticides. The suggested changes are intended as a point of reference for the development of all phases of the coffee industry. They are intended to provide substantial habitat value for birds and wildlife, as well as local economic benefits to help prevent further destruction of intact natural forests.

Traditional varieties of coffee thrive in shady, forested surroundings, naturally providing a high level of on-farm biodiversity and often adding to wildlife corridors. The move to sun-grown coffee resulted in the loss of thousands of acres of bird and wildlife habitat in Latin America. Furthermore, expansion of coffee regions into Southeast Asia is causing widespread tropical deforestation and has caused a glut in worldwide coffee supplies, resulting in depressed prices and economic difficulties. The movement to more shade-grown coffee demands changes to these conditions.

Sources and Further Reading: Dietsch, "Assessing the Conservation Value of Shade-Grown Coffee: A Biological Perspective Using Neotropical Birds"; Fridell, *Fair Trade Coffee: The Prospects and Pitfalls of Market-Driven Social Justice*; McMahon, "'Cause Coffees' Produce a Cup with an Agenda," http://www.usatoday.com/money/general/2001-07-26-coffee-usat.htm; Pendergrast, *Uncommon Grounds: The History of Coffee and How It Transformed Our World*; Smithsonian Migratory Bird Center, First Sustainable Coffee Congress overview paper; Rainforest Alliance, "Leading Conservation Groups Release Guidelines for Growing Earth-Friendly Coffee," Press Release, www.ra.org/news/2001/coffee-principles.html; Wild, *Coffee: A Dark History*.

CELEBRITY ACTIVISM IN ENVIRONMENTAL AND GLOBAL ISSUES

Time Period: 1980s to the present
In This Corner: Celebrity activists
In the Other Corner: Moral opponents
Other Interested Parties: Environmental groups, anti-poverty organizations, AIDS prevention groups
General Environmental Issue(s): Global AIDS, poverty, global warming, environment

Celebrity activism has reached new heights in recent years. Celebrities such as Angelina Jolie, Bono, Leonardo DiCaprio, Robert Redford, and Oprah Winfrey typify celebrity involvement by committing both their time and financial resources to causes such as AIDS, international poverty, global warming, and the environment. Because of the celebrities' wide range of influence and the number of people that will listen to them, their work is helping to spread the word about many of these global issues.

Many celebrities from around the world are using their celebrity clout and access to media to educate others about a variety of global causes. The celebrities are able to reach audiences either formerly uninterested or unaware of their particular issue. Angelina Jolie uses her popularity to draw attention to the plight of refugees and the environmental conditions surrounding many of the refugee camps. Working first on her own and more recently as a goodwill ambassador for the U.N. refugee committee, Jolie has committed personal time and finances, approximately one-third of her income, to these issues (CNN 2006). Others, such as Robert Redford and Leonardo DiCaprio, have long been supporters of environmental causes focusing on global warming, endangered species protection, and anti-nuclear policies. Another key figure is the band U2's frontman Bono, who is a fervent activist against AIDS and poverty in the Third World. Bono's efforts in public education, communication, and mobilization highlight his celebrity leadership. His One Campaign asks fans not for money but rather to "only" make a personal commitment to take a stand against poverty. Bono claims his efforts have resulted in over two million members to the One Campaign in the United States alone and that, by 2008, the membership of the One Campaign will surpass that of the National Rifle Association (Hicks 2007).

Celebrities are uniquely able to reach people that otherwise may "turn off" to international environmental or humanitarian causes. In light of current celebrity involvement, it appears to be a growing trend among celebrities to become involved in some charitable work. In many cases, it creates a win-win scenario in which celebrities gain media attention for themselves and their cause. For celebrities, this attention can translate into more box-office support or general interest, and the cause also gets media focus and citizen involvement.

The growing interest of celebrities to become involved with a cause led to the formation of The Creative Coalition, an organization that pairs celebrities up with causes in which they are interested. It also helps to educate celebrities on how to approach members of Congress and participate in press conferences. In this way, celebrities can be a real asset to a particular cause.

Critics of celebrity activism often challenge the authenticity of the involvement. Some accuse celebrities of taking money or gifts for their involvement, thus weakening the credibility of celebrities involved. Others point out that, at times, celebrities get involved in trendy causes during times of crisis in what appears to be simply a way to gain media attention. These celebrities move from one issue to the next, depending on what will garner them press.

Perhaps a more serious criticism is that celebrities do not always provide accurate information or reflect a true understanding of the issue du jour. An example is the effort to raise money to send to impoverished countries. Simply "throwing" money at a country fails to acknowledge the corruption of governments and the fact that much of the money does not make it into the hands of the intended recipients, thereby simply adding more to the cadre for the corrupted government to use.

There is also the concern over celebrity fatigue. The amount of motivation and commitment needed to reverse global warming or eradicate extreme poverty is extensive. What happens when a celebrity is no longer credible, interesting, or offering something new to say? If citizens invest their time in what celebrities are saying instead of directly working with the cause, at what point does public attention wane?

Finally, some opponents suggest a moral argument against celebrity activism. The wealthy celebrities who speak out for their causes often overlook the question of whether "it is morally possible to live with integrity at any level of material comfort in our industrialized society" (Hicks 2007). Although these celebrities are educating their fans, are they not also sending the message that an economically privileged lifestyle, one in which Americans can attend celebrity concerts and afford a television on which to watch the celebrities, is morally acceptable? The argument that material excess is harmful for the earth and expands rather than alleviates poverty is hard for celebrities to contest.

Despite arguments against celebrity activists, at this time, the public is coming to expect celebrities to show leadership for a variety of causes. In fact, celebrities who are not involved in political, humanitarian, or environmental causes may soon find themselves under pressure to do so. In general, people expect those who are the most privileged to give back to society, and, failing to do so could potentially affect offers, roles, and opportunities available to celebrities.

Arguably, the global AIDS epidemic and environmental concerns have benefited from the likes of Bono, Leonardo DiCaprio, and others. Although critics say that many do it for their own glorification and lack personal conviction, supporters counter that, as long as the causes are gaining support, it is not important why celebrities get involved, just that they get involved. Ultimately, the true measure of celebrity activists' success as leaders on global issues depends on whether their movements can create and maintain an international structure that delivers political, economic, and most importantly social change.

Sources and Further Reading: CNN, "One-on-one with Angelina Jolie," http://www.cnn.com/ CNN/Programs/anderson.cooper.360/blog/2006/06/one-on-one-with-angelina-jolie.html; Freydkin, "Celebrity Activists Put Star Power to Good Use," http://usatoday.com/life/people/2006-06-22-celebcharities-main-x.htm; Harris, "Celebs and Charity: Trendiness or Benevolence? The Hottest Trend in Hollywood Is Taking Up a Cause," http://abcnews.go.com/Entertainment/WNT/ story?id=2794458&page=1; Hicks, "The Limits of Celebrity Activism," http://www.religion-online.org/showarticle.asp?title=3331; Lahusen, *The Rhetoric of Moral Protest: Public Campaigns, Celebrity Endorsement, and Political Mobilization*; Quest, "On the Trail of the Celebrity Activist"; Shear, "Celebrity Activists," http://www.wiretapmag.org/stories/36850/; Voice of America, "Celebrity Activism: Publicity Stunt or Sincere Care?" http://www.voanews.com/english/archive/ 2007-02/2007-02-21-voa38.cfm?CFID=137892992&CFTOKEN=74360178.

INTERNATIONAL PEACE PARKS/TRANSNATIONAL PARKS/ TRANSNATIONAL PROTECTED AREAS

Time Period: 1932 to the present

In This Corner: Environmentalists, scientists looking at ecosystems, diplomats

In the Other Corner: Citizens concerned about sovereignty and security

Other Interested Parties: Citizens of countries sharing parks, international boundary law enforcement

General Environmental Issue(s): Protected biosphere reserves, World Heritage Site, national parks preservation

On the borders of Canada and the United States rest two national parks, Waterton Lakes Forest Park in Alberta and Glacier National Park in Montana. In 1931, members of the Rotary Clubs of Alberta and Montana proposed joining the parks to symbolize the peace and friendship between the two countries. In 1932, that plan came to fruition with the designation of the Waterton-Glacier International Peace Park, the first "International Peace Park" formed. Since that time, according to the World Conservation Union study, there are more than 188 potential transboundary conservation areas worldwide, involving 112 countries (Peace Parks Foundation 2007).

Glacier National Park Superintendent David Mihalic ponders the view from a trail in the park during a transnational hike on July 22, 1998. Glacier and Waterton Lakes' national parks straddle the U.S.-Canadian border and make up the world's first international peace park designated a World Heritage Site by the U.N. in 1995. AP Photo/Billings Gazette, Michael Milstein.

The idea for an international peace park began with the work of Rotarians from Alberta and Montana. These individuals came together in Waterton Lakes Park for the first annual international goodwill meeting. Members of the Rotary Clubs worked on their own governments, trying to get the Canadian Parliament and the U.S. Congress to pass laws setting aside Waterton-Glacier as an international peace park. By 1932, the groups had achieved their goal. The Waterton-Glacier International Peace Park was established. The two parks cooperate and collaborate on many issues, yet they maintain fiscal independence and administrative autonomy. Although the flora and fauna of the region know no boundaries, the sovereignty of the two countries is still maintained.

The success of the Waterton-Glacier International Peace Park started a movement around the globe. Peace parks, also referred to as transnational parks or transnational protected areas, are formed when neighboring countries agree to link and work together to manage bordering national parks, wildlife reserves, or other protected areas. The goal of these parks is to facilitate international cooperation and environmental protection of existing ecosystems with exchange of information and research by respective countries. Thus, many of today's transnational protected areas are situated in biologically sensitive areas and aid in the protection of these areas.

Today, about one-third of the world's more than 300 international boundaries are home to transboundary protected areas, with most of them located in Europe (Worldwatch Institute 2006). Although the majority of the transboundary protected areas span just two countries, as many as thirty-one cover three nations (Worldwatch Institute 2006). Altogether, transboundary protected areas comprise nearly 10 percent of the world's protected lands (Worldwatch Institute 2006).

Supporters of transboundary protected areas suggest that these areas serve both the natural and human communities in and around the protected areas. They note that the protected areas provide an open space along the border between or among countries, which helps to both preserve biodiversity by maintaining large areas of wilderness and encourage economic development, mainly providing an incentive for greater tourism. Furthermore, it is suggested that transboundary protected areas help to inspire cooperative efforts through research, protection, and education projects.

Those in support of transboundary protected areas also point to the benefits of facilitating peaceful relations between two neighboring countries. This is particularly helpful when the area of the transboundary protected area encompass areas where there is some level of conflict. For countries in conflict, peace parks provide a neutral ground in which to focus on positive interactions and cooperation. The hope is that, by sharing joint management, protection, and knowledge of parks, the increased communication and interaction will provide a bridge to greater cultural, social, and political understanding. Following this line of thinking, a number of international groups have proposed developing an international peace park in the demilitarized zone between North Korea and South Korea.

Critics of transboundary protected areas often focus on sovereignty and security issues. They suggest that, by creating transboundary or international parks, the line dividing where one country ends and where the next begins is less clear. Although most transboundary protected areas maintain fiscal and administrative independence, the division is still not clearly drawn. Decisions concerning what can occur on, for example, U.S. land can be determined in part by citizens of another country, such as Canada. Critics of transboundary protected areas believe these events and the efforts to "coordinate decision-making" between or among countries minimizes the power of the national government by giving away some decision-making power.

A second argument against transboundary protected areas focuses on security issues. Since September 11, 2001, countries such as the United States have increased their focus on the issue of border security. The current desire to ensure secure borders and keep unwanted elements out of a given country leaves some people questioning the logic of an open boundary such as what exists in many transboundary protected areas. Critics argue that the desire to control cross-border traffic is not compatible with the idea of an open-border transboundary protected area.

Despite critiques, the ideas of peace parks and transboundary protected areas are growing in popularity. In 1993, there were approximately seventy transboundary protected areas in sixty-five countries. By 2007, the numbers had more than doubled. In addition, there are ongoing efforts to continue to expand the number of transboundary protected areas and parks. Among the areas under discussion for transboundary protection are several locations in Southern Africa, the border between Bolivia and Paraguay, and a United States–Mexico park joining Big Bend National Park in the United States with the Maderas del Carmen and Canon de Santa Elena protected areas in Mexico.

The formation of the Waterton-Glacier International Peace Park facilitated the expansion and development of peace parks throughout the world. Its success as a peace park of unparalleled natural wonders and abundant wildlife earned the park the designation of a World Heritage Site and Biosphere Reserve by UNESCO in 1995. Today, Waterton-Glacier International Peace Park stands as an example of friendship, international agreement, and commitment to environmental protection, something many hope to duplicate in other locations around the world with other transboundary protected areas.

Sources and Further Reading: Barry and Brown, *The Challenge of Cross-border Environmentalism*; Budowski, *Peace through Parks*, http://www.unep.org/OurPlanet/imgversn/144/budowski.html; Fall, *Drawing the Line: Nature, Hybridity and Politics in Transboundary Spaces. Conservation and Biodiversity: The Social Dimension of Linking Local Level Development and Conservation through Protected Areas*; Goodale et al., *Transboundary Protected Areas: The Viability of Regional Conservation Strategies*; International Conference on Transboundary Protected Areas as a Vehicle for International Co-operation, "Parks for Peace"; Knight and Landres, *Stewardship Across Boundaries*; U.S. National Park Service, *Glacier National Park*, http://www.nps.gov/glac/historyculture.index.htm; Worldwatch Institute, "Transboundary Parks Become Popular," http://www.worldwatch.org/node/4277; Zbicz, "Transboundary Cooperation in Conservation: A Global Survey of Factors Influencing Cooperation between Internationally Adjoining Protected Areas."

WILDFIRES ON PUBLIC LANDS

Time Period: Historic, 1910–1960 and through to the present

In This Corner: Biologists, ecologists, environmentalists

In the Other Corner: Logging interests, homeowners and business owners in the wildland urban interface

Other Interested Parties: Firefighters, public lands managers, hunters

General Environmental Issue(s): Wildfire, invasive species, habitat restoration

Wildfires are common occurrences in many places around the world, particularly in climates where there is sufficient moisture to allow for the growth of trees and grasses but where there are also extended dry, hot periods. Wildfires nourish the environment yet also often threaten suburban homes and developments located in the wildland urban interface. These two effects of wildfire have created conflicting opinions on the proper management of wildfire on public lands. The 2007 fire season brought out each of these sides of the issue as firefighters combed the west to fight a record number of fires.

Particularly prevalent in the summer and fall, wildfires occur when fallen branches, leaves, grasses, and other material dry out and become highly flammable. They are most common during years of severe drought and thrive on days of strong, dry winds. Wildfires are natural and are a vital part of a healthy forest or grassland environment. Historically, Native Americans used fire for many purposes, from clearing land and hunting game to warfare. Their use, along with fires from lighting strikes and other natural causes kept fire a regular part of the ecosystem.

By the 1900s, damages to personal property and wildlife, along with threats to human life, caused the U.S. government to take action against wildfires on public lands. The result was the suppression of wildfires, whether natural or human caused. This fear of wildlife was intensified in 1910 with the historic event know as "The Big Blowup." The great fires that burned in the Northern Rockies of Idaho and Montana during 1910 burned more than three million acres and took the lives of seventy-eight firefighters. In addition to the loss of private property and loss of life, the USFS also spent millions of dollars putting out the fires (Colorado State Forest Service 2007). The Big Blowup prompted a government decision to put out wildfires on public lands as quickly as possible.

For many decades after The Big Blowup, the USFS worked under the directive to suppress all fires. This policy was epitomized by Smokey Bear, the USFS mascot, and was the basis for parts of the movie *Bambi*. It was not until the 1960s when scientists began to question the fire suppression policy. It was realized that no new Giant Sequoia has grown in the forests of California and that the cause of this situation was the lack of fire; fire is essential to the lifecycle of the Giant Sequoia. It was becoming clear that the decision to emphasize fire suppression had saved private and public property but has also led to severe changes in the forests of the United States.

Supporters of wildfire suppression on public lands argue that, on occasions, wildfires have caused large-scale damage to private and public property, causing not only the destruction of homes but also deaths. In addition, they note that the smoke from wildfires increases air pollution and that, once an area burns, the likelihood for invasive species to take hold increases. Finally, supporters also include some logging interests who would rather take the trees than see them burn.

Critics of fire suppression include many biologists and ecologists. These individuals point to how excessive first suppression has led to fuel buildup and has had negative impact on not only forest heath but also on humans and wildlife that inhabit these areas. These critics argue that today it is accepted that wildfires are a natural part of the ecosystem of public lands and that many plants depend in part on fire for the survival of their species. For example, the lodge-pole pine is partially dependent on fire to spread its seeds (Colorado State Forest Service 2007), and the process of germination is promoted in some plants when they are exposed to smoke from burning plants.

Many hunters and environmentalists were also critics of extreme fire suppression. Hunters realized that fire, at least controlled burns, reduced underbrush, making hunting easier. Environmentalists recognized that fire restored a natural cycle to the ecosystem, revitalizing habitat for some species and returning nutrients to the soil, which added to the health of many species. Finally, many managers and individuals began to realize that fire in limited amounts would actually reduce the risk of dangerous high-intensity fires caused by many years of fuel and undergrowth buildup.

As biologists, ecologists, and environmentalists make their arguments, the U.S. federal government and its agencies are changing their policies on fire suppression. Although there are still many areas that follow fire suppression, regions removed from private property are now viewed as areas that will benefit from natural burns. Furthermore, as regions such as Yellowstone National Park begin to thrive after recovering from wildfires, more people are making the connection between healthy ecosystems and fire.

The federal government reported that, by August 2007, forty large fires were actively burning, with fire activity and poor air quality conditions persisting across central Idaho and northwest Montana. According to estimates from the National Interagency Fire Center, as of September 10th, almost 70,000 wildland fires had been reported across the United States so far in 2007, with approximately 7.37 million acres burned. With evidence of the destructive force of fire, as seen in states such as Utah and Idaho during the fire season of 2007, it is clear that loss of private property and loss of life are still possibilities when dealing with wildfires on public lands. This leaves public land managers with the task of finding a balance between letting wildfires burn and extensive suppression.

Sources and Further Reading: Alverson, Waller, and Kuhlamnn, *Wild Forests: Conservation Biology and Public Policy*; Colorado State Forest Service, *About Wildfire: Introduction and History*, http://csfs.colostate.edu/wildfire.htm; Dysart and Clawson, *Managing Public Lands in the Public Interest*; Machlis et al., *Burning Questions: A Social Science Research Plan for Federal Wildland Fire Management*; Merrill, *Public Lands and Political Meaning: Ranchers, the Government, and the Property between Them*; Miyanishi and Johnson, "A Re-examination of the Effects of Fire Suppression in the Boreal Forest"; Pyne, *Tending Fire: Coping With America's Wildland Fires*; Wuerthner, *Wildfire: A Century of Failed Forest Policy*.

WOLVES IN YELLOWSTONE NATIONAL PARK

Time Period: 1990s to the present

In This Corner: Environmental organizations, National Park Service, U.S. Fish and Wildlife Service

In the Other Corner: Ranchers, Farm Bureau, logging industry

Other Interested Parties: Public

General Environmental Issue(s): Ecosystem health, Endangered Species Act, predator recovery

In early 1995, Yellowstone National Park welcomed wolves back into the region after nearly twenty-five years of confirmed absence. The wolves' absence from the park exemplified the historical battle between the large predators such as the wolf and resource extraction and ranching interests. It also signifies the changing attitudes toward predator protection and recovery in the United States, having shifted from fear and extirpation to eventual reintroduction.

In the late 1800s as more people began to seek their fortunes in the western landscape of the United States, a growing concern over large predators such as the wolf began to take hold. People feared the wolves, believing that they would harm elements of their livelihoods, specifically their livestock and the large game animals such as elk. By 1914, the U.S. government took action to address this issue. In a move to protect livestock and elk populations in and around Yellowstone National Park, the U.S. Congress appropriated funds to be used for the purpose of "destroying wolves, prairie dogs, and other animals injurious to agriculture and animal husbandry" on public lands (Chase 1986, 122). The NPS hunters carried out these orders, beginning the government sanctioned predator control policy. The NPS continued the policy until 1935, although as early as 1926 the wolves of Yellowstone National Park were virtually eliminated.

Over the decades after the wolf eradication program, few wolf packs were observed in Yellowstone. By the 1970s, there was no longer any evidence of a wolf population in the national park. Curiously, around this same time, two landmark conservation efforts were introduced. In 1972, U.S. President Nixon announced in his State of the Union address that the use of predator poisons was no longer acceptable on public lands. Then, in 1973, the U.S. Congress enacted and President Nixon signed the ESA. The wolf was one of the first mammal species listed with the ESA. These two efforts together opened the door for species recovery, and environmentalists saw this as an opportunity to work toward wolf reintroduction in Yellowstone National Park.

The culmination of government legislation, an administrative directive, and interest by environmental organizations started a roughly twenty-year battle leading to wolf recovery in Yellowstone. Arguing against these recovery efforts were ranchers, loggers, and perhaps most powerfully, the American Farm Bureau Federation, who argued that wolf recovery would have a negative impact on local ranchers and loggers. These groups argued that wolves are often elusive to rural residents and will kill livestock when given the opportunity. In addition, the American Farm Bureau Federation argued that wolf recovery granted the federal government more control over private lands and thus hindered the ability of ranchers and farmers to have economically successful businesses. Finally, many rural residents expressed concern that wolves could attack or kill them or their families if reintroduction occurred.

In 1987, the U.S. Fish and Wildlife Service, the government agency that oversees threatened and endangered species, issued the Northern Rocky Mountain Wolf Recovery Plan. This proposal reflected the U.S. Fish and Wildlife Service's evidence that there is enough contiguous protected lands in the park to establish wolf recovery and that the existing prey, including moose, mule deer, and bison, were sufficient to sustain the wolves' presence in the park. Thus, the plan proposed the reintroduction of an "experimental population" of wolves into Yellowstone National Park.

Environmental groups seized on the publication of this recovery plan as an opportunity to help facilitate wolf reintroduction efforts. In particular, one environmental organization, Defenders of Wildlife, took the helm. Defenders of Wildlife began its campaign for reintroduction of the wolf by changing its logo to the wolf and actively addressing the agricultural community's economic concerns by establishing a permanent Wolf Compensation Fund. This fund was designed to compensate ranchers financially for the loss of livestock when the rancher could demonstrate the loss was caused by wolves. The hope was to eliminate the very real concern over economic losses to ranchers and to offer some good faith gesture in their restoration.

In addition to Defenders of Wildlife efforts, other environmental organizations worked to gain public support for wolf reintroduction. These programs included educational programs for the public about wolves and wolf behavior. These programs showed the need for large predators within ecosystems and, in an attempt to address the concern about human attacks, pointed out that there has never been a documented wolf attack on a person.

Despite these campaigns, the American Farm Bureau Federation, ranchers, loggers, and many rural residents were still opposed to wolf recovery because of the possible threat to their livelihood. Even with compensation, these individuals would still need to be more diligent about protecting their livestock, deal with the nuisance of wolves killing their livestock, and take the time to collect evidence to demonstrate proof of livestock losses caused by wolves.

By the 1990s, the federal government had reversed its views on wolves. A controversial decision taken by the U.S. Fish and Wildlife Service set the stage for the reintroduction of wolves. Mackenzie Valley wolves, imported from Canada, were reintroduced into Yellowstone National Park. On March 21, 1995, the first of three groups of gray wolves were released at Crystal Creek within Yellowstone National Park. Ultimately, sixty-six wolves were reintroduced into Yellowstone National Park. These reintroductions fulfilled both a mandate under the ESA and a goal of environmental activists and wildlife managers, who view ecosystem restoration as key for preserving, protecting, and restoring complete ecosystems. These groups now argue that ecosystem restoration needs to include the reintroduction of large predators such as the wolf. Without these predators, a truly viable, complete ecosystem does not and cannot exist.

The reintroduction efforts are considered successful. By the end of 2006, at least 136 wolves in thirteen packs occupied Yellowstone National Park, with more in the surrounding ecosystem (Smith et al. 2007). Although controversy still exists among the agricultural community, support for the program exists within the environmentalist community, with many wildlife managers, and with the many tourists who come to Yellowstone National Park to see the wolves.

Sources and Further Reading: Chase, *Playing God in Yellowstone: The Destruction of America's First National Park*; Fisher, *Wolf Wars: The Remarkable Inside Story of the Restoration of Wolves to Yellowstone*; Franz, "The Yellowstone Wolves Win One"; Johnson, "Yellowstone Will Shelter Wolves Again," http://query.nytimes.com/gst/fullpage.html?res=9401E7D7173DF934A25755 C0A962958260&n=Top%2fReference%2fTimes%20Topics%2fOrganizations%2fS%2fSierra%20 Club; McNamee, *The Return of the Wolf to Yellowstone*; National Park Service, U.S. Department of the Interior, *Yellowstone National Park Wolf Restoration*, http://www.nps.gov/yell/naturscience/ wolfrest.htm; Robbins, "Resurgent Wolves Now Considered Pests by Some"; Smith and Ferguson, *Decade of the Wolf: Returning the Wild to Yellowstone*; Smith, Stahler, Guernsey, Metz, Nelson, Albers, and McIntyre, *Yellowstone Wolf Project, Annual Report 2006*; U.S. Department of the Interior, *Ecological Issues on Reintroducing Wolves into Yellowstone National Park*.

EPILOGUE

In the shadow of Hurricane Katrina in 2005 and growing attention to the implications of global climate change, fire came as never before to residential California in the fall of 2007. Only in such an arid region could small fires grow to threaten nearly two million acres, and, in the region of California near San Diego and Los Angeles where suburbs reach into the driest, most volatile areas of scrub brush, million-dollar homes go up as quickly as tall pines. The need for housing in this area has forced developers to press the envelope of the ecology of southern California.

After a very dry summer and with heavier than usual Santa Ana winds, the outcome could be horrible. Residents always knew this, but the upsides of living in dreamy California far outweighed such possibilities.

October 2007 was the first time that the worst-case scenario played out: one-half million acres burned, approximately 2,000 homes destroyed, and seven deaths. Financial losses have been estimated in excess of $1 billion. If the fires had burned in the same locations in 1980, the *New York Times* estimated that approximately 61,000 homes would have been within a mile of the fire, by 2000, the figure would be 106,000 homes, and, in 2007, more than 125,000 homes were threatened.

Although for generations reformers have called for curtailing residential patterns in such fragile ecosystems, the fires of 2007 will surely make this problem one of the next great debates in a book like this one. Already housing codes are being reconsidered and emergency plans revised. However, the fact remains that millions of people must find a way to call this delicate region home.

So many of the debates recounted in these pages pitted reformers against profit-centered business interests. Particularly during the early twentieth century, reformers, unions, journalists, and politicians sought to demand safer, healthier, and more efficient methods and products from manufacturers. In many instances, reformers have called for the involvement of a governmental authority, whether on the local, state, national, or international levels. In most of these stories, such involvement has greatly helped the plight of many consumers. In fact, in recent years, many large corporations have clearly come to feel that it is important to present an environmentally friendly image to the public.

In the energy sector, which is dominated by some of the nation's most powerful corporations, industry leaders have realized that consumer demands for change are also in the best interest of their long-term profits. Biofuels, particularly the use of ethanol, as well as increased fuel efficiency in autos and the availability of hybrid cars and a variety of sustainable products demonstrate that many companies believe green business can be good business.

Does this mean American manufacturers have responded to a century of increasing calls for reform and regulation by admitting that consumers do have some say in how they do business? Or is it another episode of "greenwashing" that only shows a temporary change? Only time will tell.

And only time will tell how well government officials work with planners to implement the lessons of the 2007 fires in California and the floodwaters of 2005 in New Orleans. It is critical, however, that each of these events and the debates and issues from which they grow be viewed in historical context. This volume seeks to contribute to this context.

BIBLIOGRAPHY

Readers are encouraged to consult any of the environmental history essays at http://nationalhumanitiescenter.org/tserve/nattrans/nattrans.htm.

Abruzzi, W. S. "The Social and Ecological Consequences of Early Cattle Ranching in the Little Colorado River Basin," *Human Ecology* 23 (1995): 75–98.

Adams, D. A. *Renewable Resource Policy: The Legal-Institutional Foundation.* Washington, DC: Island Press, 1993.

Adams, J. A. *The American Amusement Park Industry.* Boston, MA: Twayne Publishers, 1991.

Adams, S. P. *The US Coal Industry in the Nineteenth Century,* http://eh.net/encyclopedia/article/adams.industry.coal.us.

Albion, R. G. *Forests and Sea Power.* Cambridge, MA: Harvard University Press, 1926.

Albright, H. M., Cahn, R. *The Birth of the National Park Service: The Founding Years, 1913–33.* Salt Lake City, UT: Howe Brothers, 1985.

Albright, H. M., Schenck, M. A. *Creating the National Park Service: The Missing Years.* Norman, OK: University of Oklahoma Press, 1999.

Alverson, W. S., Waller, D., and Kuhlmann, W. *Wild Forests: Conservation Biology and Public Policy.* Washington, DC: Island Press, 1994.

Ambler, M. *Breaking the Bonds: Indian Control of Energy Development.* Lawrence, KS: University Press of Kansas, 1990.

Ambrose, S. E. *Undaunted Courage: Meriwether Lewis, Thomas Jefferson, and the Opening of the American West.* New York: Simon & Schuster, 1996.

American Frontiers: A Public Lands Journey. "Energy from Public Lands," http://americanfrontiers.net/energy/Energy2.php.

American State Papers. *Gallatin's Report on Roads and Canals,* http://www.union.edu/PUBLIC/ECODEPT/kleind/eco024/documents/internal/internal_callendar.htm.

Anderson, Jr., O. E. *The Health of a Nation: Harvey W. Wiley and the Fight for Pure Food.* Chicago, IL: University of Chicago Press, 1958.

Anderson, T. H. *The Movement and the Sixties.* New York, NY: Oxford University Press, 1995.

Andreas, P. *Border Games: Policing the U.S.–Mexico Divide.* Cornell, NY: Cornell University Press, 2000.

Andrews, R. N. L. *Managing the Environment, Managing Ourselves.* New Haven, CT: Yale University Press, 1999.

Annis, S. "Evolving Connectedness among Environmental Groups and Grassroots Organizations in Protected Areas of Central America," *World Development* 20, no. 4 (1992): 587–95.

Antarctic Conservation Act of 1978. 16 U.S.C. §§2401–2413, October 28, as amended 1996. Antarctic Treaty. Scientific Committee on Antarctic Research, www.scar.org/treaty/.

Anti-Defamation League. 2007. http://www.adl.org/.

APVA Preservation Virginia. *Old Cape Henry Lighthouse*, http://www.apva.org/capehenry/origin.php.

Ashworth, W. *The Late, Great Lakes: An Environmental History.* New York, NY: Knopf, 1986.

Associated Press. "PETA Claims Victory as Fashion House Drops Fur," 2006.

Athansiou, T. *Divided Planet: The Ecology of Rich and Poor.* Athens, GA: University of Georgia Press, 1998.

Atomic Archive. *Report on the Trinity Test by General Groves,* http://www.atomicarchive.com/Docs/Trinity/Groves.shtml.

Aurand, H. W. *Coalcracker Culture: Work and Values in Pennsylvania Anthracite, 1835–1935.* Harrisburg, PA: Susquehanna University Press, 2003.

Australian Center for the Moving Image. *Edison: The Invention of the Phonograph and the Electric Light Bulb,* http://www.acmi.net.au/AIC/EDISON_INVENT.html.

Bailey, A. J. *The Chessboard of War: Sherman and Hood in the Autumn Campaigns of 1864.* Lincoln, NE: University of Nebraska Press, 2000.

Bailey, A. J. *War and Ruin: William T. Sherman and the Savannah Campaign.* Wilmington, DE: Scholarly Resources, 2003.

Ballard, J. N., ed. *The History of the U.S. Army Corps of Engineers.* New York, NY: Diane Publishers, 1998.

Banerjee, N. "Tanking on a Coal Mining Practice as a Matter of Faith," *The New York Times*, October 28, 2006.

Barbalace, R. C. *Environmental Justice and the NIMBY Principle,* http://environmentalchemistry.com/yogi/hazmat/articles/nimby.html.

Barney, W. L. *The Passage of the Republic: An Interdisciplinary History of Nineteenth-Century America.* Lexington, MA: D. C. Heath, 1987.

Barry, J. *Rising Tide.* New York: Simon & Schuster, 1998.

Barry, T., and Brown, H. *The Challenge of Cross-border Environmentalism.* Albuquerque, NM: The Resource Center, 1994.

Bartram, W. *Travels.* New York, NY: Dover, 1983.

BBC News. "Iceland Violates Ban on Whaling," 2006, http://news.bbc.co.uk/2/hi/europe/6074230.stm.

Beale, E. F. The report of Edward Fitzgerald Beale to the Secretary of War concerning the wagon road from Fort Defiance to the Colorado River; April 26, 1858. 35th Congress, 1st Session, House of Representatives, Executive Document, no. 124, 137–281, in *Uncle Sam's Camels: The Journal of May Humphreys Stacey Supplemented by the Report of Edward F. Beale (1857–1858).* Glorieta, NM: Rio Grande Press, 1970.

Beilharz, E. A., and Lopez, C. U., eds. *We Were 49ers! Chilean Accounts of the California Gold Rush.* Pasadena, CA: Ward Ritchie Press, 1976.

Belasco, J. *Americans on the Road.* Cambridge, MA: MIT Press, 1979.

Benedict, R. E. *Ozone Diplomacy.* Cambridge, MA: Harvard University Press, 1991.

Benfield, F. K., Terris J., and Vorsanger, N. *Solving Sprawl: Models of Smart Growth in Communities Across America.* National Resource Defense Council. Washington, DC: Island Press, 2001.

Bennett, Jr. L. *Before the Mayflower.* New York, NY: Penguin, 1993.

Benton, T. H. *Thrilling Sketch of the Life of Col. J. C. Fremont.* London, UK: J. Field, 1850.

Bergon, F., ed. *The Journals of Lewis and Clark.* New York, NY: Penguin Books, 1989.

Berman, M. *All That Is Solid Melts into Air.* New York, NY: Penguin, 1988.

Berry, W. *The Unsettling of America: Culture & Agriculture.* Washington, DC: Sierra Club Books, 2004.

Betts, R. B. *In Search of York: The Slave Who Went to the Pacific With Lewis and Clark.* Boulder, CO: University of Colorado Press, 2000.

Billington, R. A., and Hedges, J. B. *Westward Expansion: A History of the American Frontier.* New York, NY: The Macmillan Company, 1949.

Black, B. "Organic Planning: Ecology and Design in the Landscape of TVA," in *Environmentalism in Landscape Architecture,* ed. Michel Conan, Washington DC: Dumbarton Oaks, 2000a.

Black, B. *Petrolia: The Landscape of America's First Oil Boom.* Baltimore, MD: Johns Hopkins University Press, 2000b.

Black, B. *Contesting Gettysburg: Preserving an American Shrine.* Chicago, IL: Center for American Places, University of Chicago Press. Forthcoming.

Black, B. C. "Addressing the Nature of Gettysburg: Addition and Detraction in Preserving an American Shrine." RECONSTRUCTION, online international journal of contemporary culture, Winter 2006. Available at: http://reconstruction.eserver.org/072/black.shtml.

Black, R. "Did Greens Help Kill the Whale?", 2007, http://news.bbc.co.uk/2/hi/science/nature/6659401.stm.

Blaut, J. M. *Colonizer's Model of the World: Geographic Diffusionism and Eurocentric History.* New York, NY: The Guilford Press, 1993.

Boli, J., and Thomas, G. *Constructing World Culture: International Nongovernmental Organizations since 1875.* Stanford, CA: Stanford University Press, 1999.

Bormann, F. H., Balmori, D., Geballe, G. T., and Vernegaard, L. *Redesigning the American Lawn.* New Haven, CT: Yale University Press, 1995.

Boyer, P. *By The Bomb's Early Light.* Chapel Hill, NC: University of North Carolina Press, 1994.

Brack, D. *International Trade and the Montreal Protocol.* London, UK: Earthscan/James and James Press, 1996.

Bradsher, K. *High and Mighty: SUVs: The World's Most Dangerous Vehicles and How They Got That Way.* New York, NY: Public Affairs, 2002.

Brady, L. "The Wilderness War: Nature and Strategy in the American Civil War." *Environmental History,* vol. 10, no. 3 (July 2005).

Bragg, W. H. *Griswoldville.* Macon, GA: Mercer University Press, 2000.

Brands, H. W. *The Age of Gold.* New York, NY: Doubleday, 2002.

Brehm, V. M. "Environment, Advocacy, and Community Participation: MOPAWI in Honduras," *Development in Practice* 10, no. 1 (2000): 94–98.

Brennan, T. J., Palmer, K. L., Kopp, R. J., and Krupnick, A. J. *A Shock to the System—Restructuring America's Electricity Industry.* Washington, DC: Resources for the Future, 1996.

Broder, J. M. "Rule to Expand Mountaintop Coal Mining," *The New York Times,* August 23, 2007, Late edition—final, sec. A, p. 1.

Brown, D. *Bury My Heart at Wounded Knee: An Indian History of the American West,* NY: Owl Books, 2001.

Brown, L. *Seeds of Change.* New York, NY: Praeger Publishers, 1970.

Bruegmann, R. *Sprawl: A Compact History.* Chicago, IL: University of Chicago Press, 2005.

Brinkley, D. *Wheels for the World: Henry Ford, His Company and a Century of Progress.* New York, NY: Viking, 2003.

Broad, W. *The Universe Below.* New York, NY: Simon & Schuster, 1997.

Brooks, H. A. *The Prairie School.* New York, NY: W. W. Norton, 1996.

Brower, M. *Cool Energy: Renewable Solutions to Environmental Problems,* revised ed. Cambridge, MA: MIT Press, 1992.

Bryant, B. I. *Environmental Justice: Issues, Policies and Solutions.* Washington, DC: Island Press, 1995.

Bryant, Jr., K. L., ed. *Railroads in the Age of Regulation, 1900–1980.* New York: Facts on File, 1988.

Bryson, J. M. *Strategic Planning for Public and Nonprofit Organizations.* San Francisco, CA: Jossey-Bass Publishers, 1995.

Buckley, G. L. *Extracting Appalachia: Images of the Consolidation Coal Company, 1910–1945.* Akron, OH: Ohio University Press, 2004.

Budowski, G. *Peace through Parks,* United Nations Environment Program, 2007, http://www.unep.org/OurPlanet/imgversn/144/budowski.html.

Bullard, R., Lewis, J., and Chavis, B. *Unequal Protection: Environmental Justice and Communities of Color.* San Francisco, CA: Sierra Club Books, 1994.

Burian, S. J., Nix, S. J., Pitt, R. E., and Durrans, S. R. "Urban Wastewater Management in the United States: Past, Present, and Future," *Journal of Urban Technology* 7, no. 3 (2000): 33–62.

Burke, J. F. *Mestizo Democracy: The Politics of Crossing Borders*. College Station, TX: Texas A & M University Press, 2004.

Burton, L. *American Indian Water Rights and the Limits of Law*. Lawrence, KS: University Press of Kansas, 1991.

Bush, G. W. Letter from June 13, 2001, http://www.usemb.se/Environment/letter.html.

Cagan, J., et al. *Field of Schemes: How the Great Stadium Swindle Turns Public Money into Private Profit*. New York, NY: Common Courage Press, 1998.

California Department of Water Resources. "All-American Canal Lining Agreement Signed." News Release, Sacramento, CA, January 29, 2002.

California Department of Water Resources. "Coachella Canal and All-American Canal Lining Projects," 2007, http://wwwdpla.water.ca.gov/sd/environment/canal_linings.html.

Callcott, G. H. *Maryland & America: 1940 to 1980*. Baltimore and London: The Johns Hopkins University Press, 1985.

Calloway, C. G. *First Peoples*. Boston, MA: Bedford, 1999.

Calthorpe, P. *The Next American Metropolis*. New York, NY: Princeton Architectural Press, 1993.

Cantelon, P., and Williams R. C. *Crisis Contained: Department of Energy at Three Mile Island*. Carbondale, IL: Southern Illinois University Press, 1982.

Carleson, R. R. *Paradise Paved: The Challenge of Growth in the New West*. Salt Lake City, UT: University of Utah Press, 1996.

Carlsen, L. "After NAFTA—CAFTA and AFTA." Americas Program. Center for International Policy, September 19, 2005, http://americas.irc-online.org/am/655.

Carlton, W. "New England Masts and the King's Navy," *The New England Quarterly* 12, no. 1 (1939), 4–18.

Caro, R. A. *The Power Broker: Robert Moses and the Fall of New York*. New York, NY: Alfred A. Knopf, 1970.

Carr, E. *Wilderness by Design*. Lincoln, NE: University of Nebraska Press, 1988.

Carson, R. *Silent Spring*. New York, NY: Mariner Books, 2002.

Carstensen, V. *The Public Lands: Studies in the History of the Public Domain*. Madison, WI: University of Wisconsin Press, 1963.

Catlin, L. G. *Letters and Notes*, Letter No. 2: Mouth of Yellow Stone, Upper Missouri. Can be found at http://catlinclassroom.si.edu/interviews/al-batis.html.

Catton, T. *Wonderland: An Administrative History of Mount Ranier National Park*, 1996, http://www.nps.gov/history/history/online_books/mora/adhi/adhi.htm.

Center for the State of the Parks, Park Assessments. *Waterton-Glacier International Peace Park*, 2002. http://www.npca.org/stateoftheparks/glacier/.

Cerritos College, http://www.cerritos.edu/soliver/American%20Identities/Trail%20of%20Tears/quotes.htm.

Chadwick, D. *Yellowstone to Yukon*. Washington, DC: National Geographic Society, 2007.

Chamberlain, K. P. *Under Sacred Ground: A History of Navajo Oil, 1922–1982*. Albuquerque, NM: University of New Mexico Press, 2000.

Chase, A. *Playing God in Yellowstone: The Destruction of America's First National Park*. Orlando, FL: Harcourt Brace & Company, 1986.

Chernow, R. *Titan: The Life of John D. Rockefeller, Sr.* New York, NY: Random House, 1998.

Chester, C. C. "Landscape Vision and the Yellowstone to Yukon Conservation Initiative," in *Conservation Across Borders: Biodiversity in an Interdependent World*, chap. 4. Washington, DC: Island Press, 2006, 134–216.

Child Lead Poisoning and the Lead Industry, http://www.sueleadindustry.homestead.com/.

Christianson, G. *Greenhouse: The 200-Year Story of Global Warming*. New York, NY: Penguin, 2000.

Cincinnati Children's Hospital Medical Center. *History of Lead Advertising*, http://www.cincinnatichildrens.org/research/project/enviro/hazard/lead/leadadvertising/industry-role.htm.

Clapham, P. J., and Baker, C. S. "Modern Whaling," in *Encyclopedia of Marine Mammals*, eds. W. F. Perrin, B. Wursig, and J. G. M. Thewissen. New York, NY: Academic Press, 2002, 1328–32.

Clark, A. M., Friedman E. J., and Hochstetler K. "The Sovereign Limits of Global Civil Society: A Comparison of NGO Participation in UN World Conferences on the Environment, Human Rights, and Women," *World Politics*, no. 51 (1998): 1–35.

Clark, Jr., C. E. *The American Family Home*. Chapel Hill, NC: University of North Carolina Press, 1986.

Clarke, C. G. *The Men of the Lewis and Clark Expedition: A Biographical Roster of the Fifty-one Members and a Composite Diary of Their Activities from all the Known Sources*. Glendale, CA: A. H. Clark, 1970.

Clary, D. A. *Timber and the Forest Service*. Lawrence, KS: University of Kansas Press, 1986.

Clawson, M. "The Bureau of Land Management, 1947–1953," *The Cruiser* [newsletter of the Forest History Society], 10, no. 3 (1987): 3–6.

Clearwater. *Hudson River PCB Pollution Timeline*, http://www.clearwater.org/news/timeline.html.

Clough-Riquelme, J., and Bringas, N. L., eds. *Equity and Sustainable Development: Reflections from the US–Mexico Border*. La Jolla, CA: Center for U.S.–Mexican Studies, 2006.

CNN. "Bush OKs 7000-mile Border Fence," *Politics*, October 26, 2006, http://www.cnn.com/2006/POLITICS/10/26/border.fence/index.html.

CNN. "One-on-one with Angelina Jolie," *Anderson Cooper Blog 360°*, June 19, 2006, http://www.cnn.com/CNN/Programs/anderson.cooper.360/blog/2006/06/one-on-one-with-angelina-jolie.html.

CNN. "Senate Immigration Bill Suffers Crushing Defeat," June 28, 2007, http://www.cnn.com/2007/POLITICS/06/28/immigration.congress/index.html.

Coates, P. *Trans-Alaskan Pipeline Controversy: Technology, Conservation, and the Frontier*. Anchorage: University of Alaska Press, 1993.

Cody, B. A. *CRS Report for Congress*, "Major Federal Land Management Agencies: Management of Our Nation's Lands and Resources," 1995, http://www.ncseonline.org/NLE/CRSreports/Natural/nrgen-3.cfm?&CFID=8734533&CFTOKEN=91528013.

Colignon, R. A. *Power Plays*. Albany, NY: State University of New York Press, 1997.

Collin, R. H. *Theodore Roosevelt's Caribbean: The Panama Canal, The Monroe Doctrine, and the Latin American Context*. Baton Rouge, LA: Louisiana State University Press, 1990.

Colorado Plateau Land Use History of North America. *Native Americans and the Environment: A Survey of Twentieth Century Issues with Particular Reference to Peoples of the Colorado Plateau and Southwest*, http://cpluhna.nau.edu/Research/native_americans4.htm.

Colorado State Forest Service. "About Wildfire: Introduction and History," 2007, http://csfs.colostate.edu/wildfire.htm.

Colten, C. *Transforming New Orleans and Its Environs*. Pittsburgh, PA: University of Pittsburgh Press, 2001.

Colton, H. S. "Some Notes on the Original Condition of the Little Colorado River: A Side Light on the Problem of Erosion," *Museum Notes of the Museum of Northern Arizona* 10 (1937): 17–20.

Columbian Exposition. *The World's Columbian Exposition: Idea, Experience, Aftermath*, http://xroads.virginia.edu/~MA96/WCE/title.html.

Columbus, C. *The Four Voyages*. New York, NY: Penguin, 1969.

Commission for Environmental Cooperation. *Who We Are*, 2007, http://www.cec.org/who_we_are/index.cfm?varlan=english.

Commission for Labor Cooperation. *Objectives, Obligations, and Principles*, 2007, http://www.naalc.org/english/objective.shtml.

Connolly, J. A. *Three Years in the Army of the Cumberland: The Letters and Diary of Major James A. Connolly*, ed. P. M. Angle. 1928. Reprint, Bloomington, IN: Indiana University Press, 1996.

Conservation Law Foundation. *Early History of CLF's Fight to Clean up Boston Harbor 1983–1986*, http://www.clf.org/programs/cases.asp?id=188.

Contra Costa Times, 2003. "Rodeo Residents' Beef with PETA Must Wait." http://www.beefusa.org/newsrodeoresidentsbeefwithpetamustwait13689.aspx.

Conway, G. *The Doubly Green Revolution*. Ithaca, NY: Cornell University Press, 1998.

Conzen, M., ed. *The Making of the American Landscape*. Boston, MA: Unwin Hyman Publishers, 1990.

Cooke, M. *Report of the Great Plains Drought Area Committee*, http://newdeal.feri.org/hopkins/hop27.htm.

Cooper, G. *Air-Conditioning America*. Baltimore, MD: Johns Hopkins University Press, 2002.

Cornelius, W. *Death at the Border: The Efficacy and the "Unintended" Consequences of U.S. Immigration Control Policy 1993–2000.* San Diego, CA: The Center for Comparative Immigration Studies, University of California, 2001.

Cowdrey, A. *This Land, This South.* Lexington, KY: University Press of Kentucky, 1983.

Cowles, H. C. *Ecology and the American Environment,* http://memory.loc.gov/ammem/award97/icuhtml/aepsp4.html.

Creese, W. L. *TVA's Public Planning.* Knoxville, TN: University of Tennessee Press, 1990.

Creighton, M. S. *Rites and Passages: The Experience of American Whaling, 1830–1870.* Cambridge, 1995.

Creighton, M. S. *The Colors of Courage: Gettysburg's Forgotten History: Immigrants, Women, And African Americans in the Civil War's Defining Battle.* NY: Basic Books, 2006.

Cronon, W. *Changes in the Land.* New York, NY: W. W. Norton, 1991a.

Cronon, W. *Nature's Metropolis.* New York, NY: W. W. Norton, 1991b.

Cronon, W., ed. *Uncommon Ground: Rethinking the Human Place in Nature.* New York, NY: Norton, 1996.

Crosby, A. *Ecological Imperialism.* New York, NY: Cambridge University Press, 1986.

Crosby, A. *America's Forgotten Pandemic: The Influenza of 1918.* New York, NY: Cambridge University Press, 1990.

Culp, P. W. *Restoring the Colorado Delta with the Limits of the Law of the River.* Tucson, AZ: Udall Center for Studies in Public Policy, The University of Arizona, 2000.

Cunfer, G. *On the Great Plains: Agriculture and the Environment.* College Station, TX: Texas A & M University, 2005.

Custer, G. A. *My Life on the Plains: Or Personal Experiences with the Indians.* Norman: University of Oklahoma Press, 1977.

Cutright, P. *Theodore Roosevelt: The Making of a Conservationist.* Urbana, IL: University of Illinois Press, 1985.

Dana, C. W. *The Great West, or the Garden of the World.* Boston, MA: Wentworth and Co., 1857.

Dana, S. T., and Fairfax, S. K. *Forest and Range Policy: Its Development in the United States.* 2nd ed. New York, NY: McGraw-Hill, 1980.

Dangerfield, G. *The Awakening of American Nationalism, 1815–1828.* New York, NY: Harper and Row, 1965.

Darrah, W. C. *Pithole, the Vanished City.* Gettysburg, PA: 1964.

Darst, R. G. *Smokestack Diplomacy: Cooperation and Conflict in East-West Environmental Politics.* Cambridge, MA: MIT Press, 2001.

Darwin, C. *Origin of Species.* New York, NY: Signet Classics, 2003.

Dary, D. *The Buffalo Book: The Full Saga of the American Animal.* Athens, OH: Swallow Press/Ohio University Press, 1989.

Davis, D. *When Smoke Ran Like Water.* New York, NY: Basic Books, 2002.

Davis, J., ed. *The Earth First! Reader: Ten Years of Radical Environmentalism.* Salt Lake City, UT: Peregrine Smith Books, 1991.

Davis, S. G. *Spectacular Nature.* Berkeley, CA: University of California Press, 1997.

Davison, C. *White Pines for the Royal Navy,* http://www.nhssar.org/essays/Whtpines.html.

Day, D. *The Whale War.* San Francisco, CA: Sierra Club Books, 1987.

De Tocqueville, A. *Democracy in America,* http://xroads.virginia.edu/~HYPER/DETOC/home.html.

Demarest, Jr. D. P. *"The River Ran Red": Homestead, 1892.* Pittsburgh, PA: University of Pittsburgh Press, 1992.

DeVoto, B., ed. *The Journals of Lewis and Clark.* Boston, MA: Houghton Mifflin Co., 1997.

Diamond, J. *Guns, Germs and Steel: The Fates of Human Societies.* New York, NY: W. W. Norton, 1997.

Dietsch, T. V. "Assessing the Conservation Value of Shade-Grown Coffee: A Biological Perspective Using Neotropical Birds," *Endangered Species Update* 17 (2000): 122–30, University of Michigan, School of Natural Resources.

Dolin, E. J. *Political Waters: The Long, Dirty, Contentious, Incredibly Expensive but Eventually Triumphant History of Boston Harbor—A Unique Environmental Success Story.* Cambridge, MA: University of Massachusetts Press, 2004.

Domer, D., ed. *Lawrence on the Kaw: A Historical and Cultural Anthology.* Lawrence, KS: University Press of Kansas, 2000.

Donahue, D. L. *The Western Range Revisited: Removing Livestock from Public Lands to Conserve Native Biodiversity.* Legal History of North America Series, vol. 5. Norman, OK: University of Oklahoma Press, 1999. http://ipl.unm.edu/cwl/fedbook/taylorgr.html.

Dorsey, K. *The Dawn of Conservation Diplomacy.* Seattle: University of Washington Press, 1994.

Douglass, F. *My Bondage and My Freedom.* Urbana, IL: University of Illinois Press, 1987.

Dowie, M. *American Foundations: An Investigative History.* Cambridge, MA: MIT Press, 2001.

Downing, A. J. *A Treatise on the Theory and Practice of Landscape Gardening,* 9th ed. New York: Orange Judd, 1875. Reprint Little Compton, RI: Theophrastus Publishers, 1977.

Doyle, J. *Taken for a Ride: Detroit's Big Three and the Politics of Air Pollution.* New York, NY: Four Walls Eight Windows, 2000.

Dreyer, E. L. *Early Ming China: A Political History 1355–1435.* Stanford, CA: Stanford University Press, 1982.

Duany, A., and Plater-Zyberk, E. *Suburban Nation: The Rise of Sprawl and the Decline of the American Dream.* New York, NY: North Point Press, 2000.

Dunlop, B., and Scully, V. *Building a Dream: The Art of Disney Architecture.* New York, NY: Harry N. Abrams, 1996.

Dysart, B. C., and Clawson, M. *Managing Public Lands in the Public Interest.* New York, NY: Praeger Publishers, 1988.

Earth First! 2007. http://www.earthfirst.org/.

Eaton, D. J., and Anderson, J. M. *The State of the Rio Grande/Rio Bravo: A Study of Water Resource Issues along the Texas/Mexico Border.* Tucson, AZ: University of Arizona Press, 1986.

Edwards, M., and Hulme, D. *Beyond the Magic Bullet: NGO Performance and Accountability in the Post-Cold War World.* West Hartford, CT: Kumarian Press, 1996.

Egan, T. "New Fight in Old West: Farmers vs. Condo City." *The New York Times,* October 3, 1989.

Eichstaedt, P. H. *If You Poison Us: Uranium and Native Americans.* Sante Fe, NM: Crane Books, 1994.

El Nasser, H., and Overberg, P. "A Comprehensive Look at Sprawl in America," *USA Today,* February 22, 2001. http://www.usatoday.com/news/sprawl/main.htm.

Elkington, J., and Fennell, S. "Partners for Sustainability: Business-NGO Relations and Sustainable Development," *Greener Management International,* no. 24 (1998): 48–61.

Emerson, R. W. *Nature.* http://oregonstate.edu/instruct/phl302/texts/emerson/nature-contents.html.

Environmental Literacy Council. *Superfund,* http://www.enviroliteracy.org/article.php/329.html.

Erikson, K. *A New Species of Trouble—The Human Experience of Modern Disasters.* New York, NY: W. W. Norton, 1994.

Etheridge, E. W. *Sentinel for Health.* Berkeley, CA: University of California Press, 1992.

Eureka Country, Nevada Nuclear Waste Office. 2007. http://www.yuccamountain.org/new.htm.

Everhart, W. C. *The National Park Service.* Boulder, CO: Westview Press, 1983.

Ewers, J. C. *The Blackfeet: Raiders on the Northwestern Plains.* Norman, OK: University of Oklahoma Press, 1958.

Ewing, F. E. *America's Forgotten Statesman: Albert Gallatin.* New York, NY: Vantage Press, 1959.

Fairmont Water Works Interpretive Center. http://www.fairmountwaterworks.com.

Fall, J. *Drawing the Line: Nature, Hybridity and Politics in Transboundary Spaces.* Burlington, VT: Ashgate Publishing Company, 2005.

Fattig, P. "Forest Speaker Says He's No Terrorist," *Mail Tribune: Southern Oregon's News Source,* January 23, 2003, Local page. http://archive.mailtribune.com/archive/2003/0123/local/stories/07local.htm.

Federal Land and Policy Management Act. http://www.blm.gov/flpma/ and http://www.blm.gov/flpma/snapshot.htm.

Federal Wildlife Laws Handbook. http://ipl.unm.edu/cwl/fedbook/taylorgr.html.

Federation of American Scientists. *NSC-68,* http://www.fas.org/irp/offdocs/nsc-hst/nsc-68.htm.

Feller, D. *The Jacksonian Promise: America, 1815–1840.* Baltimore, MD: Johns Hopkins University Press, 1995.

Finkel, M. "From Yellowstone to Yukon," *Audubon* (1999): 44–53.

Fisher, H. *Wolf Wars: The Remarkable Inside Story of the Restoration of Wolves to Yellowstone.* Missoula, MT: Fischer Outdoor Discoveries, 2003.

Fishman, R. *Bourgeois Utopias: The Rise and Fall of Suburbia.* New York, NY: Basic Books, 1989.

Fixico, D. L. *The Invasion of Indian Country in the Twentieth Century: American Capitalism & Tribal Natural Resources.* Niwot, CO: University Press of Colorado, 1998.

Flink, J. J. *The Automobile Age.* Cambridge, MA: MIT Press, 1990.

Flores, D. "Bison Ecology and Bison Diplomacy: The Southern Plains from 1800–1850," *Journal of American History* 78 (1991): 2.

Flores, D. "The West That Was, and the West That Can Be," *High Country News.* August 18, 1997. http://www.hcn.org/servlets/hcn.Article?article_id=3560.

Floudas, D. A., and Rojas, L. F. "Some Thoughts on NAFTA and Trade Integration in the American Continent," *International Problems* LII (2000): 371–89.

Food and Agriculture Organization of the United Nations. "Women and the Green Revolution," 2007. http://www.fao.org/FOCUS/E/Women/green-e.htm.

Foreman, D. *Confessions of an Eco-Warrior.* NY: Three River Press, 1993.

Foreman, D. *Ecodefense: A Field Guide to Monkeywrenching.* 3rd ed. Chico, CA: Abbzugg Press, 1994.

Foresta, R. A. *America's National Parks and Their Keepers.* Washington, DC: Resources for the Future, 1985.

Formwalt, L. W. "Benjamin Henry Latrobe and the Revival of the Gallatin Plan of 1808," *Pennsylvania History* 48, no. 1 (19:1) 99–128.

Fox, S. *The American Conservation Movement.* Madison, WI: University of Wisconsin Press, 1981.

Franz, D. "The Yellowstone Wolves Win One," *E Magazine* 11 (2000): 14.

Freedmen's Act. http://www.multied.com/documents/Freedman.html.

Freese, B. *Coal: A Human History.* New York: Penguin Books, 2004.

Freydkin, D. "Celebrity Activists Put Star Power to Good Use," *USA Today.* June 22, 2006. http://usatoday.com/life/people/2006-06-22-celebcharities-main-x.htm.

Fri, R. "The Corporation as Nongovernment Organization." *Columbia Journal of World Business* 27, no. 3/4 (1992): 90–96.

Fridell, G. *Fair Trade Coffee: The Prospects and Pitfalls of Market-Driven Social Justice.* Studies in Comparative Political Economy and Public Policy. Toronto, Canada: University of Toronto Press, 2007.

Frisvold, G. B., and Caswell, M. F. "Trans-boundary Water Management: Game-Theoretic Lessons for Projects on the U.S.–Mexico Border," *Agricultural Economics* 24 (2000): 101–11.

Furze, B., de Lacy, T., and Birckhead, J. *Culture, Conservation and Biodiversity: The Social Dimension of Linking Local Level Development and Conservation through Protected Areas.* Hoboken, NJ: John Wiley & Sons, 1996.

Gadd, B. "The Yellowstone to Yukon Landscape," in *A Sense of Place: Issues, Attitudes and Resources in the Yellowstone to Yukon Ecoregion,* A. Harvey, ed. Canmore, Canada: Yellowstone to Yukon Conservation Initiative, 1998, 9–18.

Gadsden Purchase Treaty. http://www.yale.edu/lawweb/avalon/diplomacy/mexico/mx1853.htm.

Gallery of Lead Pollution Promotions. http://www.uwsp.edu/geo/courses/geog100/lead-ads.htm.

Gardner, J. S., and Sainato, P. "Mountaintop Mining and Sustainable Development in Appalachia," *Mining Engineering* 48 (2007): 48–55.

Garner, J. S., ed. *The Midwest in American Architecture.* Chicago, IL: University of Illinois Press, 1991.

Garwin, R. L., and Charpak, G. *Megawatts and Megatons: A Turning Point in the Nuclear Age.* New York, NY: Knopf, 2001.

Gatell, O., and Goodman, P. *Democracy and Union: The United States, 1815–1877.* New York, NY: Holt, Rinehart, and Winston, 1972.

Gates, P. W. *History of Public Land Law Development.* Washington, DC: U.S.G.P.O. for the Public Land Law Review Commission, 1968.

Gates, Jr., H. L., ed. *The Classic Slave Narratives.* New York, NY: Penguin, 1987.

Gelbspan, R. *The Heat Is On: The Climate Crisis, the Cover-Up, the Prescription.* NY: Perseus Books, 1998.

Gelletly, L. *Mexican Immigration: The Changing Face of North America.* Broomall, PA: Mason Crest Publishers, 2004.

Geraint, J., and Sheard, R. *Stadia: A Design and Development Guide*. Boston, MA: Architectural Press, 1997.

Gerard, D. *1872 Mining Law: Digging A Little Deeper*, PERC Policy Series, PS-11. Bozeman, MT: Political Economy Research Center, 1997.

German MAB National Committee, ed. *Full of Life: UNESCO Biosphere Reserves—Model Regions for Sustainable Development. The German Contribution to the UNESCO Programme Man and the Biosphere (MAB)*. New York, NY: Springer, 2005.

Giddens, P. *Early Days of Oil*. Gloucester, MA: Peter Smith, 1964.

Gifford, J. *The Exceptional Interstate Highway System*, http://onlinepubs.trb.org/onlinepubs/trnews/trnews244newvision.pdf.

Gilman, C. *The Poetry of Traveling in the United States*. New York: S. Colman, 1838.

Ginsburg, S. *Nuclear Waste Disposal: Gambling on Yucca Mountain*. Walnut Creek, CA: Aegean Park Press, 1994.

Glatthaar, J. T. *The March to the Sea and Beyond: Sherman's Troops in the Savannah and Carolinas Campaign*. New York, NY: New York University Press, 1985.

Glave, D. "A Garden So Brilliant with Colors, So Original in Its Design," *Environmental History* 8, no. 3 (2003): 395–411.

Goldish, M. *Gray Wolves: Return to Yellowstone*. New York, NY: Bearport Publishing Company, 2008.

Goodale, U. M., Lanfer A. G., Stern, M. J., Margoluis, C., and Fladeland, M., eds. *Transboundary Protected Areas: The Viability of Regional Conservation Strategies*. Binghamton, NY: The Haworth Press, 2003.

Gordon, R. B., and Malone, P. M. *The Texture of Industry*. New York, NY: Oxford, 1994.

Gordon, R., and VanDorn, P. *Two Cheers for The 1872 Mining Law*. Washington, DC: CATO Institute, 1998.

Gorman, H. *Redefining Efficiency: Pollution Concerns, Regulatory Mechanisms, and Technological Change in the U.S. Petroleum Industry*. Akron, OH: University of Akron Press, 2001.

Gottleib, R. *Forcing the Spring: The Transformation of the American Environmental Movement*. Washington, DC: Island Press, 1993.

Gowda, M., Rajeev, V., and Easterling, D. "Nuclear Waste and Native America: The MRS Sitting Exercise," *Risk: Health, Safety & Environment* 229 (1998): 229–58.

Graebner, N. B., ed. *Manifest Destiny*. New York, NY: Bobbs Merrill, 1968.

Graham, F. *The Adirondack Park: A Political History*. New York, NY: Random House, 1978.

Grinde, D. A., and Johansen, B. E. *Ecocide of Native America: Environmental Destruction of Indian Lands and Peoples*. Sante Fe, NM: Clear Light, 1995.

Griswold, D. T. *NAFTA at 10: An Economic and Foreign Policy Success*, Cato Institute, 2002. http://www.freetrade.org/node/87.

Gruen, V., and Smith, L. *Shopping Towns USA: The Planning of Shopping Centers*. New York, NY: Van Nostrand Reinhold Company, 1960.

Gumprecht, B. "Transforming the Prairie: Early Tree Planting in an Oklahoma Town," *Historical Geography* 29 (2001): 116–34.

Gura, P. F., and Myerson, J., eds. *Critical Essays on American Transcendentalism*. London: G. K. Hall, 1982.

Gutfreund, O. D. *20th Century Sprawl: Highways and the Reshaping of the American Landscape*. New York, NY: Oxford University Press, 2005.

Hage, W. *Storm Over Rangelands: Private Rights in Federal Lands*. Bellevue, WA: Free Enterprise Press, 1989.

Hague, J. A., ed. *American Character and Culture*. New York: Eveett Edwards Press, 1964.

Haines, F. *The Buffalo: The Story of American Bison and Their Hunters from Prehistoric Times to the Present*. New York, NY: Crowell, 1970. Reprint Norman, OK: University of Oklahoma Press, 1995.

Hales, P. B. *William Henry Jackson and the Transformation of the American Landscape*. Philadelphia, PA: Temple University Press, 1988.

Hampton, W. *Meltdown: A Race against Nuclear Disaster at Three Mile Island: A Reporter's Story*. Cambridge, MA: Candlewick Press, 2001.

Hanford Downwinders Litigation Information Resource. http://www.downwinders.com/index.html.

Hardin, G. "The Tragedy of the Commons," *Science* 162 (1968): 1243–48.

Hargrove, E. C. *Prisoners of Myth*. Princeton, NJ: Princeton University Press, 1994.

Hargrove, E. C., and Conkin, P. K., eds. *TVA: Fifty Years of Grass-Roots Bureaucracy*. Knoxville, TN: University of Tennessee Press, 1984.

Harris, D. "Celebs and Charity: Trendiness or Benevolence? The Hottest Trend in Hollywood is Taking Up a Cause," *ABC News*, January 14, 2007. http://abcnews.go.com/Entertainment/WNT/story?id=2794458&page=1.

Harris, T. "Texas Patrols the Mexican Border—Virtually," *ABC News*, June 6, 2006. http://abcnews.go.com/US/story?id=2044968&page=1.

Hart, J. F., ed., *Our Changing Cities*. Baltimore, MD: Johns Hopkins University Press, 1991.

Hartzog, Jr., G. B. *Battling for the National Parks*. Mt. Kisco, NY: Moyer Bell, 1988.

Harvey, M. *Wilderness Forever: Howard Zahniser and the Path to the Wilderness Act*. Seattle, WA: University of Washington Press, 2005.

Haycox, Jr., E. "Building The Transcontinental Railroad, 1864–1869," *Montana: The Magazine of Western History*, 45 (2001).

Hayes, D. L. "The All-American Canal Lining Project: A Catalyst for Rational and Comprehensive Groundwater Management on the United States–Mexico Border," *Natural Resources Journal* 31 (1991): 806–15.

Hays, S. P. *Beauty, Health, and Permanence: Environmental Politics in the United States, 1955–85*. New York, NY: Cambridge University Press, 1993.

Hays, S. P. *Conservation and the Gospel of Efficiency*. Pittsburgh, PA: University of Pittsburgh Press, 1999.

Heacox, K. "Antarctica. The Last Continent," *National Geographic Destinations*. April 1, 1999.

Helms, D. "Conserving the Plains: The Soil Conservation Service in the Great Plains," Reprinted from *Agricultural History* 64 (1990): 58–73. http://www.nrcs.usda.gov/about/history/articles/ConservingThePlains.html.

Henderson, H. L., and Woolner, D. B., eds. *FDR and the Environment*. New York, NY: Palgrave, 2004.

Herzog, L. A., ed. *Shared Space: Rethinking the U.S.–Mexico Border Environment*. La Jolla, CA: Center for U.S.–Mexican Studies, University of California, San Diego, 2000.

Hickman and Eldredge. http://www.forester.net/msw_9909_brief_history.html.

Hicks, D. A. *The Limits of Celebrity Activism*. 2007. http://www.religion-online.org/showarticle.asp?title=3331.

Hietala, T. R. *Anxiety and Aggrandizement: The Origins of American Expansion in the 1840s*. Ann Arbor, MI: University Microfilms International, 1981.

Higgs, E. *Nature by Design: People, Natural Process, and Ecological Restoration*. Cambridge, MA: Massachusetts Institute of Technology Press, 2003.

Higham, J. *Critical Issues in Ecotourism: Understanding a Complex Tourism Phenomenon*. Burlington, MA: Elsevier, 2007.

Hine, R. V., and Faragher, J. M. *The American West: A New Interpretive History*. New Haven, CT: Yale University Press, 2000.

Hines, T. S. "The Imperial Mall: The City Beautiful Movement and the Washington Plan of 1901–1902," in *The Mall in Washington, 1791–1991*. Washington, DC: National Gallery of Art, 1991.

Hirschhorn, J. S. *Sprawl Kills: How Blandburbs Steal Your Time, Health, and Money*. New York: Sterling & Ross, 2005.

Hirt, P. W. *A Conspiracy of Optimism: Management of the National Forests since World War Two*. Lincoln, NE: University of Nebraska Press, 1994.

Hohman, E. *The American Whaleman*. Clifton, NJ: Augustus M. Kelley, 1928.

Holliday, J. S. *The World Rushed In*. New York, NY: Simon & Schuster, 1981.

Honey, M. *Ecotourism and Sustainable Development: Who Owns Paradise?* Washington, DC: Island Press, 1999.

Hopkins, J. "'Slow Food' Movement Gathers Momentum," *USA Today*. November 25, 2003.

Horgan, P. *Great River: The Rio Grande in North American History: Volume 1, Indians and Spain. Volume 2, Mexico and the United States*. Hanover, NH: Wesleyan University Press, 1984.

Hornaday's report. http://etext.lib.virginia.edu/railton/roughingit/map/figures3/bufhornaday.html.

Horowitz, D. *Jimmy Carter and the Energy Crisis of the 1970s.* New York, NY: St. Martin's Press, 2005.

Horsman, R. *Race and Manifest Destiny.* Cambridge: Harvard University Press, 1981.

Hughes, T. *Networks of Power: Electrification in Western Society, 1880–1930.* Baltimore, MD: Johns Hopkins University Press, 1983.

Hughes, T. *American Genesis.* New York, NY: Penguin, 1989.

Hunter, L. C., and Bryant, L. *A History of Industrial Power in the United States, 1780–1930,* vol 3, The Transmission of Power. Cambridge, MA: MIT Press, 1991, 207–8.

Hurley, A. *Environmental Inequalities: Class, Race, and Industrial Pollution in Gary, Indiana, 1945–1980.* Chapel Hill, NC: University of North Carolina Press, 1995.

Immigrant Solidarity Network. *2006 Report on Migrant Deaths at the US–Mexico Border.* November 18, 2006. http://www.immigrantsolidarity.org/cgi-bin/datacgi/database.cgi?file=Issues&report=SingleArticle&ArticleID=0646.

Imperial Irrigation District. *Water Department.* 2007. http://www.iid.com/Water_Index.php.

International Boundary and Water Commission. *International Boundary & Water Commission: United States and Mexico, United States Section,* 2007. http://www.ibwc.state.gov/home.html.

International Conference on Transboundary Protected Areas as a Vehicle for International Co-operation. *Parks for Peace.* Conference Proceedings. September 16–18, 1997. Somerset West, South Africa.

Inventory of Conflict and Environment. *The Buffalo Harvest,* http://www.american.edu/TED/ice/buffalo.htm.

Irwin, W. *The New Niagara.* University Park, PA: Pennsylvania State University Press, 1996.

Isaacs, J. C. "The Limited Potential of Ecotourism to Contribute to Wildlife Conservation," *The Ecologist* 28 (2000): 61–69.

Ise, J. *Our National Park Policy: A Critical History.* Baltimore, MD: Johns Hopkins University Press, 1961.

Isenberg, A. *The Destruction of the Bison.* New York, NY: Cambridge University Press, 2001.

Jackson, D. C. *Building the Ultimate Dam.* Lawrence, KS: University of Kansas Press, 1995.

Jackson, K. T. *Crabgrass Frontier: The Suburbanization of the United States.* New York, NY: Oxford University Press, 1985.

Jackson, W. H. *Time Exposure: The Autobiography of William Henry Jackson.* Tucson, AZ: The Patrice Press, 1994.

Jackson, W. H., and Fielder, J. *Colorado, 1870–2000.* New York, NY: Westcliffe Publishing, 1999.

Jacobs, J. *The Death and Life of Great American Cities.* New York, NY: Vintage Books, 1961.

Jefferson, T. "Observations on the Whale-Fishery," in *Public and Private Papers.* New York, NY: Vintage Books, 1990.

Jenkins, V. S. *The Lawn.* Washington, DC: Smithsonian Institution Press, 1994.

Johnson, D. "Yellowstone Will Shelter Wolves Again," *The New York Times.* June 17, 1994. http://query.nytimes.com/gst/fullpage.html?res=9401E7D7173DF934A25755C0A962958260&n=Top%2fReference%2fTimes%20Topics%2fOrganizations%2fS%2fSierra%20Club.

Johnson, P. M., and Beaulieu, A. *The Environment and NAFTA: Understanding and Implementing the New Continental Law.* Washington, DC: Island Press, 1996.

Jones, H. R. *John Muir and the Sierra Club: The Battle for Yosemite.* San Francisco, CA: Sierra Club, 1965.

Jones, L. C. "Assessing Transboundary Environmental Impacts on the U.S.–Mexican and U.S.–Canadian Borders," *Journal of Borderlands Studies* 12 (1997): 81.

Jones, L. Y., ed. *The Essential Lewis and Clark.* New York, NY: Ecco Press, 2000.

Jordan, T., and Kaups, M. *The American Backwoods Frontier.* Baltimore, MD: Johns Hopkins University Press, 1986.

Jorgensen, J. G. *Native Americans and Energy Development II.* Boston, MA: Anthropology Resource Center & Seventh Generation Fund, 1984.

Josephson, P. R. *Red Atom: Russia's Nuclear Power Program from Stalin to Today.* New York, NY: W. H. Freeman, 2000.

Kamauro, O. *Ecotourism: Suicide or Development? Voices from Africa #6: Sustainable Development, UN Non-Governmental Liaison Service.* United Nations News Service, 1996.

Karliner, J. *A Brief History of Greenwash.* CorpWatch, http://www.greenwashing.net/.

Kasson, J. *Amusing the Million: Coney Island at the Turn of the Century.* New York, NY: Hill and Wang, 1978.

Kay, J. H. *Asphalt Nation.* Berkeley, CA: University of California Press, 1997.

Keller, C. *Philanthropy Betrayed.* www.apspub.com/proceedings/1441/Keller.pdf.

Kellogg, P. U. *The Pittsburgh Survey.* http://www.clpgh.org/exhibit/stell30.html.

Kelly, G. "Grazing Act Still at Work to Protect Grasslands," *Denver Rocky Mountain News.* November 2, 1999.

Kennedy, R. F. *Crimes Against Nature: How George W. Bush and His Corporate Pals Are Plundering the Country and Hijacking Our Democracy.* New York, NY: Harper Collins, 2004.

Kennett, L. B. *Marching through Georgia: The Story of Soldiers and Civilians during Sherman's Campaign.* New York, NY: Harper Collins, 1995.

Kern, S. *The Culture of Time and Space, 1880–1918.* Cambridge, MA: Harvard University Press, 1983.

Kinder, C. *Preventing Lead Poisoning in Children,* http://www.yale.edu/ynhti/curriculum/units/1993/5/93.05.06.x.html#b.

Kirby, A. *Extinction Nears for Whales and Dolphins.* 2003. http://news.bbc.co.uk/2/hi/science/nature/3024785.stm.

Kirby, J. T. *The American Civil War: An Environmental View.* http://www.nhc.rtp.nc.us:8080/tserve/nattrans/ntuseland/essays/amcwar.htm.

Klinkenborg, V. "The New Range Wars," *Mother Jones.* 2003. http://www.motherjones.com/news/featurex/2003/09/nrw.pdf.

Klobuchar, A. *Uncovering the Dome.* New York, NY: Waveland Press, 1986.

Knight, R., and Landres, P. *Stewardship Across Boundaries.* Washington, DC: Island Press, 1998.

Kraut, A. M. *Silent Travelers: Germs, Genes, and the Immigrant Menace.* New York, NY: Basic Books, 1994.

Krech III, S. *Buffalo Tales: The Near Extermination of the American Bison,* http://www.nhc.rtp.nc.us/tserve/nattrans/ntecoindian/essays/buffalo.htm.

Krech III, S. *The Ecological Indian: Myth and History.* New York, NY: W. W. Norton, 1999.

Kremen, C., Merenlender, A. M., and Murphy, D. D. "Ecological Monitoring: A Vital Need for Integrated Conservation and Development Programs in the Tropics," *Conservation Biology* 8, no. 2 (1994): 388–97.

Kunstler, J. H. *The Geography of Nowhere: The Rise and Decline of America's Man-Made Landscape.* New York, NY: Touchstone Books, 1993.

Kuppenheimer, L. B. *Albert Gallatin's Vision of Democratic Stability: An Interpretive Profile.* Westport, CT: Praeger Publishers, 1996.

Kushida, H. "Searching for Federal Aid: The Petitioning Activities of the Chesapeake and Delaware Canal Company," *Japanese Journal of American Studies* 14 (2003): 87–103. http://www.soc.nii.ac.jp/jaas/periodicals/JJAS/PDF/2003/No.14-087.pdf.

Labaree, B. *America and the Sea.* Mystic, CT: Mystic Seaport, 1999.

Lahusen, C. *The Rhetoric of Moral Protest: Public Campaigns, Celebrity Endorsement, and Political Mobilization.* New York, NY: Walter de Gruyter and Company, 1996.

Lamar, H., ed. *The New Encyclopedia of the American West.* New Haven, CT: Yale University Press, 1998.

Lavathes, L. *When China Ruled the Seas: The Treasure Fleet of the Dragon Throne 1405–1433.* Oxford, UK: Oxford University Press, 1994.

Leahy, P. Senate Speech 2004. http://leahy.senate.gov/press/200404/042604a.html

Lear, L. *Rachel Carson: Witness for History.* New York, NY: Owl Books, 1997.

Lederman, D., Maloney, W., and Serven, L. *Lessons from NAFTA for Latin America and the Caribbean.* Palo Alto, CA: Stanford University Press, 2005.

Lee, M. *Earth First!: Environmental Apocalypse.* Syracuse, NY: Syracuse University Press, 1995.

Leffler, W. L. *Deepwater Petroleum Exploration and Production.* New York, NY: PennWell, 2003.

Legler, D., and Korab, C. *Prairie Style: Houses and Gardens by Frank Lloyd Wright and the Prairie School.* New York, NY: Stewart, Tabori and Chang, 1999.

Legrain, P. *Immigrants: Your Country Needs Them*. Princeton, NJ: Princeton University Press, 2007.

Lemann, N. *The Promised Land: The Great Black Migration and How It Changed America*. New York, NY: Vintage Books, 1992.

Lemlich, C. *Life in the Shop*, http://www.ilr.cornell.edu/trianglefire/texts/stein_ootss/ootss_cl.html?location=Sweatshops+and+Strikes.

Leopold, A. *A Sand County Almanac, and Sketches Here and There*. [1948] New York, NY: Oxford University Press, 1987.

Leshy, J. D. *The Mining Law: A Study in Perpetual Motion*. Washington, DC: Resources For The Future, 1987.

Letts, C. W., Ryan, W. P., and Grossman, A. *High Performance Nonprofit Organizations: Managing Upstream for Greater Impact*. New York, NY: John Wiley and Sons, 1999.

Levesque, S. M. "The Yellowstone to Yukon Conservation Initiative: Reconstructing Boundaries, Biodiversity, and Beliefs," in *Reflections on Water: New Approaches to Transboundary Conflicts and Cooperation*, eds. J. Blatter and H. Ingram. Cambridge, MA: MIT Press, 2001, 123–62.

Lewis, D. R. *Neither Wolf nor Dog: American Indians, Environment, and Agrarian Change*. New York, NY: Oxford University Press, 1994.

Lewis, J. *Lead Poisoning: A Historical Perspective*, http://www.epa.gov/history/topics/perspect/lead.htm.

Lewis, M. J. "The First Design for Fairmount Park," *The Pennsylvania Magazine of History and Biography* 130.3 (2006): 33 pars. http://www.historycooperative.org/journals/pmh/130.3/lewis.html.

Lewis, T. *Divided Highways*. New York, NY: Penguin Books, 1997.

Library of Congress. *The Extermination of the American Bison*, http://memory.loc.gov/learn/features/timeline/riseind/west/bison.html.

Library of Congress. *Treaty for Russion for the Purchase of Alaska*, http://www.loc.gov/rr/program/bib/ourdocs/Alaska.html.

Liebs, C. H. *Main Street to Miracle Mile*. Baltimore, MD: Johns Hopkins University Press, 1995.

Liftin, K. T. *Ozone Discourses*. New York, NY: Columbia University Press, 1984.

Limerick, P. N. *Legacy of Conquest*. New York, NY: W. W. Norton, 1987.

Little, C. *Jefferson and Nature*. Baltimore: Johns Hopkins, 1988.

Livernash, R. "The Growing Influence of NGOs in the Developing World," *Environmental Conservation* 34, no. 5 (1992): 12–43.

Loeb, P. *Moving Mountains: How One Woman and Her Community Won Justice from Big Coal*. Lexington, KY: The University Press of Kentucky, 2007.

Los Angeles Department of Water and Power. *The Story of the Los Angeles Aqueduct*, http://www.ladwp.com/ladwp/cms/ladwp001006.jsp.

Lorey, D. E. *The U.S.–Mexican Border in the Twentieth Century*. Washington, DC: SR Books, 1999.

Lovins, A. *Soft Energy Paths*. New York, NY: Harper Collins, 1979.

Low, N., and Gleeson, B. *Justice, Society and Nature: An Exploration of Political Ecology*. New York, NY: Routledge, 1998.

Lowenthal, D. *George Perkins Marsh: Prophet of Conservation*. Seattle, WA: University of Washington Press, 2000.

Lower East Side Tenement Museum. *Health and Disease*, chap. 4. http://www.tenement.org/encyclopedia/diseases_cholera.htm.

Lowitt, R. *The New Deal and the West*. Norman, OK: University of Oklahoma Press, 1984.

MacArthur, J. R. *The Selling of "Free Trade": NAFTA, Washington, and the Subversion of American Democracy*. Berkeley, CA: University of California Press, 2001.

Macfarlane, A. M., and Ewing, R. C., eds. *Uncertainty Underground: Yucca Mountain and the Nation's High Level Nuclear Waste*. Cambridge, MA: MIT Press, 2006.

Machlis, G. E., Kaplan, A. B., Tuler, S. P., Bagby, K. A., and McKendry, J. E. *Burning Questions: A Social Science Research Plan for Federal Wildland Fire Management*. Report to the National Wildfire Coordinating Group. Contribution Number 943 of the Idaho Forest, Wildlife and Range Experiment Station. College of Natural Resources, University of Idaho, Moscow, 2002.

Mackintosh, B. *The National Parks: Shaping the System*. Washington, DC: National Park Service, 1991.

Magoc, C. *Yellowstone*. Santa Fe, NM: University of New Mexico Press, 1999.

Maher, N. *Nature's New Deal*. New York: Oxford University Press, 2007.

Maher, N. "Neil Maher on Shooting the Moon," *Environmental History* 9.3 (2004): 12 pars.

Makah: http://www.historylink.org/essays/output.fm?file_id=5301.

Malin, *The Grassland of North America*, http://www.kancoll.org/books/malin/mgchap02.htm.

Marcello, P. *Ralph Nader: A Biography*. Westport, CT: Greenwood.

Markell, D. L., and Knox, J. H., eds. *Greening NAFTA: The North American Commission for Environmental Cooperation*. Stanford Law and Politics. Stanford, CA: Stanford University Press, 2003.

Markowitz, G., Rosner, D. *Deceit and Denial: The Deadly Politics of Industrial Pollution*. Berkeley, CA: University of California Press, 2002.

Marks, R. B. *Origins of the Modern World*. Oxford, UK: Rowan and Littlefield, 2002.

Marsh, G. P. *Lectures on the English Language*. New York, NY: Charles Scribner's Sons, 1859, 1884, 1887.

Marsh, G. P. *The Camel—His Organization, Habits and Uses*. Boston, MA: Gould and Lincoln, 1856 (for material on importing the camel to America, see Chapters XVII, XVIII, and Appendix D).

Marsh, G. P. *Man and Nature*. Cambridge, MA: The Harvard University Press, 1965 (notated reprint of the original 1864 edition).

Marsh, G. P. *The Earth as Modified by Human Action: Man and Nature*. New York: Scribner, Armstrong, and Co., 1976 (straight reprint of the 1874 edition).

Marston, E. "The Old West Is Going Under," *High Country News*. April 27, 1998. http://www.hcn.org/servlets/hcn.Article?article_id=4105.

Martin, A. *Railroads Triumphant: The Growth, Rejection and Rebirth of a Vital American Force*. New York, NY: Oxford University Press, 1992.

Martin, R. "New Urban Islands Dot the West," *Monday Business Roundup*. September 24, 2007. New West-Boulder. http://www.newwest.net/city/article/new_urban_islands_dot_the_west/C94/L94/.

Marx, L. *The Machine in the Garden*. New York, NY: Oxford University Press, 1964.

Massachusetts Turnpike Authority. *The Big Dig*, http://www.masspike.com/bigdig/background/index.html.

Massey, D. *Smoke and Mirrors: U.S. Immigration Policy in the Age of Globalization*. New York, NY: Russell Sage Foundation Publications, 2001.

Matthiessen, P. *Arctic National Wildlife Refuge: Seasons of Life and Land*. Seattle, WA: The Mountaineers Books, 2005.

May, E. R. *American Cold War Strategy*. Boston, MA: Bedford Books, 1993.

May, E. T. *Homeward Bound*. New York, NY: Basic Books, 1988.

Mayer, F. W. *Interpreting NAFTA: The Science and Art of Political Analysis*. New York, NY: Columbia University Press, 1998.

Mayr, E. *The Growth of Biological Thought*. Cambridge, MA: Harvard University Press, 1982.

McCullough, D. *Path Between the Seas*. New York, NY: Touchstone, 1977.

McCullough, D. *The Johnstown Flood*. New York, NY: Simon & Schuster, 1968.

McCullough, D. *Great Bridge*. New York, NY: Simon & Schuster, 1983.

McGreevy, P. V. *Imagining Niagara*. Amherst, MA: University of Massachusetts Press, 1994.

McGrory, K. C. *Who Controls Public Lands?: Mining, Forestry, and Grazing Policies, 1870–1990*. Chapel Hill, NC: University of North Carolina Press, 1996.

McGuire, T., Lord, W. B., and Wallace, M. G., eds. *Indian Water in the New West*. Tucson, AZ: University of Arizona Press, 1993.

McHarg, I. *Design with Nature*. New York, NY: John Wiley and Sons, 1992.

McHugh, T. *The Time of the Buffalo*. New York, NY: Knopf, 1972.

McKee, B. "As Suburbs Grow, So Do Waistlines," *The New York Times*. September 4, 2003.

McKinsey, E. *Niagara Falls: Icon of the American Sublime*. Cambridge, MA: Cambridge University Press, 1985.

McLaren, D. *Rethinking Tourism and Ecotravel: The Paving of Paradise and What You Can Do to Stop It*. West Hartford, CT: Kamarian Press, 1998.

McMahon, P. "'Cause Coffees' Produce a Cup with an Agenda," *USA Today*. July 25, 2001. http://www.usatoday.com/money/general/2001-07-26-coffee-usat.htm.

McNamee, T. *The Return of the Wolf to Yellowstone*. New York, NY: Henry Holt and Company, 1997.

McNeil, J. R. *Something New Under the Sun: An Environmental History of the Twentieth-Century World*. New York, NY: W. W. Norton, 2001.

McPhee, J. *Assembling California*. New York, NY: Farrar, Straus and Giroux, 1993.

McShane, C. *Down the Asphalt Path*. New York, NY: Columbia University Press, 1994.

Medicine Crow, J. *From the Heart of the Crow Country: The Crow Indians' Own Stories*. New York, NY: Orion Books, 1992.

Meikle, J. L. *American Plastic: A Cultural History*. New Brunswick, NJ: Rutgers University Press, 1997.

Meilander, P. C. *Towards a Theory of Immigration*. New York, NY: Palgrave Macmillian, 2001.

Melosi, M. *Coping with Abundance*. New York, NY: Knopf, 1985.

Melosi, M. *Sanitary City*. Baltimore, MD: Johns Hopkins University Press, 1999.

Melville, H. *Moby Dick* (originally published as *The Whale*). New York, NY: Harper and Brothers, 1851.

Merchant, C. *Green versus Gold*. New York, NY: Island Press, 1998.

Merchant, C. *Major Problems in American Environmental History*. New York, NY: Heath, 2005.

Merrill, K. *Public Lands and Political Meaning: Ranchers, the Government, and the Property between Them*. Los Angeles, CA: University of California Press, 2002.

Merrill, K. *The Oil Crisis of 1973–1974: A Brief History with Documents*. New York, NY: Bedford Books, 2007.

Metlar, G. W. *Northern California, Scott and Klamath Rivers, Their Inhabitants and Characteristics, Historical Features, Arrival of Scott and his Friends, Mining Interests*. Yreka, CA: Yreka Union Office, 1856.

Millennium Whole Earth Catalog. San Francisco, CA: Harper, 1998.

Miller, B. *Coal Energy Systems*. Burlington, MA: Elsevier Academic Press, 2005.

Miller, C. *Gifford Pinchot and the Making of Modern Environmentalism*. New York, NY: Shearwater Books, 2004.

Miller, C. E. *Jefferson and Nature*. Baltimore, MD: Johns Hopkins University Press, 1988.

Miller, J. *My Life amongst the Indians*. Chicago, IL: Morril, Higgins, and Co., 1892.

Miller, J. *Germs: Biological Weapons and America's Secret War*. New York, NY: Simon & Schuster, 2001.

Miller, J. B. *An Evolving Dialogue: Theological and Scientific Perspectives on Evolution*. London: Trinity Press International, 2001.

Miller, P., ed. *The Transcendentalists: An Anthology*. Cambridge: Harvard University Press, 1950.

Miller, J. *An Evolving Dialogue: Theological and Scientific Perspectives on Evolution*. London: Trinity Press International, 2001.

Milligan, S. "US Senate Passes Bill to Build Mexican Border Fence," *The Boston Globe*, Nation, September 30, 2006. http://www.boston.com/news/nation/articles/2006/09/30/us_senate_passes_bill_to_build_mexican_border_fence/.

Mitchell, J. G. "When Mountains Move," *National Geographic*. March 2006. http://www7.nationalgeographic.com/ngm/0603/feature5/index.html.

Miyanishi, K., and Johnson, E. A. "A Re-examination of the Effects of Fire Suppression in the Boreal Forest," *Conservation Biology* 16 (2001): 1177–78.

Montrie, C. *To Save the Land and People: A History of Opposition to Surface Coal Mining in Appalachia*. Chapel Hill, NC: University of North Carolina Press, 2003.

Moorhouse, J. C., ed. *Electric Power: Deregulation and the Public Interest*. San Francisco, CA: Pacific Research Institute for Public Policy, 1986.

Morris, E. *Theodore Rex*. New York, NY: Random House, 2001.

Morrison, E. *J. Horace McFarland*. Harrisburg, PA: Pennsylvania Historical and Museum Commission, 1995.

Moss, S., and deLeiris, L. *Natural History of the Antarctic Peninsula*. New York, NY: Columbia University Press, 1988.

Motavalli, J. *Forward Drive: The Race to Build "Clean" Cars for the Future*. San Francisco, CA: Sierra Club Books, 2001.

"Mr. Coal's Story." http://www.sip.ie/sip019B/Mr_%20Coal's%20Story_files/Mr_%20Coal's%20Story. htm, Ohio State University.

Muhn, J., and Hanson R. S., eds. *Opportunity and Challenge: The Story of BLM.* Washington, DC: U.S. Printing Office, 1988.

Muir, J. *Travels in Alaska,* http://www.sierraclub.org/john_muir_exhibit/writings/travels_in_alaska/.

Mulvaney, K. *The Whaling Season: An Inside Account of the Struggle to Stop Commercial Whaling.* Washington DC: Island Press, 2003.

Mumford, L. *Technics and Civilization.* New York, NY: Harcourt, 1963.

Nabhan, G. P. *Coming Home to Eat: The Pleasures and Politics of Local Foods.* New York, NY: W. W. Norton, 2002.

Naess, A. "The Shallow and the Deep, Long-Range Ecology Movement," *Inquiry* 16 (1973): 95–100.

Nash, R. *Wilderness and the American Mind.* New Haven, CT: Yale University Press, 1982.

Nash, Roderick, ed. *American Environmentalism.* New York, NY: Cambridge University Press, 2001.

National Conservation Training Center. *Origins of the U.S. Fish and Wildlife Service,* http://training. fws.gov/history/origins.html.

National Council for Science and the Environment. *Chippewa Treaty Rights: History and Management in Minnesota and Wisconsin,* http://ncseonline.org/nae/docs/chippewa.html and http://ncseonline. org/NAE/fishing.html.

National Humanities Center. http://www.nhc.rtp.nc.us/pds/amerbegin/contact/contact.htm.

National Interagency Fire Center. www.nifc.gov.

National Park Service. *Flood Witnesses,* http://www.nps.gov/archive/jofl/witness.htm.

National Park Service. *Fort Raleigh,* http://www.nps.gov/history/history/online_books/hh/16/hh16d2.htm.

National Park Service. *Mount Rainier,* http://www.nps.gov/archive/mora/adhi/adhit.htm.

National Park Service. "Reading 1: Flour Milling," in *Wheat Farms, Flour Mills, and Railroads: A Web of Interdependence.* http://www.cr.nps.gov/nr/twhp/wwwlps/lessons/106wheat/106facts1.htm.

National Park Service. "Reading 1: The Work at Hopewell Furnace," in *Hopewell Furnace: A Pennsylvania Iron-Making Plantation.* http://www.cr.nps.gov/nr/twhp/wwwlps/lessons/97hopewell/97facts1.htm.

National Park Service. "Reading 2: The Bonanza Farms of North Dakota," in *Wheat Farms, Flour Mills, and Railroads: A Web of Interdependence.* http://www.cr.nps.gov/nr/twhp/wwwlps/lessons/106wheat/106facts2.htm.

National Park Service. *Yellowstone National Park Wolf Restoration.* 2007. http://www.nps.gov/yell/naturscience/wolfrest.htm.

National Research Council. *Hardrock Mining on Federal Lands, Committee on Hardrock Mining on Federal Lands.* Washington, DC: National Academy Press, 1999.

Natural Resources Defense Commission. *Historic Hudson River Cleanup to Begin After Years of Delay, But Will General Electric Finish the Job?,* http://www.nrdc.org/water/pollution/hhudson.asp

New Jersey Lighthouse Society. *An Act for the Establishment and Support of Lighthouse, Beacons, Buoys, and Public Piers,* http://www.njlhs.org/historicdocs/act.htm.

Newkirk, I. "The ALF: Who, Why, and What?" in *Terrorists or Freedom Fighters? Reflections on the Liberation of Animals,* S. Best and A. J. Nocella, eds. Seattle, WA: Lantern Press, 2004, 341.

Niven, J. *The Coming of the Civil War, 1837–1861.* Arlington Heights, IL: Harlan Davidson, 1990.

Norman, J., MacLean, H. L., and Kennedy, C. A. "Comparing High and Low Residential Density: Life Cycle Analysis of Energy Use and Greenhouse Gas Emissions," *Journal of Urban Planning and Development* 132 (2006): 10–21.

Norrell, B. "Yucca Mountain Lawsuit Filed," *Indian Country Today,* March 11, 2005. http://www.indiancountry.com/content.cfm?id=1096410530.

Norris, F. *The Octopus.* New York, NY: Penguin, 1986.

North American Development Bank. 2007. http://www.nadbank.org/.

Novak, B. *Nature and Culture.* New York, NY: Oxford University Press, 1980.

NPR. *Teaching Evolution: A State-by-State Debate,* http://www.npr.org/templates/story/story. php?storyId=4630737.

Nugent, W. "Western History, New and Not So New," Reprinted from the *OAH Magazine of History* 9 (1994).

Numbers, R. L. *The Creationists: The Evolution of Scientific Creationism*. Berkeley, CA: University of California Press, 1991.

Nye, D. *Technological Sublime*. Boston, MA: MIT Press, 1996.

Nye, D. *Electrifying America: Social Meanings of a New Technology*. Boston, MA: MIT Press, 1999.

Oblinger, U. W. *Letter from Uriah W. Oblinger to Mattie V. Oblinger and Ella Oblinger*, April 6, 1873. Courtesy of the Nebraska State Historical Society, Oblinger Family Collection.

Ohlemacher, S. "Number of Immigrants Hits Record 37.5 Million," Associated Press. September 12, 2007. http://news.yahoo.com/s/ap/20070912/ap_on_go_ot/census_demographics.

Oliens, R. M., and Davids, D. *Oil and Ideology: The American Oil Industry, 1859–1945*. Chapel Hill, NC: University of North Carolina Press, 1999.

Olmsted, F. L. "Draft of Preliminary Report upon the Yosemite and Big Tree Grove" and "Letter on the Great American Park of the Yosemite." Typed transcriptions. Frederick Law Olmsted Papers. Manuscript Division, Library of Congress, Washington, DC.

Olmsted, F. L. Letter, *New York Evening Post*, June 18, 1868.

Olmsted, F. L. "Yosemite and the Mariposa Grove: A Preliminary Report, 1865," *Landscape Architecture* 43, no. 1 (1952).

Olmsted, F. L. *The Papers of Frederick Law Olmsted, Volume Five: The California Years, 1863–1865*, ed. V. P. Ranney. Baltimore, MD: Johns Hopkins University Press, 1990.

Olmsted, F. L. *Yosemite and the Mariposa Grove: A Preliminary Report, 1865*. Yosemite, CA: Yosemite Association, 1995. http://www.yosemite.ca.us/history/olmsted/report.html.

Opie, J. *Nature's Nation*. New York, NY: Harcourt Brace, 1998.

Oster, S. M. *Strategic Management for Nonprofit Organizations*. New York, NY: Oxford University Press, 1995.

Oxfam. "Dumping without Borders: How U.S. Agricultural Policies are Destroying the Livelihoods of Mexican Corn Farmers." 2003. http://www.oxfam.org/en/files/pp030827_corn_dumping.pdf.

Painter, N. I. *Exodusters*. New York, NY: W. W. Norton, 1992.

Parrington, V. L. *The Romantic Revolution in America, 1800–1860*. Norman, OK: University of Oklahoma Press, 1987.

Patton, contained in Soderlund. *William Penn, and the Founding of Pennsylvania.*

PBS. *Timeline: Life and Death of the Electric Car*, http://www.historynet.com/exploration/great_migrations/3036611.html.

Peace Parks Foundation. "What Is the International Status of Peace Parks?" 2007. http://www.peace-parks.org/faq.php?mid=451&pid=302.

Peffer, W. A. *The Farmer's Side*. New York, 1891.

Pendergrast, M. *Uncommon Grounds: The History of Coffee and How It Transformed Our World*. New York, NY: Basic Books, 1999.

People for the Ethical Treatment of Animals. 2007. http://www.peta.org/.

Perkins, J. H. *Geopolitics and the Green Revolution: Wheat, Genes, and the Cold War*. New York, NY: Oxford University Press, 1997.

Peschard-Sverdrup, A. *U.S.–Mexico Transboundary Water Management: The Case of the Rio Grande/Rio Bravo*. CSIS Monograph. Washington, DC: Center for Strategic & International Studies, 2003.

Peterson, C. S. "A Portrait of Lot Smith—Mormon Frontiersman," *The Western Historical Quarterly* 1 (1970): 393–414.

Peterson, C. S. *Take Up Your Mission: Mormon Colonizing along the Little Colorado River 1870–1900*. Tucson, AZ: University of Arizona Press, 1973.

Petrikin, J. S. *Environmental Justice*, New York, NY: Greenhaven, 1995.

Petrini, C., and Padovani, G. *Slow Food Revolution: A New Culture for Eating and Living*. New York, NY: Rizzoli, 2006.

Petroski, H. *The Evolution of Useful Things*. New York, NY: Knopf, 1992, 100–101.

Pfanz, H. *Gettysburg*. Chapel Hill, NC: University of North Carolina Press, 1993.

Pinchot, G. *Breaking New Ground.* New York, NY: Island Press, 1998.

Pinchot, G. *The Use of the National Forests.* New York: Intaglio Press, 1907.

Pinchot, G. *A Primary of Forestry,* http://www.forestry.auburn.edu/sfnmc/class/pinchot.html.

Pitt, J. "Two Nations, One River: Managing Ecosystem Conservation in the Colorado River Delta," *Natural Resources Journal* 40 (2000): 855–57.

Pollan, *Omnivore's Dilemma.* New York: Penguin Books, 2007.

Pollan, M. *Second Nature.* New York, NY: Delta, 1992.

Poole, Jr., R. W., ed. *Unnatural Monopolies: The Case for Deregulating Public Utilities.* Lexington, MA: Lexington Books, 1985.

Posewitz, J. "Yellowstone to the Yukon (Y2Y): Enhancing Prospects for a Conservation Initiative," *International Journal of Wilderness* 4 (1998): 25–27.

Pratt, J. *Offshore Pioneers: Brown & Root and the History of Offshore Oil and Gas.* Houston, TX: Gulf Professional Publishing, 1997.

Preston, J. "U.S. Farmers Go Where Workers Are: Mexico," *International Herald Tribune.* September 4, 2007. http://www.iht.com/articles/2007/09/04/america/export.php.

Price, J. *Flight Maps.* New York, NY: Basic Books, 2000.

Price, M. "Ecopolitics and Environmental Nongovernmental Organizations in Latin America," *Geographic Review* 84, no. 1 (1994): 42–59.

Price, M. F. *Mountain Research in Europe: Overview of MAB Research from the Pyrenees to Siberia.* United Nations Educational, Scientific and Cultural Organization. New York, NY: The Parthenon Publishing Group, 1995.

Priest, T. *The Offshore Imperative: Shell Oil's Search for Petroleum in Postwar America.* Houston, TX: Texas A & M University Press, 2007.

Princen, T., and Finger, M. *Environmental NGOs in World Politics.* London, UK: Routledge, 1994.

Pyne, S. *Fire in America: A Cultural History of Wildland and Rural Fire.* Princeton, NJ: Princeton University Press, 1982.

Pyne, S. J. *Vestal Fire: An Environmental History, Told through Fire, of Europe and Europe's Encounter with the World,* 1997.

Pyne, S. J. *Tending Fire: Coping With America's Wildland Fires.* Washington, DC: Island Press, 2004.

Pyne. S. J. *History with Fire in Its Eye: An Introduction to Fire in America,* http://www.nhc.rtp.nc.us/tserve/nattrans/ntuseland/essays/fire.htm.

Quest, R. "On the Trail of the Celebrity Activist," CNN. September 1, 2005. http://edition.cnn.com/2005/WORLD/europe/08/11/quest/.

Rabe, B. *George Beyond Nimby: Hazardous Waste Siting in Canada and the United States,* Brookings Institute, November 1994.

Raber, P., ed. *The Archaic Period in Pennsylvania.* Harrisburg, PA: Pennsylvania Historical and Museum Commission, 1998.

Rainforest Alliance. "Leading Conservation Groups Release Guidelines for Growing Earth-Friendly Coffee," Press Release. May 30, 2001. www.ra.org/news/2001/coffee-principles.html.

Raustalia, K. "States, NGOs, and International Environmental Institutions," *International Studies Quarterly* 41 (1997): 719–40.

Rediker, M. *Between the Devil and the Deep Blue Sea.* New York, NY: Cambridge University Press, 1987.

Redstone Arsenal. *Women and War,* http://www.redstone.army.mil/history/women/welcome.html.

Reece, E. "Moving Mountains: The Battle for Justice Comes to the Coal Fields of Appalachia," *Orion Magazine.* January/February 2006. http://www.orionmagazine.org/index.php/articles/article/166/.

Reece, E. *Lost Mountain: A Year in the Vanishing Wilderness: Radical Strip Mining and the Devastation of Appalachia.* New York, NY: Penguin Group, 2006.

Reid, J. *Rio Grande.* Austin, TX: University of Texas Press, 2004.

Reiger, J. *American Sportsmen and the Origins of Conservation.* Norman, OK: University of Oklahoma Press, 1988.

Reisner, M. *Cadillac Desert.* New York, NY: Penguin, 1993.

Relph, E. *The Modern Urban Landscape.* Baltimore, MD: Johns Hopkins University Press, 1987.

Rettie, D. F. *Our National Park System: Caring for America's Greatest Natural and Historic Treasures.* Urbana, IL: University of Illinois Press, 1995.

Reuss, M. *Water Resources Administration in the United States: Policy, Practice, and Emerging Issues.* Ann Arbor, MI: Michigan State University Press, 1993.

Rhodes, R. *Audubon.* New York, NY: Knopf, 2004.

Rice, O. K. *The Allegheny Frontier.* Lexington, KY: University of Kentucky Press, 1970.

Ridenour, J. M. *The National Parks Compromised: Pork Barrel Politics and America's Treasures.* Merrillville, IN: ICS Books, 1994.

Riegel, R. E. *Young America, 1830–1840.* Norman, OK: University of Oklahoma Press, 1949.

Ries, L. A., and Stewart, J. S. *This Venerable Document,* http://www.phmc.state.pa.us/bah/dam/charter/charter.html.

Rifkin, J. *The Hydrogen Economy.* New York, NY: Penguin, 2003.

Riis, J. *How the Other Half Lives,* http://www.cis.yale.edu/amstud/inforev/riis/title.html.

Ringholz, R. C. *Uranium Frenzy, Boom and Bust on the Colorado Plateau,* Albuquerque, NM: University of New Mexico Press, 1989.

Risser, P. G., and Cornelison, K. D. *Man and the Biosphere.* Norman, OK: University of Oklahoma Press, 1979.

Rivera, J. A. *Acequia Culture: Water, Land, and Community in the Southwest.* Albuquerque, NM: University of New Mexico Press, 1998.

Robb, J., and Riebsame, W. E., eds. *Atlas of the New West: Portrait of a Changing Region.* New York, NY: W. W. Norton, 1997.

Robbins, J. "Resurgent Wolves Now Considered Pests by Some," *Wyoming Journal.* March 7, 2006.

Robbins, R. M. *Our Landed Heritage: The Public Domain, 1776–1970.* Lincoln, NE: University of Nebraska Press, 1976.

Robbins, W. G. *Colony and Empire: The Capitalist Transformation of the American West.* Lawrence, KS: University Press of Kansas, 1995.

Roberts, P. *Anthracite Coal Communities.* 1904. Reprint, Greenwood Publishers, 1970.

Rohrbough, M. J. *The Land Office Business: The Settlement and Administration of America's Public Lands, 1789–1837.* New York, NY: Oxford University Press, 1968.

Rohrbough, M. J. *Days of Gold: The California Gold Rush and the American Nation.* Berkeley, CA: University of California Press, 1997.

Rohter, L. "Canal Project Sets Off U.S.–Mexico Clash Over Water for Border Regions," *The New York Times,* page 2, October 2, 1989.

Rome, A. *The Bulldozer in the Countryside: Suburban Sprawl and the Rise of American Environmentalism.* New York: Cambridge University Press, 2001.

Roosevelt, T. *The Roosevelt Corollary to the Monroe Doctrine.* 1904. http://www.theodore-roosevelt.com/trmdcorollary.html.

Roper, L. W. *FLO: A Biography of Frederick Olmsted.* Baltimore, MD: John Hopkins University Press, 1973.

Rosenberg, C. *The Cholera Years: The United States in 1832, 1849, and 1866.* Chicago, IL: University of Chicago Press, 1962.

Rosensweig, R., and Blackmar, E. *The Park and the People: A History of Central Park.* Ithaca, NY: Cornell University Press, 1998.

Rosner, D. "The Living City: Engineering Social and Urban Change in New York City, 1865 to 1920," *Bulletin of the History of Medicine* 73 (1999): 124–29.

Rosner, D., ed. "Introduction: Hives of Sickness," in *Hives of Sickness: Public Health and Epidemics in New York City.* Piscataway, NJ: Rutgers University Press, 1995.

Rosove, M. H. *Let Heroes Speak: Antarctic Explorers, 1772–1922.* Annapolis, MD: Naval Institute Press, 2000.

Roth, L. M. *A Concise History of American Architecture.* New York, NY: Harper and Row, 1970.

Rothman, H. K. *Preserving Different Pasts: The American National Monuments.* Urbana, IL: University of Illinois Press, 1989.

Rothman, H. K. *The Greening of a Nation.* New York, NY: Harcourt, 1998.

Rothman, H. K. *Saving the Planet: The American Response to the Environment in the 20th Century*. Chicago, IL: Ivan R. Dee, 2000.

Rottenberg, D. *In the Kingdom of Coal: An American Family and the Rock That Changed the World*. New York, NY: Routledge, 2003.

Rowley, W. D. *U.S. Forest Service Grazing and Rangelands: A History*. College Station, TX: Texas A & M University Press, 1985.

Roy, A. *The Coal Mines*. New York, NY: Robison, Savage & Co., 1876.

Royster, J. "Water Quality and the Winters Doctrine." http://www.ucowr.siu.edu/updates/pdf/V107_A9.pdf.

Runte, A. *National Parks: The American Experience*. 3rd ed. Lincoln, NE: University of Nebraska Press, 1997.

Rupp, L. *Mobilizing Women for War*. Princeton: Princeton University Press, 1978.

Rush, E. *Annexing Mexico: Solving the Border Problem through Annexation and Assimilation*. Jamul, CA: Level 4 Press, 2007.

Russell, E. *War and Nature: Fighting Humans and Insects with Chemicals from World War I to Silent Spring*. New York, NY: Cambridge University Press, 2001.

Russell, E., and Tucker, R. P., eds. *Natural Enemy, Natural Ally: Toward an Environmental History of War*. Corvallis, OR: Oregon State University Press, 2005.

Rybczynski, W. "Suburban Despair," *Slate*. November 7, 2005. www.slate.com/id/2129636/?nav=tap3.

Sabin, P. *Crude Politics: The California Oil Market, 1900–1940*. Berkeley, CA: University of California Press, 2005.

Salm, J. "Coping with Globalization: A Profile of the Northern NGO Sector." *Nonprofit and Voluntary Sector Quarterly* 28, no. 4 (1999): 87–103.

Sanchea Munguia, V., ed. *El Revestimiento del Canal Todo Americano: Competencia o cooperacion por el agua en la frontera Mexico–Estados Unidos?* Tijuana, Baja California, Mexico: El Colegio de la Frontera Norte, 2004.

Savage, W. S. *Blacks in the West*. Westport, CT: Greenwood Press, 1976.

Scarce, R. *Eco-Warriors: Understanding the Radical Environmental Movement*. Chicago, IL: Nobel Press, 2006.

Scharff, V. *Taking the Wheel: Women and the Coming of the Motor Age*. New York: Free Press; Toronto: Collier Macmillan Canada, 1991.

Schiffer, M. B., Butts, T. C., and Grimm, K. K. *Taking Charge: The Electric Automobile in America*. Washington, DC: Smithsonian Institution Press, 1994.

Schlebecker, J. T. *Whereby We Thrive: A History of American Farming, 1607–1972*. Ames, IA: Iowa State University Press, 1975.

Schlosser, E. *Fast Food Nation: The Dark Side of the All-American Meal*. New York, NY: Houghton Mifflin Company, 2001.

Schuyler, D. *Apostle of Taste: Andrew Jackson Downing, 1815–1852*. Baltimore, MD: Johns Hopkins University Press, 1996.

Schweder, T. "Protecting Whales by Distorting Uncertainty: Non-precautionary Mismanagement?" *Fisheries Research* 52 (2001): 217–25.

Science Education Resource Center. *Impacts of Resource Development on Native American Lands*, http://serc.carleton.edu/research_education/nativelands/index.html.

Sears, S. *Sacred Places*. New York, NY: Oxford University Press, 1989.

Sellars, R. W. *Preserving Nature in the National Parks: A History*. New Haven, CT: Yale University Press, 1997.

Sellers, C. C. *Mr. Peale's Museum*. New York, NY: Regina Ryan Publishing Enterprises, 1979.

Semonin, P. *American Monster: How the Nation's First Prehistoric Creature Became a Symbol of National Identity*. New York, NY: New York University Press, 2000.

Shankland, R. *Steve Mather of the National Parks*. 3rd ed. New York, NY: Knopf, 1976.

Shaw, R. E. *Canals for a Nation: The Canal Era in the United States, 1790–1860*. Lexington, KY: University of Kentucky Press, 1990.

Shear, L. "Celebrity Activists," *WireTap Magazine*. May 2006. http://www.wiretapmag.org/stories/36850/.

Sheehan, B. *Seeds of Extinction; Jeffersonian Philanthropy and the American Indian*. New York, NY: Norton, 1973.

Sheppard, M. *Cloud by Day: The Story of Coal and Coke and People*. Pittsburgh, PA: University of Pittsburgh Press, 2001.

Sheriff, C. *The Artificial River: The Erie Canal and the Paradox of Progress, 1817–1862*. New York: Hill and Wang, 1997.

Shiva, V. *The Violence of the Green Revolution: Ecological Degradation and Political Conflict in Punjab*. New Delhi, India: Zed Press, 1992.

Sierra Nevada Earth First! "History of Earth First!" 2007. http://www.sierranevadaearthfirst.org/.

Singer, P. *Animal Liberation*. New York, NY: Harper Collins, 1975.

Singer, P. *In Defense of Animals: The Second Wave*. Malden, MA: Blackwell Publishing, 2006.

Sixeas, V. M. "Saving the Rio Grande," *Environment* 42 (2000): 7.

Sloane, D. C. *The Last Great Necessity*. Baltimore, MD: Johns Hopkins University Press, 1995.

Slotkin, R. *The Fatal Environment: The Myth of the Frontier in the Age of Industrialization, 1800–1890*. Norman: University of Oklahoma Press, 1998.

Slow Food International. 2007. http://www.slowfood.com/.

Slow Food U.S.A. 2007. http://www.slowfoodusa.org/.

Smil, V. *Energy in China's Modernization: Advances and Limitations*. Armonk, NY: M. E. Sharpe, 1988.

Smil, V. *Energy in World History*. Boulder, CO: Westview Press, 1994.

Smith, D. *Mining America: The Industry and the Environment, 1800–1980*. Lawrence, KS: Kansas University Press, 1987.

Smith, D. W., and Ferguson, G. *Decade of the Wolf: Returning the Wild to Yellowstone*. Guilford, CT: The Lyons Press, 2005.

Smith, D. W., Stahler, D. R., Guernsey, D. S., Metz, M., Nelson, A., Albers, E., and McIntyre, R. Smithsonian Migratory Bird Center, First Sustainable Coffee Congress overview paper. *Yellowstone Wolf Project, Annual Report 2006*. National Park Service, U.S. Department of the Interior. Yellowstone Center for Resources. Yellowstone National Park, Wyoming, 2007.

Smith, H. N. *Virgin Land: The American West as Symbol and Myth*. Cambridge, MA: Harvard University Press, 1978.

Smith, T. *Making the Modern*. Chicago, IL: University of Chicago Press, 1993. http://www.smokeybear.com/vault/wartime_prevention.asp.

Smith, T. G. *Green Republican: John Saylor and Preservation of America's Wilderness*. Pittsburgh, PA: University of Pittsburgh Press, 2006.

Sobel, R. *Conquest and Conscience: The 1840's*. New York, NY: Crowell, 1971.

Society of Plactics Industry. *History of Plastics*, http://www.plasticsindustry.org/industry/history.htm.

Soderlund, J. R. ed. *William Penn and the Founding of Pennsylvania, 1680–1684: A Documentary History*. Philadelphia, PA: University of Pennsylvania Press, 1983.

Solnit, R. *Savage Dreams: A Journey into the Hidden Wars of the American West*. San Francisco, CA: Sierra Club Books, 1994.

Soule, M. E., and Terborgh, J. *Continental Conservation: Scientific Foundations of Regional Reserve Networks*. Washington, DC: Island Press, 1999.

Spence, C. "The Golden Age of Dredging," *Western Historical Quarterly* (1980): 403–14.

Spence, M. D. *Dispossessing the Wilderness: Indian Removal and the Making of the National Parks*. New York, NY: Oxford University Press, 2000.

Stanford University. *The Influenza Pandemic of 1918*. http://www.stanford.edu/group/virus/uda/.

State Library of North Carolina. *First English Settlement in the New World*, http://statelibrary.dcr.state.nc.us/nc/ncsites/english1.htm.

Steen, H. K. *The U.S. Forest Service: A History*. Seattle, WA: University of Washington Press, 1976.

Stegner, W. *The American West as Living Space*. Ann Arbor, MI: University of Michigan Press, 1987.

Stegner, W. *Mormon Country*. Lincoln, NE: Bison Books, 2003.

Steinberg, T. *Nature Incorporated: Industrialization and the Water of New England*. New York, NY: Cambridge University Press, 1991.

Steinberg, T. *Down to Earth*. New York: Oxford University Press, 2002.

Stephanson, A. *Manifest Destiny: American Expansionism and the Empire of Right*. New York: Hill and Wang, 1995.

Stern, W. E., and Long, D. W. "U.S. Supreme Court Upholds 1995 Department of the Interior Grazing Regulations." 2000. http://www.modrall.com/articles/article_68.html#.

Stevens, D. L. *A Homeland and a Hinterland*. 1991. http://www.cr.nps.gov/history/online_books/ozar/hrs4.htm.

Stevens, J. E. *Hoover Dam*. Norman, OK: University of Oklahoma Press, 1988.

Stevenson, E. *Park Maker: A Life of Frederick Law Olmsted*. New York, NY: Macmillan, 1999.

Stewart, M. *"What Nature Suffers to Groe": Life, Labor and Landscape on the Georgia Coast, 1680–1920*. Athens, GA: University of Georgia Press, 1996.

Stiles, J. *Brave New West: Morphing Moab at the Speed of Greed*. Tucson, AZ: University of Arizona Press, 2007.

Stilgoe, J. *Common Landscapes of America*. New Haven, CT: Yale University Press, 1982.

Stilgoe, J. *Metropolitan Corridor: Railroads and the American Scene*. New Haven, CT: Yale University Press, 1983.

Stilgoe, J. R. *Borderland*. New Haven, CT: Yale University Press, 1988.

Stilgoe, J. *Alongshore*. New Haven: Yale University Press, 1998.

Stine, J. K. *Mixing the Waters: Environment, Politics, and the Building of the Tennessee-Tombigbee Waterway*. Akron, OH: University of Akron Press, 1993.

Stokes, K. M. *Man and the Biosphere: Toward a Coevolutionary Political Economy*. Armonk, NY: M. E. Shape, 1992.

Stonehouse, B. *The Last Continent: Discovering Antarctica*. New York, NY: W. W. Norton, 2000.

Stradling, D. *Smokestacks and Progressives: Environmentalists, Engineers, and Air Quality in America, 1881–1951*. Baltimore, MD: Johns Hopkins University Press, 1999.

Stratton, D. *Tempest over Teapot Dome: The Story of Albert B. Fall*. Norman, OK: University of Oklahoma Press, 1998.

Strohmeyer, J. *Extreme Conditions: Big Oil and the Transformation of Alaska*. Anchorage, AK: University of Alaska Press, 1997.

Sutter, P. S. *Driven Wild: How the Fight Against Automobiles Launched the Modern Wilderness Movement*. Seattle, WA: University of Washington Press, 2002.

Swain, D. C. *Wilderness Defender: Horace M. Albright and Conservation*. Chicago, IL: University of Chicago Press, 1970.

Tansley. *Ecology and the American Environment*, http://memory.loc.gov/ammem/award97/icuhtml/aepsp6.html.

Tarbell, I. *All in the Day's Work: An Autobiography*. Champaign, IL: University of Illinois Press, 2003.

Tarr, J. *The Search for the Ultimate Sink*. Akron, OH: University of Akron Press, 1996.

Tarr, J., ed. *Devastation and Renewal*. Pittsburgh, PA: University of Pittsburgh Press, 2003.

Taylor Grazing Act 43 U.S.C. §§315-316o, June 28, 1934, as amended 1936, 1938, 1939, 1942, 1947, 1948, 1954, and 1976. http://www4.law.cornell.edu/uscode/html/uscode43/usc_sup_01_43_10_8A_20_I.html.

Taylor, B., Chait, R., and Holland, T. "The New Work of the Nonprofit Board," *Harvard Business Review* (1996): 36–46.

Taylor, J. M. *Bloody Valverde: A Civil War Battle on the Rio Grande, February 21, 1862*. Albuquerque, NM: 1999.

Teacher Oz's Kingdom of History. *Women and the Home Front during World War II*, http://www.teacheroz.com/WWIIHomefront.htm.

Tempest-Williams, T. *Refuge*. New York: Vintage, 1992. http://www.ratical.org/radiation/inetSeries/TTW_C1-BW.html.

Terrie, P. *Forever Wild: A Cultural History of Wilderness in the Adirondacks*. Syracuse, NY: Syracuse University Press, 1994.

Teysott, G., ed. *The American Lawn*. Princeton, NJ: Princeton Architectural Press, 1999.

Thoreau, H. D. *Walden*. 1854. http://eserver.org/thoreau/walden00.html.

Thoreau, H. D. *The Maine Woods*. 1864. http://eserver.org/thoreau/mewoods.html.

Tomes, N. *The Gospel of Germs: Men, Women, and the Microbe in American Life*. Cambridge, MA: Harvard University Press, 1998.

Trachtenberg, A. *Brooklyn Bridge*. Chicago, IL: University of Chicago Press, 1979.

Trachtenberg, A. *Incorporation of America*. New York, NY: Hill and Wang, 1982.

Transboundary Parks. http://www.eoeartth.org/article/Transboundary_protected_areas.

Travis, W. R. *New Geographies of the American West: Land Use and the Changing Patterns of Place*. Washington, DC: Island Press, 2007.

Tremble, M. *The Little Colorado River*, in *Riparian Management: Common Threads and Shared Interests*, eds. B. Tellman, H. J. Cortner, M. G. Wallace, L. F. DeBano, R. H. Hamre. USDA Forest Service, Rocky Mountain Forest and Range Experiment Station, Fort Collins, CO, 1993, 283–89.

Trimble, S. *The People: Indians of the American Southwest*. Santa Fe, NM: School of American Research Press, 1993.

Turner, F. J. *The United States, 1830–1850*. New York, NY: Henry Holt, 1934. http://xroads.vir ginia.edu/~HYPER/TURNER/.

Twain, M. *Roughing It*. London, UK: George Routledge, 1871.

Tyler, A. F. *Freedom's Ferment: Phases of American Social History to 1860*. New York, NY: Harper and Row, 1962.

U.N. Educational Scientific and Cultural Organization. "The World Network of Biosphere Reserves," *UNESCO Courier*. May 1997.

U.N. Educational Scientific and Cultural Organization. *UNESCO's Man and the Biosphere Programme (MAB)*. 2007a. http://www.unesco.org/mab/mabProg.shtml.

U.N. Educational Scientific and Cultural Organization. *UNESCO Man and Biosphere Programme FAQs*. 2007b. http://www.gov.mb.ca/conservation/wno/status-report/fa-8.19.pdf.

UNEP, Ozone Secretariat. *The 1987 Montreal Protocol on Substances that Deplete the Ozone Layer (as agreed in 1987)*. 2004. http://ozone.unep.org/Ratification_status/montreal_protocol.shtml.

UNEP, Ozone Secretariat. *Evolution of the Montreal Protocol*. 2007. http://ozone.unep.org/Ratification_status/index.shtml.

United Farm Workers. *Address by Cesar Chavez*, http://www.ufw.org/_page.php?menu=research&inc=history/10.html.

U.S. Antarctic Program External Panel of the National Science Foundation. *Antarctica—Past and Present*, www.nsf.gov/pubs/1997/antpanel/antpan05.pdf.

U.S. Army Corps of Engineers. *Hydrogeomorphic Approach to Assessing Wetland Functions: Guidelines for Developing Regional Guidebooks*, http://el.erdc.usace.army.mil/wetlands/pdfs/trel02–3.pdf.

U.S. Army Corps of Engineers. *Navigational Improvements before the Civil War*, http://www.usace.army.mil/inet/usace-docs/eng-pamphlets/ep870-1-13/c-1.pdf.

U.S. Central Intelligence Agency. "Antarctica," *The World Factbook*. https://www.cia.gov/library/publications/the-world-factbook/geos/ay.html.

U.S. Department of Energy. "DOE Awards $3 Million Contract to Oak Ridge Associated Universities for Expert Review of Yucca Mountain Work." Press Release. March 31, 2006. www.energy.gov/news/3418.htm.

U.S. Department of Health and Human Services. *Health*. http://www.nlm.nih.gov/exhibition/phs_history/intro.html.

U.S. Department of State. *Purchase of Alaska, 1867*, http://www.state.gov/r/pa/ho/time/gp/17662.htm.

U.S. Department of the Interior. *Ecological Issues on Reintroducing Wolves into Yellowstone National Park*. National Park Service, SuDoc I 29.80:22, 1993.

U.S. Environmental Protection Agency. *Mountaintop Mining and Valley Fills in Appalachia: Final Programmatic Environmental Impact Statement*. 2006. www.epa.gov/region3/mtntop/index.htm.

U.S. Environmental Protection Agency. *Milestones in Garbage*. http://www.epa.gov/msw/timeline_alt.htm.

U.S. Environmental Protection Agency. *The Birth of Superfund.* http://www.epa.gov/superfund/20years/ch2pg2.htm.

U.S. Environmental Protection Agency. *Record of Decision.* http://www.epa.gov/hudson/d_rod.htm#record.

U.S. Fish and Wildlife Service. *A Guide to the Laws and Treaties of the United States for Protecting Migratory Birds.* http://www.fws.gov/migratorybirds/intrnltr/treatlaw.html.

U.S.–Canadian Wildlife Protection Treaties in the Progressive Era. Seattle: Univ. of Washington Press, 1998.

U.S. Geological Survey. *Wetlands of the United States.* http://www.npwrc.usgs.gov/resource/wetlands/uswetlan/century.htm.

U.S. Global Change Research Program. *U.S. National Assessment of the Potential Consequences of Climate Variability and Change,* http://www.usgcrp.gov/usgcrp/nacc/background/meetings/forum/greatplains_summary.html.

U.S. Government Accountability Office. *Illegal Immigration: Border-Crossing Deaths Have Doubled Since 1995.* GAO-06-770. Washington, DC: U.S. Government Accountability Office, 2006.

University at Buffalo Libraries. *Love Canal Collection.* http://ublib.buffalo.edu/libraries/projects/lovecanal/.

University of Virginia. *The Diseased City,* Benjamin Henry, Notes, http://xroads.virginia.edu/~MA96/forrest/WW/fever.html.

Unrau, H. *Gettysburg Administrative History.* Washington, DC: National Park Service, 1991.

Utley, R. *Cavalier in Buckskin: George Armstrong Custer and the Western Military Frontier.* Norman, OK: University of Oklahoma Press, 1988.

Valavenes, P. *Hysplex.* Berkeley, CA: University of California Press, 1999.

Vandenbosch, R., and Vandenbosch, S. E. *Nuclear Waste Stalemate: Political and Scientific Controversies.* Salt Lake City, UT: University of Utah Press, 2007.

Vergara, R. "NGOs: Help or Hindrance for Community Development in Latin America?" *Community Development Journal* 29, no. 4 (1994): 322–28.

Vieyra, D. I. *Fill 'Er Up: An Architectural History of America's Gas Stations.* New York, NY: Macmillan, 1979.

Virginia Places. *Canals of Virginia,* http://www.virginiaplaces.org/transportation/canals.html.

Vivanco, L. "Ecotourism, Paradise Lost—A Thai Case Study." *The Ecologist* 32 (2002): 28–30.

Voice of America. "Celebrity Activism: Publicity Stunt or Sincere Care?" Washington, February 21, 2007. http://www.voanews.com/english/archive/2007-02/2007-02-21-voa38.cfm?CFID=137892992&CFTOKEN=74360178.

Voight Jr. W. *Public Grazing Lands: Use and Misuse by Industry and Government.* New Brunswick, NJ: Rutgers University Press, 1976.

Wall, D. *Earth First! And the Anti-Roads Movement: Radical Environmentalism and Comparative Social Movements.* London: Routledge, 2001.

Wallace, *Report of the Great Plains Drought Area,* http://newdeal.feri.org/hopkins/hop27.htm.

Walters, R. *Albert Gallatin: Jeffersonian Financier and Diplomat.* New York, NY: Macmillan, 1957.

Ward, J. A. *Railroads and the Character of America.* Knoxville, TN: University of Tennessee Press, 1986.

Washington Post. "As Border Crackdown Intensifies, A Tribe Is Caught in the Crossfire." September 15, 2006.

Waste of the West. http://www.wasteofthewest.com/Chapter1.html.

Weaver, D. *Ecotourism.* Milton, Australia: John Wiley and Sons Australia, 2002.

Weber, S. J. "In Mexico, U.S. and Canada, Public Support for NAFTA Surprisingly Strong, Given Each Country Sees Grass as Greener on the Other Side." *World Public Opinion.* January 23, 2006. http://www.worldpublicopinion.org/pipa/articles/brlatinamericara/161.php?nid=&id=&pnt=161&lb=brla.

Weeks, J. *Gettysburg: Memory, Market, and an American Shrine.* Princeton: Princeton University Press, 2003.

Weiner, D. R. *Models of Nature: Ecology, Conservation, and Cultural Revolution in Soviet Russia.* Bloomington, IN: Indiana University Press, 1988.

Weiss, T. G., and Gordenker, L. *NGOs, the UN, and Global Governance*. Boulder, CO: Lynne Rienner Publishers, 1996.

Wellman, P. I. *The House Divides: The Age of Jackson and Lincoln*. Garden City, NY: Doubleday, 1966.

Wendell Cox Consultancy. "Demographia," *The Public Purpose*. 2002. www.demographia.com/dbx-intlair.htm.

Wennberg, R. N. *God, Humans, and Animals: An Invitation to Enlarge Our Moral Universe*. Grand Rapids, MI: Eerdmans Publishing Co., 2003.

Wertime, R. *Citadel on the Mountain*. New York, NY: Farrar, Straus, and Giroux, 2000.

West, E. *The Contested Plains: Indians, Goldseekers, and the Rush to Colorado*. Lawrence, KS: University of Kansas Press, 2000.

West, E. *A New Look at the Great Plains*, http://www.historynow.org/09_2006/historian2.html.

Wheeler, H. W. *Buffalo Days: Forty Years in the Old West: The Personal Narrative of a Cattleman Indian Fighter, and Army Officer Colonel Homer W. Wheeler*. New York, NY: The Bobbs-Merrill Company, 1925.

Wheelwright, J. *Degrees of Disaster: Prince William Sound, How Nature Reels and Rebounds*. New Haven, CT: Yale University Press, 1996.

White, R. *It's Your Misfortune and None of My Own*. Norman, OK: University of Oklahoma Press, 1991.

White, R. *Organic Machine*. New York, NY: Hill and Wang, 1996.

White House. "President Signs Yucca Mountain Bill." Press Release. July 23, 2002. http://www.whitehouse.gov/news/releases/2002/07/20020723-2.html.

Whitman, W. "Passage to India," in *Leaves of Grass*. 1855. http://www.bartleby.com/142/183.html.

Wild, A. *Coffee: A Dark History*. New York, NY: W. W. Norton, 2004.

Wilderness Society. *Yellowstone to Yukon*. 2007. http://www.wilderness.org/WhereWeWork/Montana/y2y.cfm?TopLevel=Y2Y.

Wiley, P., and Gottlieb, R. *Empires in the Sun: The Rise of the New American West*. Tucson, AZ: University of Arizona Press, 1985.

Williams, M. *Americans and Their Forests*. New York, NY: Cambridge University Press, 1992.

Wilson, A. *The Culture of Nature*. Cambridge, MA: Blackwell, 1992.

Wirth, C. L. *Parks, Politics, and the People*. Norman, OK: University of Oklahoma Press, 1980.

Wise Uranium Project. *Uranium Mining and Indigenous People*, http://www.wise-uranium.org/uip.html.

Wishart, D. *The Fur Trade of the American West*. Lincoln, NE: University of Nebraska Press, 1979.

Wong, P. "McCormick's Revolutionary Reaper," *Illinois History*. December 1992. http://www.lib.niu.edu/ipo/1992/ihy921205.html.

Wooster, R. *The Military and United States Indian Policy 1865–1903*. New Haven CT: Yale University Press, 1988.

Workman, D. P. *PETA Files: The Dark Side of the Animal Rights Movement*. Bellevue, WA: Merril Press, 2003.

Worster, D. *Dust Bowl: The Southern Plains in the 1930s*. New York, NY: Oxford University Press, 1979.

Worster, D. *Nature's Economy*. New York, NY: Cambridge University Press, 1994.

Worster, D. *A River Running West: Life of John Wesley Powell*, New York, NY: Oxford University Press, 2000.

Wright, A. *The Death of Ramon Gonzalez: The Modern Agricultural Dilemma*. Austin, TX: University of Texas Press, 1992.

Wright, G. *Building the Dream*. Cambridge, MA: MIT Press, 1992.

Wuerthner, G. *Wildfire: A Century of Failed Forest Policy. Foundation for Deep Ecology*. Washington, DC: Island Press, 2006.

Yellowstone National Park. Forty-Second Congress. Session II Ch. 21–24. 1872. March 1, 1872. Chap. XXIV.

Yergin, D. *The Prize: The Epic Quest for Oil, Money & Power*. New York, NY: Free Press, 1993.

Young, J. H. *The Medical Messiahs: A Social History of Health Quackery in Twentieth-Century America.* Princeton, NJ: Princeton University Press, 1967.

Young, J. H. *Pure Food: Securing the Federal Food and Drugs Act of 1906.* Princeton, N.J.: Princeton University Press, 1989.

Zbicz, D. C. *Transboundary Cooperation in Conservation: A Global Survey of Factors Influencing Cooperation between Internationally Adjoining Protected Areas.* PhD thesis, Duke University, 1999.

Zimmerman, M. E., Callicott, J. B., Sessions, G., Warren, K. J., and Clark, J., eds. *Environmental Philosophy: From Animal Rights to Radical Ecology.* Englewood Cliffs, NJ: Prentice-Hall, 1993.

Zuniga, V., and Hernandez-Leon, R., eds. *New Destinations: Mexican Immigration in the United States.* New York, NY: Russell Sage Foundation Publications, 2006.

OTHER RESOURCES

Dams, GeoGuide Online, National Geographic, http://www.nationalgeographic.com/resources/ngo/education/geoguide/dams/.http://environment.about.com/library/weekly/blrenew4.htm.

INDEX

About the Authors

BRIAN BLACK is Associate Professor in the departments of history and environmental studies at Penn State University, Altoona. He is the author of *Nature and the Environment in Nineteenth-Century American Life* and *Nature and the Environment in Twentieth-Century American Life*, each with Greenwood. In addition, Black is the author of *Petrolia: The Landscape of America's First Oil Boom* (2003) and the forthcoming *Contesting Gettysburg: Preserving a Cherished American Landscape*.

DONNA L. LYBECKER is an Assistant Professor of Political Science at Idaho State University. She received her M.A. from Tulane University and her Ph.D. from Colorado State University, where she focused on environmental politics and Latin America. Her research interests include decentralization of environmental policy, water policy in the West and along the U.S.–Mexico border, and cross-border environmental politics.